Modern Engineering 2019

IOP Conference Series: Materials Science and Engineering
Volume 589

Moscow, Russia
16 May 2019

ISBN: 978-1-5108-9462-4
ISSN: 1757-8981

Printed from e-media with permission by:

Curran Associates, Inc.
57 Morehouse Lane
Red Hook, NY 12571

Some format issues inherent in the e-media version may also appear in this print version.

Institute of Physics
Dirac House, Temple Back
Bristol BS1 6BE UK

Phone: 44 1 17 929 7481
Fax: 44 1 17 920 0979

techtracking@iop.org

Additional copies of this publication are available from:

Curran Associates, Inc.
57 Morehouse Lane
Red Hook, NY 12571 USA
Phone: 845-758-0400
Fax: 845-758-2633
Email: curran@proceedings.com
Web: www.proceedings.com

Table of contents

Volume 589

Modern Engineering

16 May 2019, Bauman Moscow State Technical University, Russian Federation

Accepted papers received: 9 July 2019
Published online: 16 August 2019

Preface

Papers

IOP Conf. Series: Materials Science and Engineering **589** (2019) 011001 doi:10.1088/1757-899X/589/1/011001

Modern Engineering-2019

The conference "Modern Engineering" is held annually in May at the Bauman Moscow State Technical University. The Department of Fluid Mechanics, Hydraulic Machines and Hydraulic and Pneumatic Automation is the conference organizer. The best conference reports are published as articles in the journal "Hydraulics", indexed in the Russian Science Citation Index. Every year, representatives of various scientific, educational and industrial organizations of Russia and other countries come to us with reports and speeches. The main topics of the reports are hydromechanics, hydraulic machines, and hydraulic drives. A part of the reports is devoted to industrial ecology issues. We are pleased to see at our conference scientists from any countries and from any organizations devoting their work to industrial hydraulics, ecology and solving fundamental issues in these areas.

Sincerely, A. Protopopov

IOP Conf. Series: Materials Science and Engineering **589** (2019) 011002 doi:10.1088/1757-899X/589/1/011002

Peer review statement

All papers published in this volume of *IOP Conference Series: Materials Science and Engineering* have been peer reviewed through processes administered by the proceedings Editors. Reviews were conducted by expert referees to the professional and scientific standards expected of a proceedings journal published by IOP Publishing.

IOP Conf. Series: Materials Science and Engineering **589** (2019) 012001 doi:10.1088/1757-899X/589/1/012001

Optimization of the output device of a disk pump for high viscous fluid

V Cheremushkin[1,2] and A Polyakov[1]

[1]Bauman Moscow State Technical University, 5 Second Baumanskaya Street, Moscow, 105005, Russian Federation

[2]Email: vcheremushkin@bmstu.ru

Abstract
In this paper the problem of designing an output device for dynamic pumps operating at a high fluid viscosity on the example of a disk pump was considered. The main objective of the study was to increase the pump efficiency by optimizing the geometry of the wet part of its volute. The dependences of the efficiency on the geometrical parameters of the output device estimated section were obtained. The results of the study are represented in a form of graphs and scalar field distributions.

Introduction

A disk pump is a friction pump which moving elements make a fluid move via viscosity forces. Thus, the mechanical energy of the fluid increases due to the viscous interaction[1].

A disk pump (Fig. 1) consists of an impeller 1 located in housing with a discharge device 3. The impeller 1 consists of a set of flat disks, which are connected to each other along the periphery.

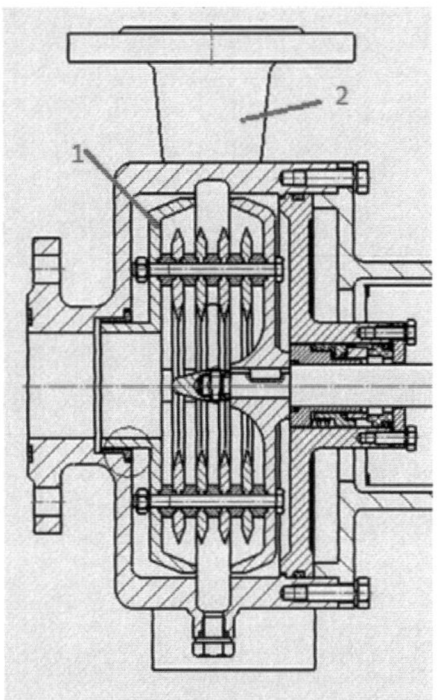

Figure1.Disk pump construction.

Disc pumps are capable to pump highly viscous liquids, liquids containing abrasive inclusions, and also may provide gentle pumping, i.e. not damage the particles within the pumped medium. These qualities give advantages to disc pumps in comparison with more common centrifugal pumps in some industries [1-3].

The methods of calculation available at current stage of technical development are not quite comprehensive. In addition, experimental data on operation of disk pumps at high viscosity are few and do not provide a complete description and explanation of the transfer process itself [4,5]. This problem leads to difficulties in the designing process and characteristics prediction of disk pumps[6].

In this work, the question of designing a volute output for a disk pump operating at a high viscosity (3000 cSt) is considered. The existing methods of diverting devices designing for centrifugal pumps are focused on the turbulent flow of a fluid, where thickness of a boundary layer is small compared to the channel width and the friction force is not significant[7,8].

In the case of a highly viscous fluid, this condition is not satisfied, and a disk pump with such a discharge device has unsatisfactory efficiency. A possible solution to the problem is to increase the output devise flow capacity.

Mathematical model and methods

The task of designing the diverting device is solved by means of CFD (computational fluid dynamics). The implementation of the numerical simulation method is based on solving the basic hydrodynamic equations discrete analogs [9-11]. In case of an incompressible fluid model, when the density is constant, these are:

Continuityequation

$$\frac{\partial \tilde{u}_j}{\partial x_j} = 0,$$

where $\tilde{u}_j$ - j axis projection of an averaged value of the fluid velocity (j=1,2,3);

The momentum conservation equation (Reynolds averaged):

$$\rho\left[\frac{\partial U_i}{\partial t} + U_j\frac{\partial U_i}{\partial x_j}\right] = -\frac{\partial P}{\partial x_i} + \frac{\partial}{\partial x_i}\left[T_{ij}^{(v)} - \rho u_i u_j\right];$$

where U, P - averaged velocity and pressure;

$\tilde{T}_{ij}^{(v)} = 2\mu\tilde{s}_{ij}$ - viscous stress tensor for incompressible fluid;

$\tilde{s}_{ij} = \frac{1}{2}\left[\frac{\partial \tilde{u}_i}{\partial x_j} + \frac{\partial \tilde{u}_j}{\partial x_i}\right]$ - instant strain rate tensor;

$\rho u_i u_j$ – Reynolds stresses.

The introduction of the Reynolds averaged Navier-Stokes equation makes the equations system not closed, as additional unknowns, Reynolds stresses, appear.

To solve this system of equation, the k-ω SST turbulence model was used. It introduces the necessary additional equations: the transport equations for the turbulent kinetic energy and the relative dissipation rate of this energy:

$$\frac{\partial k}{\partial t} + U_j\frac{\partial k}{\partial x_j} = P_k - \beta^* k\omega + \frac{\partial}{\partial x_j}\left[(\nu + \sigma_k \nu_T)\cdot\frac{\partial k}{\partial x_j}\right]$$

$$\frac{\partial \omega}{\partial t} + U_j\frac{\partial \omega}{\partial x_j} = \alpha\cdot S^2 - \beta\cdot\omega^2 + \frac{\partial}{\partial x_j}\left[(\nu + \sigma_\omega \nu_T)\cdot\frac{\partial \omega}{\partial x_j}\right] + 2\cdot(1 - F_1)\cdot\sigma_{\omega 2}\cdot\frac{1}{\omega}\cdot\frac{\partial k}{\partial x_i}\cdot\frac{\partial \omega}{\partial x_i}$$

Despite the fact that the expected type of fluid motion in a disk pump with specified parameters is laminar, the k-ω SST model can be used. This is because of the fact that the generation of turbulent kinetic energy is directly related to flown nature, the magnitude of the velocity and its derivatives. That is, in case the flow regime is essentially laminar, the generation of turbulent kinetic energy and the value of the turbulent viscosity are automatically reduced to very small values. The paper further confirms the possibility of using the k-ω SST turbulence model in the calculation.

When selecting the optimal geometric parameters of the diverting device, a combination of global (stochastic) and local (directional) search was used [12]. The range of parameters change is given in table 1.

Table1. Geometric parameters change ranges

Parameter	Minimum value, м	Maximum value , м
WidthW	0,0165	0,0760
Height B	0,0370	0,0754

To achieve maximum efficiency of the output, losses in the diffuser of the output are necessary to be minimal. That is achieved at a diffuser angle of 10°. Thus, the output diameter of the pump is defined as dependent on the parameters of the design section of volut.

$$D_{out} = \sqrt{\frac{4 \cdot W \cdot B}{\pi}} + 2 \cdot L_d \cdot tg(5°),$$

where D_{out} — pump outlet diameter;

L_d — diffuser length.

Figure 2 shows the computational grid, which has polyhedral cells in the flow core, and prismatic cells near the solid walls [13].

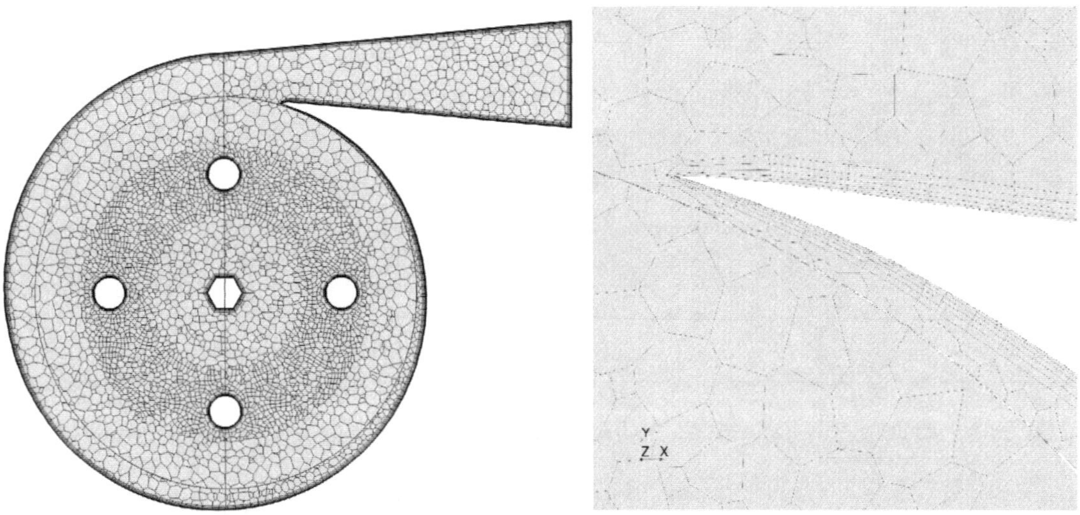

Figure2. Computational grid.

IOP Conf. Series: Materials Science and Engineering **589** (2019) 012001 doi:10.1088/1757-899X/589/1/012001

Discussion and results

Research of the pump characteristics was carried out with the following parameters:

$Q = 20\frac{m^3}{hr}$ – pump volumetric flow,

$n = 2900\,rpm$ – rotation speed,

$\nu = 3000\,cSt$ – fluid viscosity.

Table 2 shows the values of the geometric parameters of the models, as well as the characteristics obtained from the results of numerical simulation.

Table2. Parameters and characteristics of the calculated models.

№	W, m	B, m	D_out, m	Efficiency	Head, m	Area, m^2	Volumetric capacity, m
1	0,0190	0,0600	0,1112	0,2725	26,8702	0,0011	0,0094
2	0,0378	0,0600	0,0892	0,2719	26,7614	0,0023	0,0188
3	0,0378	0,0600	0,0850	0,2717	26,5076	0,0023	0,0188
4	0,0384	0,0600	0,0888	0,2713	26,7810	0,0023	0,0190
5	0,0618	0,0600	0,0970	0,2713	26,8395	0,0037	0,0306
6	0,0384	0,0639	0,0900	0,2699	26,7666	0,0025	0,0199
7	0,0578	0,0600	0,0879	0,2699	26,7367	0,0035	0,0287
8	0,0652	0,0639	0,1136	0,2689	26,7594	0,0042	0,0339
9	0,0207	0,0600	0,1014	0,2687	26,6817	0,0012	0,0103
10	0,0749	0,0600	0,1106	0,2678	26,7769	0,0045	0,0371
11	0,0652	0,0524	0,0852	0,2677	26,5665	0,0034	0,0291
12	0,0378	0,0754	0,1141	0,2673	27,0433	0,0029	0,0222
13	0,0372	0,0524	0,0848	0,2671	26,5835	0,0019	0,0166
14	0,0327	0,0408	0,0826	0,2640	26,5906	0,0013	0,0120
15	0,0760	0,0370	0,0772	0,2629	26,1659	0,0028	0,0257
16	0,0435	0,0370	0,0775	0,2617	26,0790	0,0016	0,0147
17	0,0367	0,0370	0,0889	0,2603	26,4486	0,0014	0,0124
18	0,0372	0,0408	0,0932	0,2601	26,5151	0,0015	0,0136
19	0,0760	0,0408	0,0678	0,2367	24,1646	0,0031	0,0279
20	0,0503	0,0370	0,0649	0,2219	23,1522	0,0019	0,0170

For better results clarity Figure 3 shows the efficiency dependence on the geometric parameters.

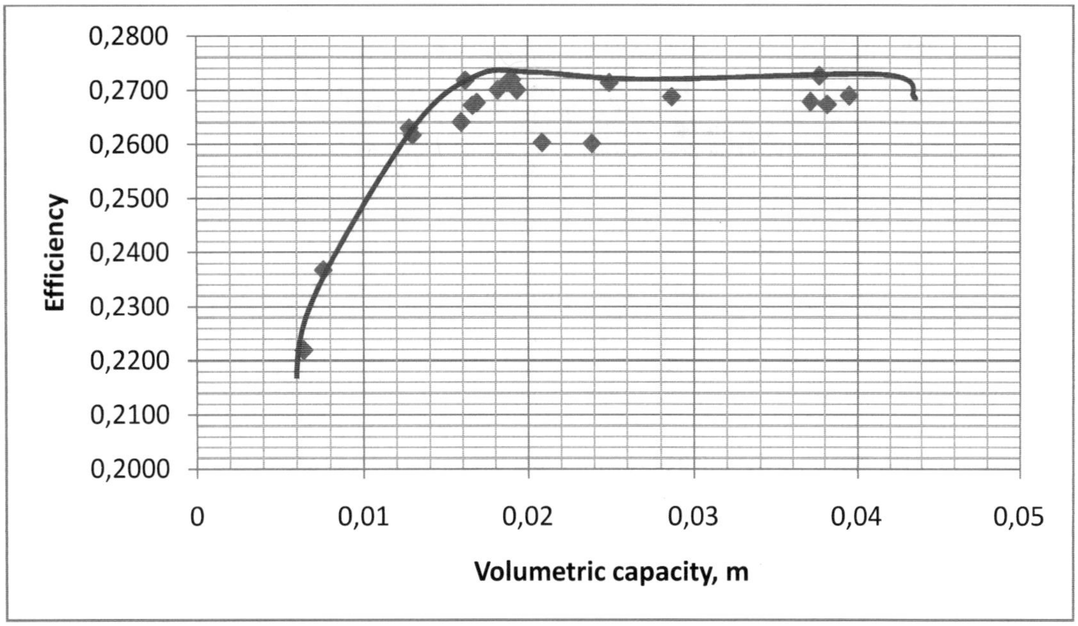
0,2800
0,2700
0,2600
0,2500
0,2400
0,2300
0,2200
0,2100
0,2000
Efficiency
0
0,01
0,02
0,03
0,04
0,05
Volumetric capacity, m

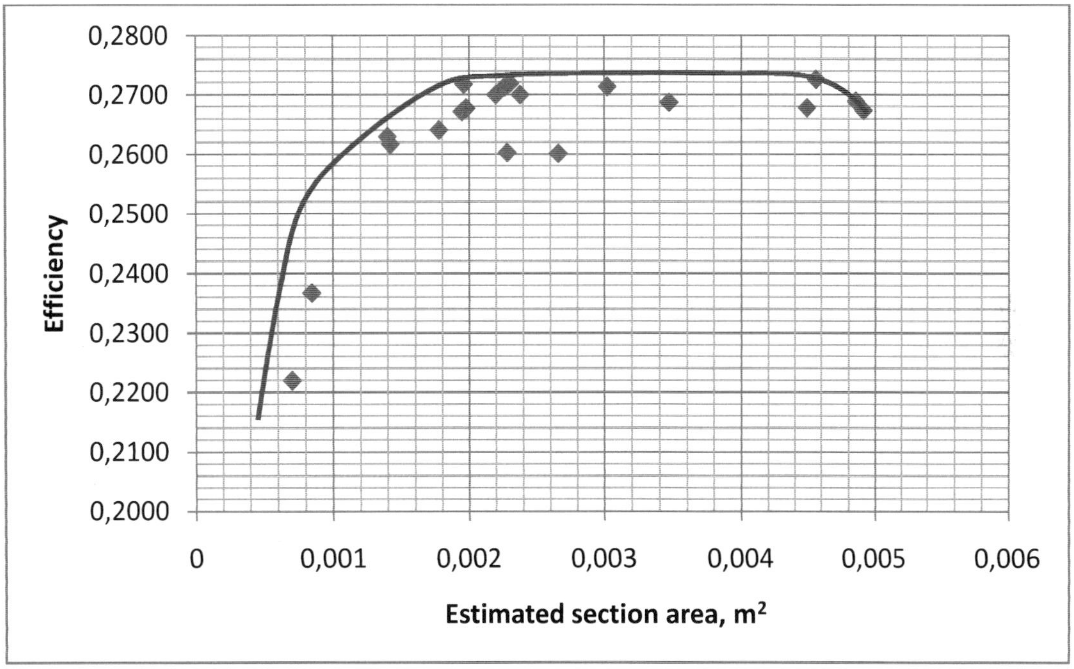
0,2800
0,2700
0,2600
0,2500
0,2400
0,2300
0,2200
0,2100
0,2000
Efficiency
0
0,001
0,002
0,003
0,004
0,005
0,006
Estimated section area, m²

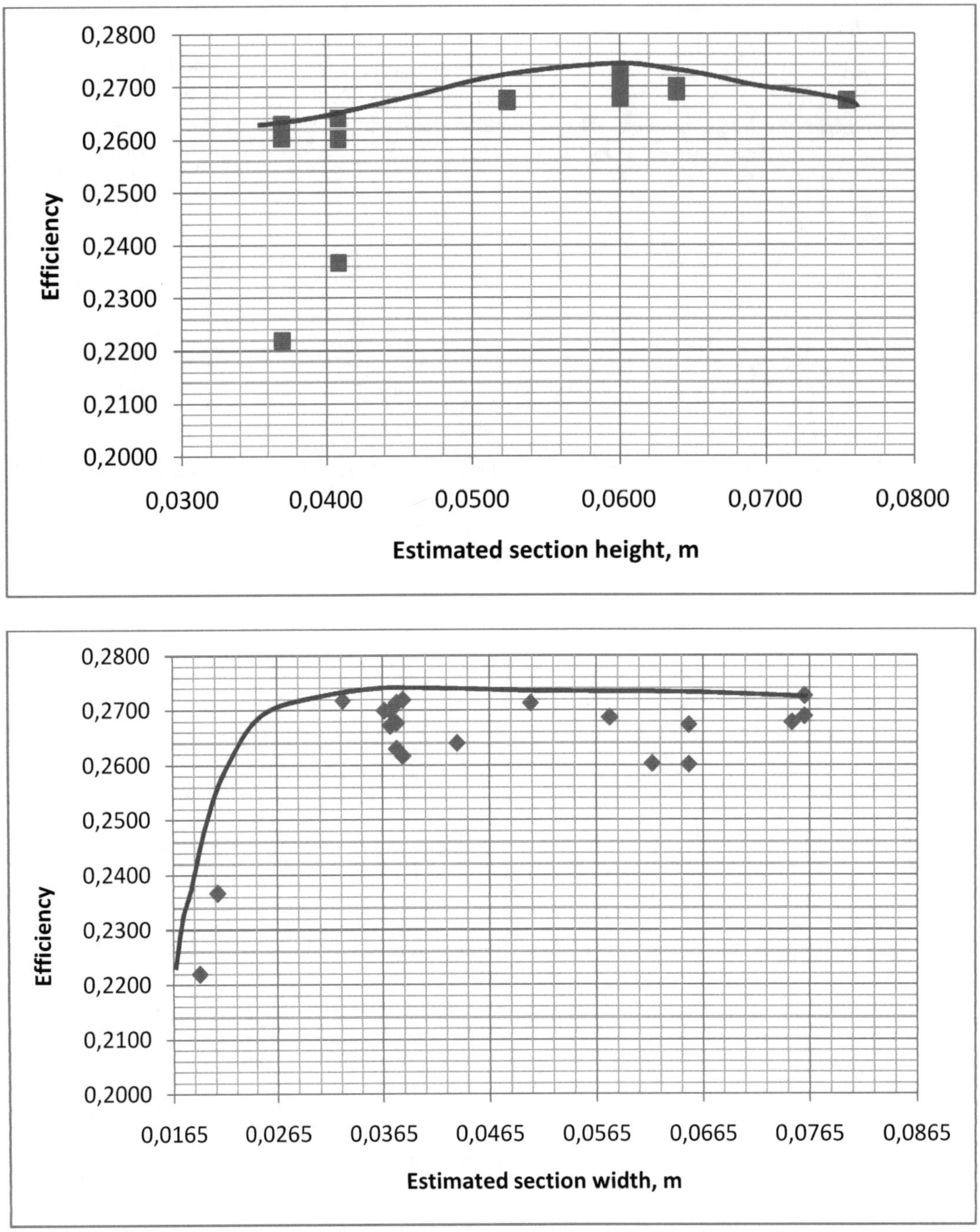

Figure3. Efficiency dependences on the geometric parameters.

In order to visualize the flow inside the wet part some scalar field functions distributions are shown on fig. 4 and 5 for two volute variations: initial one designed for water by an existing method and the one with enlarged estimated section.

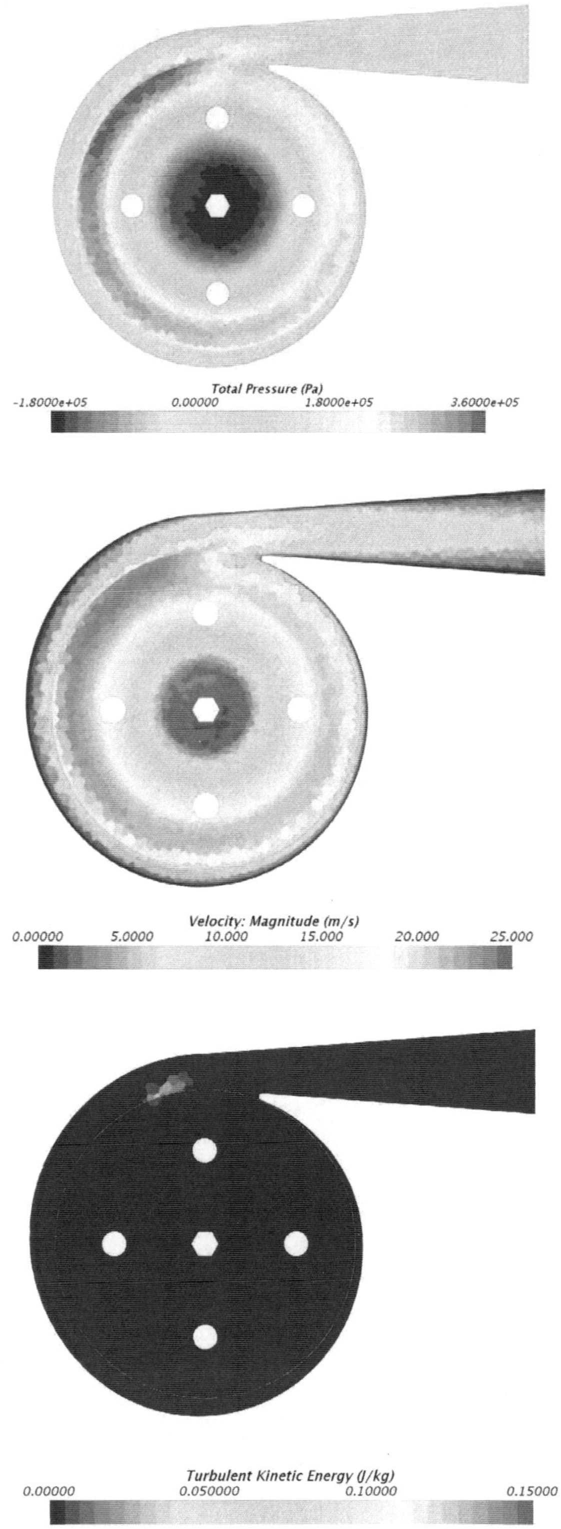

Figure 4. Scalar fields for the pump with initial volut.

IOP Conf. Series: Materials Science and Engineering **589** (2019) 012001 doi:10.1088/1757-899X/589/1/012001

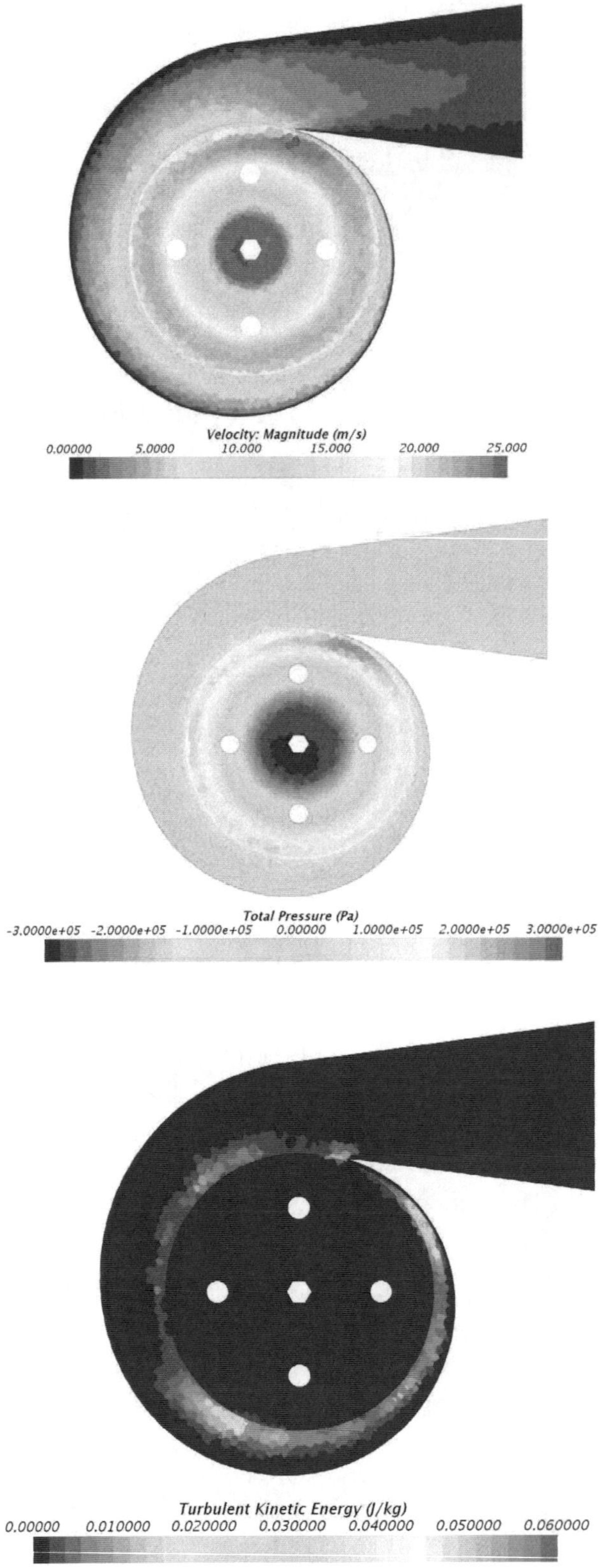

Figure5. Scalar fields for the pump with extended volute.

Table 3 shows a comparison of the characteristics of the pump with the initial output device and with the extended one. In addition, to confirm the possibility of using the k-ω SST turbulence model for laminar flows, the data obtained via the calculation of the optimized wet part in the laminar mode are presented.

Table3. Models comparison.

№	Model name	Mode	Efficiency, %	Head, m	Radial force, N
1	Initial model	Turbulent, k- ω SST	14	15,67	293,52
2	Optimized model	Turbulent, k- ω SST	27,26	26,87	78,36
3	Optimized model	Laminar	27,4	26,89	77,95

Figure 6 shows the distribution of scalar values for an extended diverter, calculated in laminar mode.

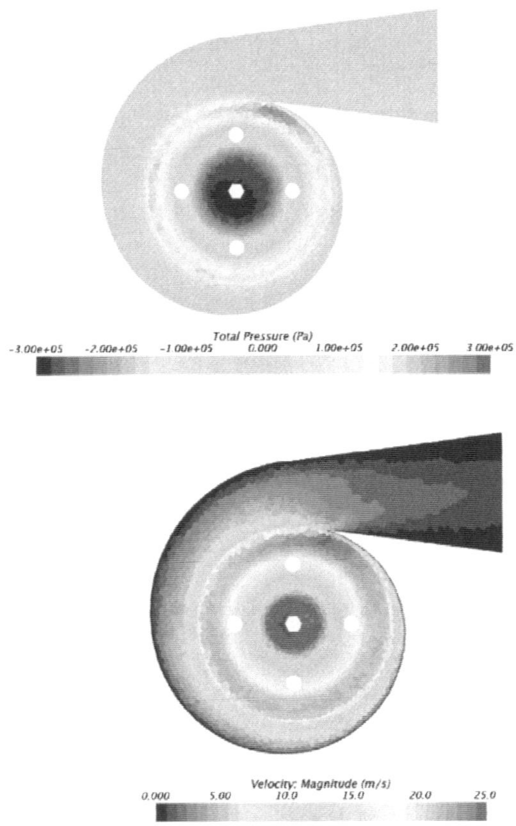

Figure6. Scalar fields distribution for laminar mode.

For a better understanding of the results of the work, characteristics of the pump with initial output device and optimized volute are shown on Figure 7. Red dots mark the efficiency dependence on the pumps flow and the blue ones - the head dependence.

IOP Conf. Series: Materials Science and Engineering **589** (2019) 012001 doi:10.1088/1757-899X/589/1/012001

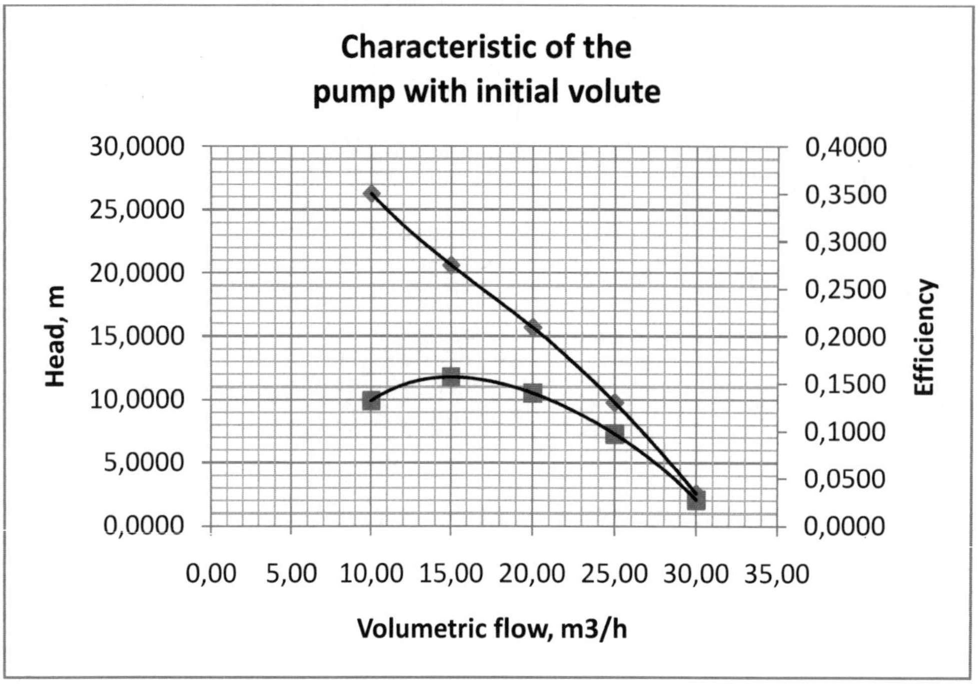

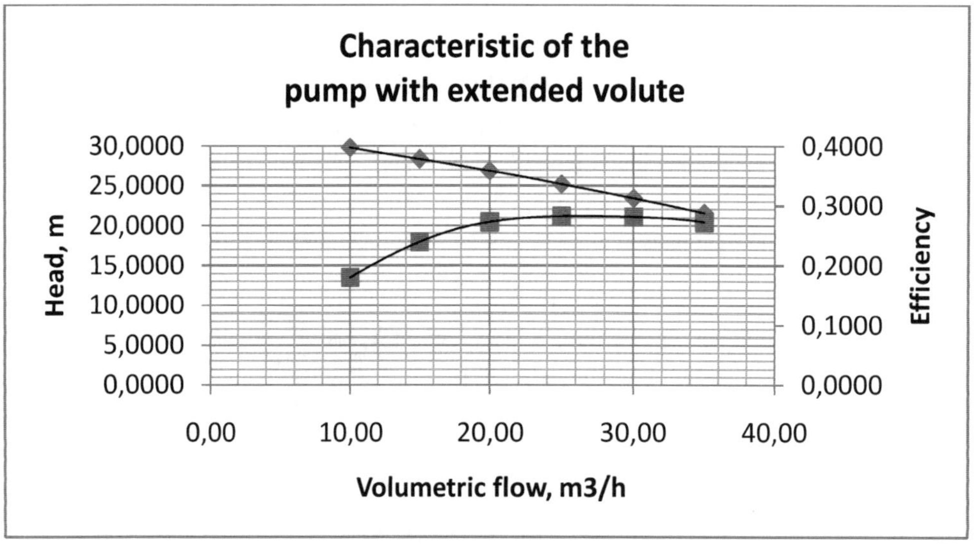

Figure7. Pump characteristics.

Conclusions

During the work it was shown that the volute designing according to classical methods (mainly for water) leads to a non - optimal design when dealing with high viscosities. This is apparently connected with the fact of speed momentum conservation hypothesis violation. The flow mode inside the wet part of the pump at a viscosity of 3000 cSt is laminar, as it was shown in the work, which causes difference of the velocity profile from that adopted in classical methods of volute designing.

The output device was optimized via two geometrical parameters variation - the width and height of the estimated section with the maximum efficiency criterion. According to the results of the calculations, the head was increased by 71.5 % and the efficiency - by 94.7 %. According to the presented graphsan obvious efficiency dependence on the height of the estimated section can be observed. The dependences on the width and estimated section volumetric capacity are not so evident, possibly due to the limitation of the varied geometrical parameters.

In the future, it is planned to expand the range of numerical research by adding the geometrical dimensions of the impeller too ther optimization parameters.

References

[1] Misyura V I, Ovsyannikov B V, Prisnyakov V F 1986 Discovyenasosy [Disc pumps] (Moscow, Mashinostroenie Publ.)

[2] Miller G E and Sidhu Amrita 1992 Multiple disk centrifugal pump as an artificial ventricle Proceedings of Engineering Mechanics pp 976–9

[3] Miller G E, Etter B D and Dorsi J M 1990 A multiple Dick Centrifugal Pump as a Blood Flow Device IEEE Transactions on Biomedical Engineering37
DOI: 10.1109/10.46255

[4] Crawford M E and Rice W 1974 Calculated Design Data For The Multiple-Disk Pump Using Incompressible Fluid *J Eng Power Trans ASME96*Ser A (3) pp 274–82

[5] ZehevTadmor, Pradip S. Mehta, Lefteris N. Valsamis and Jan-Chin Yang. *Corotating Disk Pumps for viscous Liquids* Farrel Company, Division of USM, Ansonia, Connecticut, 06401.

[6] Cheremushkin V A, Lomakin V O, Development and research of hydraulic disk pump 2019 *IOP Conf.Ser.:Mater.Sci.Eng.* 492 012039

[7] Kozlov S N, Petrov A I, Raschetiproektirovanieotvodyashchihustrojstvcentrobeznyxnasosov (BMSTU Publ.) pp 12-17

[8] Mihajlov A K, Malyushenko V V, Lopastnyenasosy.Teoriya, raschetikonstruirovanie. (Moscow, Mashinostroenie Publ.) 1997

[9] Lomakin V O 2015 Investigation of two-phase flow in axial-centrifugal impeller byhydrodynamic modeling methods *Proceedings of 2015 International Conference on FluidPower and Mechatronics,* FPM 2015 no 7337302 pp 1204–6

[10] Lomakin V. Complex optimization of the wet part of the hermetic pump by the LP-Tau search method // Pumps. Turbines. Systems. 2016, no. 1 (18), pp. 55-61

[11] Lomakin V, CheremushkinV and Chaburko P 2018 Investigation of vortex and hysteresis effects in the inlet device of a centrifugal pump 2018 Global Fluid Power Society PhD Symposium, GFPS, no 8472374 DOI: 10.1109/GFPS.2018.8472374

[12] Tesch K, and Kaczorowska-Ditrich K 2018 Discrete-continuous optimization of an axial flow blood pump *23rd Fluid Mechanics Conference, FMC 2018 (Zawiercie, Poland)* 1101 DOI: 10.1088/1742-6596/1101/1/012044

[13] Lomakin, V.O., Chaburko, P.S., Kuleshova, M.S. Multi-criteria Optimization of the Flow of a Centrifugal Pump on Energy and Vibroacoustic Characteristics // Procedia Engineering. Vol. 176, 2017, pp 476-482. DOI: 10.1016/j.proeng 2017.02.347.

IOP Conf. Series: Materials Science and Engineering **589** (2019) 012002 doi:10.1088/1757-899X/589/1/012002

Optimization of Liquid Rocket propulsion system for Reusable Orbital Transfer Vehicle

I Borovik[1], D Borovikov[1] and A Protopopov[2,3]

[1]Moscow Aviation Institute (National Research University), Moscow, Russian Federation
[2]Bauman Moscow State Technical University, 5 Second Baumanskaya Street, Moscow, 105005, Russian Federation

[3]Email: proforg6@yandex.ru

Abstract
The work is devoted to the investigation of possible ways to increase the efficiency of reusable orbital transfer vehicle (R-OTV) by using a multi-criteria optimization procedure of liquid rocket propulsion system (LRPS) design parameters. The "Federation" spacecraft is considered as an example payload of R-OTV for orbital transfer mathematical modeling, due to its known initial mass of 16,500 kg and other design parameters. Payload mass and specific cost of payload transfer are chosen as optimality criteria.
Multiple tasks are solved during the work: mathematical model and corresponding software for determining the optimal design parameters of R-OTV LRPS by mass and cost criteria are developed, R-OTV and LRPS calculation and optimization are conducted, the technical appearance of LRPS is designed.
Mathematical modeling is used to solve the presented problem. A developed mathematical model has a modular structure, in particular, it has modules for liquid rocket engine (LRE) mass calculating, OTV (Orbital Transfer Vehicle) mass calculating module, payload mass, design cost and manufacturing cost calculating modules for LRE and OTV and module for calculating the specific cost of payload transfer to target orbit. The intermediate result is used by LRPS project parameters design optimization module, then by technical appearance design module and LRPS units design module.
Multiple calculations are conducted with a developed mathematical model. Verification of the model is carried out by developing the technical appearance of real OTVLRPS. Several variants of LRPS for spacecraft "Federation" type OTV are designed for low lunar orbit, OTV parameters are compared with optimized LRPS and with existing LRPS.
Keywords: Liquid rocket engine, optimization, reusable orbital transfer vehicle.

Introduction

Traditionally, LRPS [1-5] design is based on LRE properties as a heat machine; this approach does not take into account close connection of the OTV main engine, which is the main unit of the system, with neighboring units and systems. When designing an optimal LRE, one should take into account the features of its functioning as part of the OTV. This approach will allow performing mission tasks with maximum efficiency.

The problem of reusability of the OTVLRPS in the environment of outer space arose at the dawn of the space age. At that time the primary task was to perform guaranteed launch of a spacecraft into a target orbit, the spacecraft was launched only one at a time, and there was no need for multiple uses of OTV. In the early 1980s, the creation of OTV was seriously involved in the United States [6-9]. In 1987, the company "Martin Marietta" proposed a project of a reusable OTV [8]. Also, the OTV project was developed by the Massachusetts Institute of Technology in 1994 [9].Several LRPS were developed on a competitive basis by Aerojet, Rocket dyne and Pratt&Whitney [10].

In Russia, two oxygen-hydrogen engines for R-OTV were developed. The first engine for OTV on oxygen-hydrogen propellant was the RD-56 developed by the Isaev`s Design Bureau (KBHM). A later modification of the RD-56 called KVD-1M had passed the full cycle of tests and is in normal operation as part of the last stage of the GSLV rocket (India) [11], [12]. The second engine for OTV on oxygen-hydrogen fuel is the RD-0146, developed in the 90s by Kosberg`s Design Bureau (KBKhA). The main difference from KVD-1M is that it uses expander-cycle of operation.

From the past year's projects analysis, it can be concluded that all projects of single-use and reusable oxygen-hydrogen OTV used as the main engine are expander cycle rocket engine.

Compared with R-OTV, OTV are quite expensive, butal though well-developed and reliable. According to experts [13], [14], an R-OTV will have a right to exist if its application for launching spacecraft to various orbits (including GSO) would be more cost-effective than using an OTV. Achieving this economic efficiency is possible only with sufficiently large payload quantity of transportation to the target orbit. The value of this payload quantity was predicted from hundreds to thousands of tons for the entire duration of the transportation program.

The modern stage of space exploration is characterized by the large-scale implementation of targeted integrated programs for spacecraft removal using inter orbital transport vehicles. For the period from 1993 to 2006, with the help of OTV, more than 20 spacecraft a year were launched only at the GSO. Most of them were implemented using the ESC-A (Ariane 4 and 5) and Centaur (Atlas 2, 3, 5, and Delta 2, 3, 4) rocket stages[15].

Nowadays interest in exploitation the Moon and building a long-term orbital station and an inhabited station on its surface and its orbit has increased all over the world again. [16], [17]. To ensure this project, it will be necessary to regularly transport into the orbit of the Moon both large-sized station modules and transport supply ships. In total, to ensure such a program, according to forecasts, it will be necessary to transport about 435 tons annually and about 740 tons at the deployment stage. One of the ways to significantly reduce the cost of cargo delivery is the orbital spaceport Figure 1, the construction project of which is being discussed by large aerospace companies [18].

Figure 1. An orbital spaceport in the S7 company's view.

The task of building such a structure can be made easier by reusable rocket stages for low orbit transfer. [19]. The use of such systems can reduce the unit cost of breeding by 3 times or to 1,000 $ / kg and lower. The use of reusable spaceplane figure 2 to supply the orbital refueling station with propellant and launch payloads for subsequent transportation to the target orbit will allow cargo to be moved even to high-energy orbits using reusable OTV, even taking into account their substantial loss in the mass of payloads in comparison with OTV.

Figure 2. Spaceplane.

Method

It had been theoretically proved that, for the transportation of cargo in space, it is most expedient to create R-OTV with the maximum mass of output payload with the minimum cost of launching from a low earth orbit into the target orbit. To fulfill these conditions, a wide range of factors should be taken into account when designing LRPS: propellant type; features of the LRPS design; characteristic velocity costs for orbital transition for which the R-OTV is created; many economic indicators; mass characteristics; reliability characteristics.

To take into account such a wide list of factors, a mathematical model consisting of several modules is developed, the general structure of which is shown in Figure 3.

Since the main criteria for optimizing the parameters of an OTV LRPS are the mass of the payload and the specific cost of transfer the payload into the target orbit for the OTV LRPS. In addition to the general requirements for the rocket engines fuel to ensure that these criteria are met.

Figure 4 shows a dependence of OTV payload mass as a function of the characteristic velocity of transfer into the target orbit, calculated using a developed mathematical model. It follows from this graph that the use of oxygen-hydrogen fuel, despite all its disadvantages, will be most effective for the OTV LRPS.

When choosing an OTV LRPS design, preference in the scheme should be given to expander-cycle liquid rocket engines. The advantages of this scheme are mainly in increased reliability, operating life and higher economic indicators associated with the manufacturing of LRE and are described in detail in [20]. For R-OTV, one of the main requirements is high reliability and resource of LRPS with a small mass of construction, it is the expander-cycle LRE design that is most preferable in comparison with other schemes.

The details of developed mathematical model work are described in detail in PhD thesis [21].

To carry out the verification of the model, calculations are made and the design of the expander-cycle oxygen-hydrogen fuel rocket engine is formed using design parameters of the RD-0146 engine taken from sources [22], [23], [20]. A comparison of the results showed good convergence, the error was not more than 10%

The main task of this work is the design of the technical appearance of R-OTV LRPS for "Federation" type space ship figure 5, with an initial mass of 16,500 kg, to bring the payload to low moon orbit.

IOP Conf. Series: Materials Science and Engineering **589** (2019) 012002 doi:10.1088/1757-899X/589/1/012002

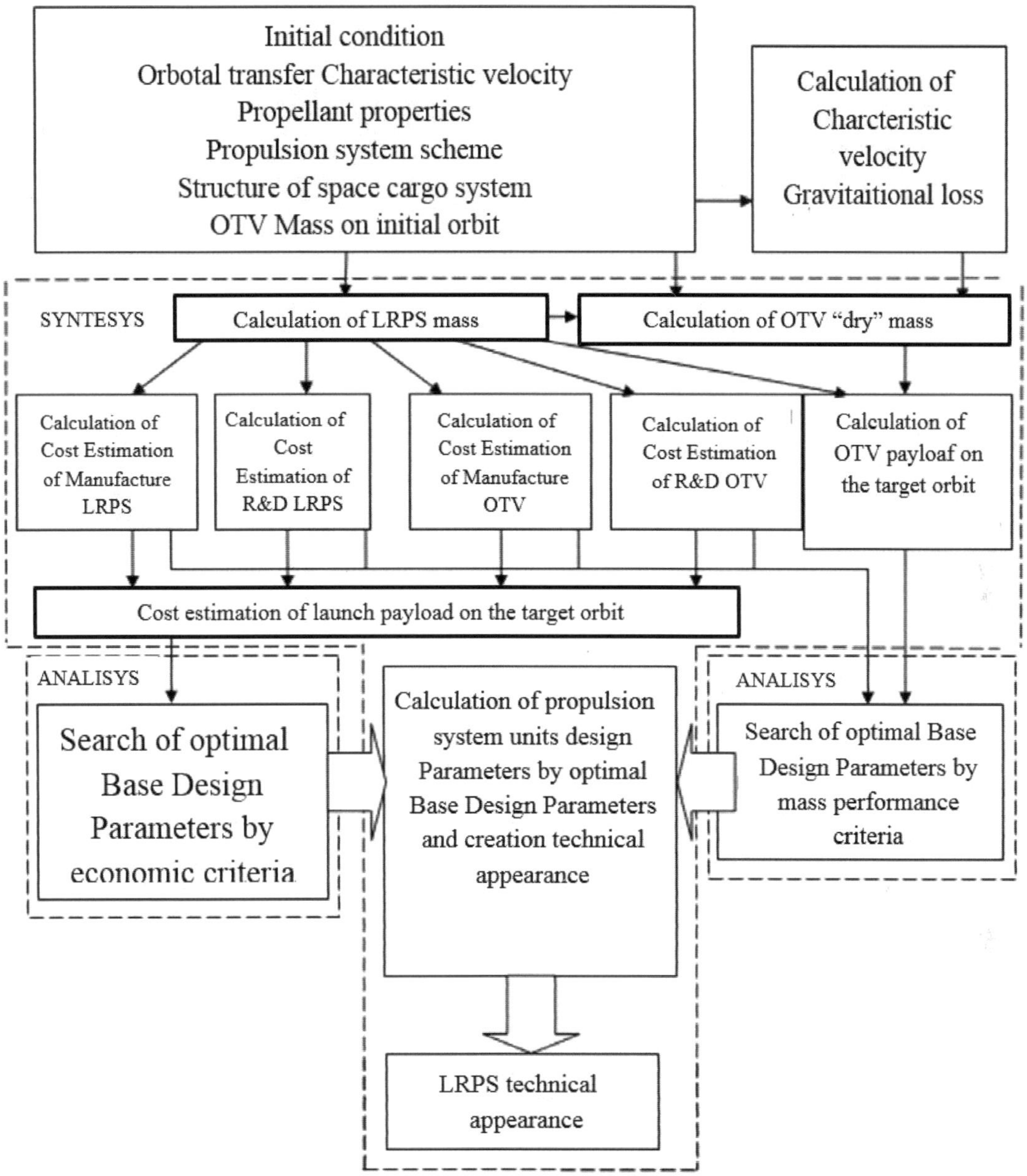

Figure 3. R-OTV LRPS optimal parameters search mathematical model structure.

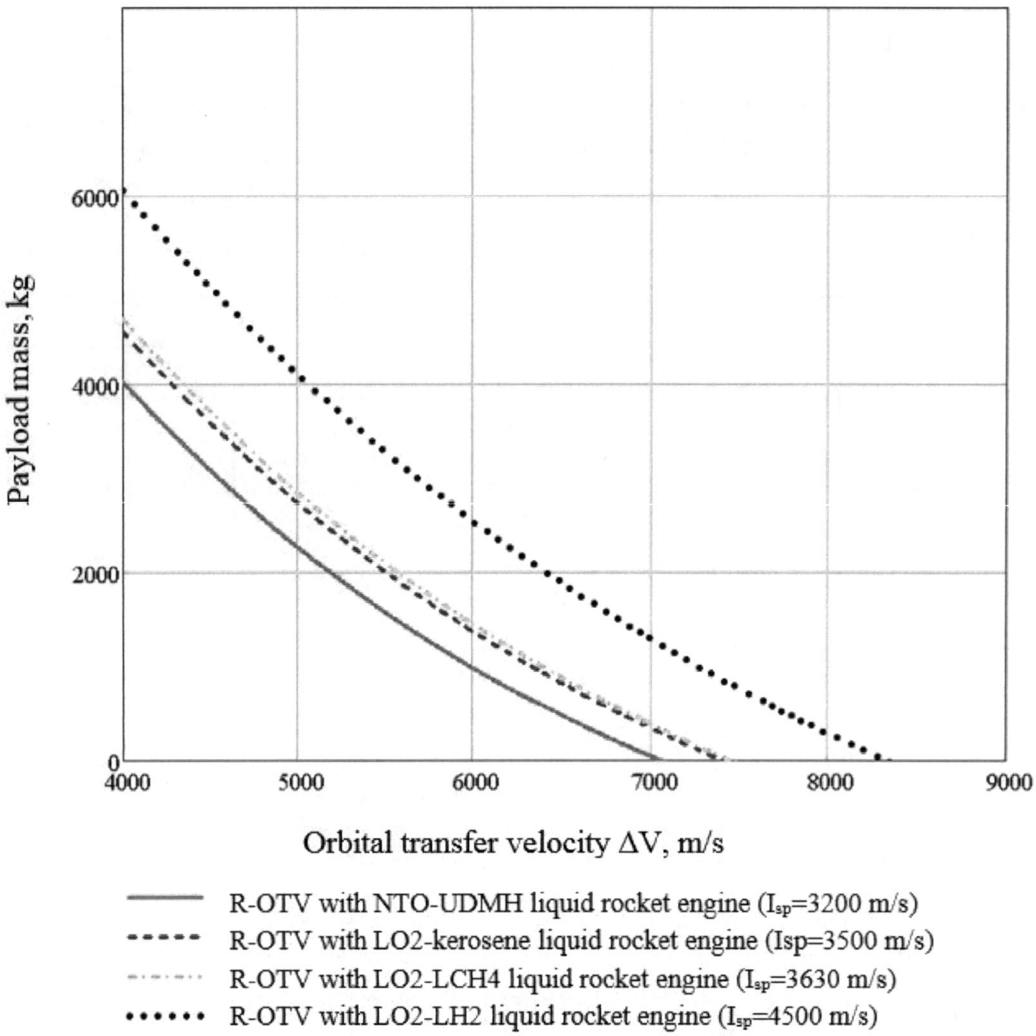

Figure 4. Comparison of the effectiveness of various propellant for the OTV LRPS, depending on mission energy requirements. The mass of OTV in the initial orbit is 22,000 kg.

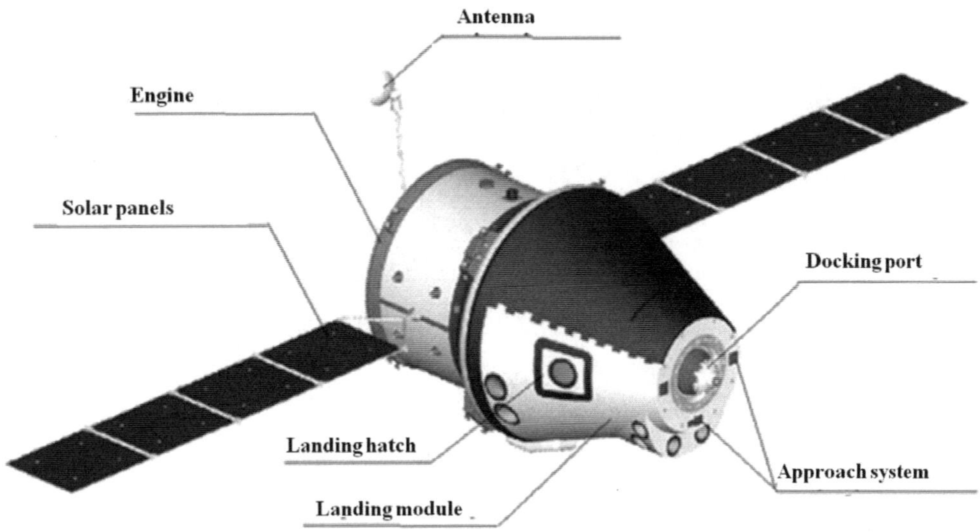

Figure 5."Federation" spaceship. The following parameters and assumptions are taken as the initial data:

- the value of the required characteristic velocity of the orbital transition is 4600 m / s;
- target orbit - low circumlunar orbit 110 km high;
- initial orbit - the height of 200 km, the inclination of the orbital plane of 51.6°;
- the diameter of fuel tanks and oxidizer is 4.1 m;

To facilitate the mode of operation of the turbo pump unit (TP) of hydrogen and provide a large resource and probability of failure-free operation, the rotational speed of the hydrogen TP rotor was set as a functional limitation - not more than 90000 rpm. To ensure the long-lasting durability of the nozzle and its resource, a functional limit was imposed on the maximum diameter of the nozzle outlet section equal to 2.2 m.

Results

The calculation results of the basic design parameters (BDP) of LRPS optimized for this R-OTV and their comparison with the LRP Son the basis of RD-0146 are shown in Table 1

Two variants of LRPS arrangement were considered - with one engine and annex tendable nozzle, and with four engines in an assembly. Obtained BDP allows to form the technical appearance of LRPS and calculate the design parameters of the propulsion system units.

In case of application of already developed LRE RD-0146, the mass of the payload transferred into the target orbit will be 9196 kg with 4 RD-0146 engines assembly and 2888 kg for one engine with an extendable nozzle, it is about 60% and 20% relatively optimized LRPS respectively. As a result, a significant increase in the R-OTV mass in the initial orbit will be required for the orbital transfer of spacecraft of a given mass for the OTV variants with RD-0146. Another disadvantage of the already developed RD-0146 is the high rotational speed of the rotor of the hydrogen turbopumps, 123000 rpm, 54% higher than in the optimized LRPS, which leads to a decrease in the reliability of the LRPS.

Main reasons for the low value of payload mass for R-OTV versions with RD-0146 are the low level of thrust-to-weight ratio (for the version with one RD-1046) and the large mass of the propulsion system with 4 RD-0146 (for the version with 4 RD- 0146). According to the calculated data, it is clear that a higher specific impulse of anassembly of 4 engines (4523 m / s) is not enough to compensate an increased mass of the propulsion system (974 kg), even despite the level of thrust close to optimal (392.8 kN).

IOP Conf. Series: Materials Science and Engineering **589** (2019) 012002 doi:10.1088/1757-899X/589/1/012002

Table 1. Basic design parameters (BDP) of the liquid propellant rocket engine optimized for the OTV, transferring a 16,500 kg of payload on low lunar orbit.

Parameter	Value for OTV with optimized LRPS	Value for OTV with optimized LRPS with expandable 2.2 m diameter nozzle	Value for OTV with LRPS based on 4 RD-0146	Value for OTV with LRPS based on RD-0146 with expandable 2.2 m diameter nozzle
Oxidizer excess ratio (oxidizer to fuel ratio)	0.917 (7.27)	0.923 (7.32)	0.743 (5.89)	0.743 (5.89)
Nozzleareaexpansionratio	121	157	200	600
Combustionchamberpressure, Pa	58.8 10^5	76 10^5	80 10^5	80 10^5
Propellant mass flow, kg / s	85	84.11	86.68	21.67
Hydrogen temperature before TP, K	296	260	300	300
The rotational speed of hydrogen TP, rpm	70868	79978	123000	123000
Combustion chamber and nozzle cooling jacket pressure drop, Pa	$30 \cdot 10^5$	$25 \cdot 10^5$	$30 \cdot 10^5$	$30 \cdot 10^5$
Thrust in vacuum, kN	374	373	392.8	99.8
Specific impulsein vacuum, m / s	4400	4441	4532	4605
Nozzleexit diameter, m	2.2	2.2	1.252	2.163
LRPSreliabilityat each start	0.993	0.996	0.996	0.996
LRPS maximum summary operation time, s	25530	25210	24390	101400
LRPS maximum on/off sequences during exploitation time	114	120	120	120
Guaranteed on/off sequences	1728	2748	2748	2748
Averageignitiontime, s	224	210	202.6	844.7
Cost of 1 kg of payload transfer into target orbit excluding development costs, $ / kg	11092	11062	18425	61436
Cost of 1 kg of payload transfer into target orbit taking into account development costs, $ / kg	13353	16560	20900	63728
Total cargo traffic to target orbit for the entire time of R-OTV operation, kg	913200	957600	95760	95760
Maximum payload mass transferred on the target orbit per launch, kg	16750	16510	9196	2888
Mass of "dry" OTV, kg	10570	9930	11170	10610
Propellanttanksmass, kg	6443	5946	6951	7579

Окончаниетабл. 1

Parameter	Value for OTV with optimized LRPS	Value for OTV with optimized LRPS with expandable 2.2 m diameter nozzle	Value for OTV with LRPS based on 4 RD-0146	Value for OTV with LRPS based on RD-0146 with expandable 2.2 m diameter nozzle
Mass of other systems, kg	2922	2791	2852	2991
Instrumentcompartmentmass, kg	549	516	516	516
LRE mass, kg	654	677	974	318
Oxidizer TP systems mass, kg	26.2	25.9	4.27	4.27
Fuel TP systems mass, kg	37	36.2	7.2	7.2
Mass of oxidizer TP, kg	28.8	39.64	18.5	18.5
Mass of fuel TP, kg	17.2	17.7	15.5	15.5
Mass of combustion chamber and nozzle with an extension mechanism, kg	348	361	117.5	190
Engine frames, automation units, and cable network mass, kg	197	196.56	80.5	82.5
Mass of OTV on the initial orbit, kg	141000	132000	132000	132000
Cost of developing OTV million. $.	2324	2392	2380	2196
Cost of development LRPS, million. $.	780	896	923	623
Cost of manufacturing OTV, million. $.	40	38	52	367
Cost of manufacturing LRE, million. $.	6.2	6.3	4	4
Number of OTV required for the implementation of the total cargo traffic	3	3	3	11
Number of OTV flights required for the implementation of the total cargo traffic	19	20	20	20

Although, the fact that cost of researching and developing the RD-0146 engine can be disregarded, the specific cost of orbital transferring the R-OTV versions with RD-0146 (18428 $ / kg for 4 RD-0146 and 61436 $ / kg for single RD-0146) more than R-OTV with optimized LRPS (16560 $ / kg), even if we take into account the cost of its development.

The cost of developing an R-OTV is approximately equal for the variants with optimized LRPS and with 4 RD-0146, and for the variant with one RD-0146, it is 10% less, which is within the model error. The cost of development for an optimized LRPS and LRPS with 4 RD-0146 is about the same, and for the version with one RD-0146, is about 30% less.

The cost of manufacturing R-OTV with 4 RD-0146 is more than the other two options due to costs of manufacturing and installation of 4 engines, their integration into R-OTV design and increased dry weight of R-OTV. An optimized LRPS manufacturing cost is higher than the cost of manufacturing RD-0146 due to the greater mass of the engine.

In case of using anassembly of 4 RD-0146, the mission may be completed only by increasing the sizes of R-OTV and propellant supply, which will entail an increase in developing and manufacturing costs of R-OTV due to increased dry mass, and require the use of a heavier-class rocket to transfer R-OTV into orbit.

Orbital transfers with a large value of characteristic velocity require a high energy efficiency of LRPS and reliable operation, despite the multiple on/off sequences and high level of heat and vibration protection for LRPS units. Entered functional limitations allowed to find BDP at which such level is reached - the rotational speed of hydrogen TP rotor was less than 80,000 rpm and required a temperature of hydrogen after the nozzle cooling jacket is 260 K.

Although, the relatively low heating temperature of hydrogen before of TP turbines, the pressure in the combustion chamber will be $76 \cdot 10^5$ Pa. This is due to the high value of the required fuel turbine efficiency 0.718. Obtaining such efficiency in a real construction may cause difficulties, in already existing hydrogen turbines with a lower mass flow efficiency level is 0.68-0.69. To eliminate these shortcomings, an optimal BDP are re-calculated with a new functional limitation - the efficiency of hydrogen TP turbine should not exceed 0.7.

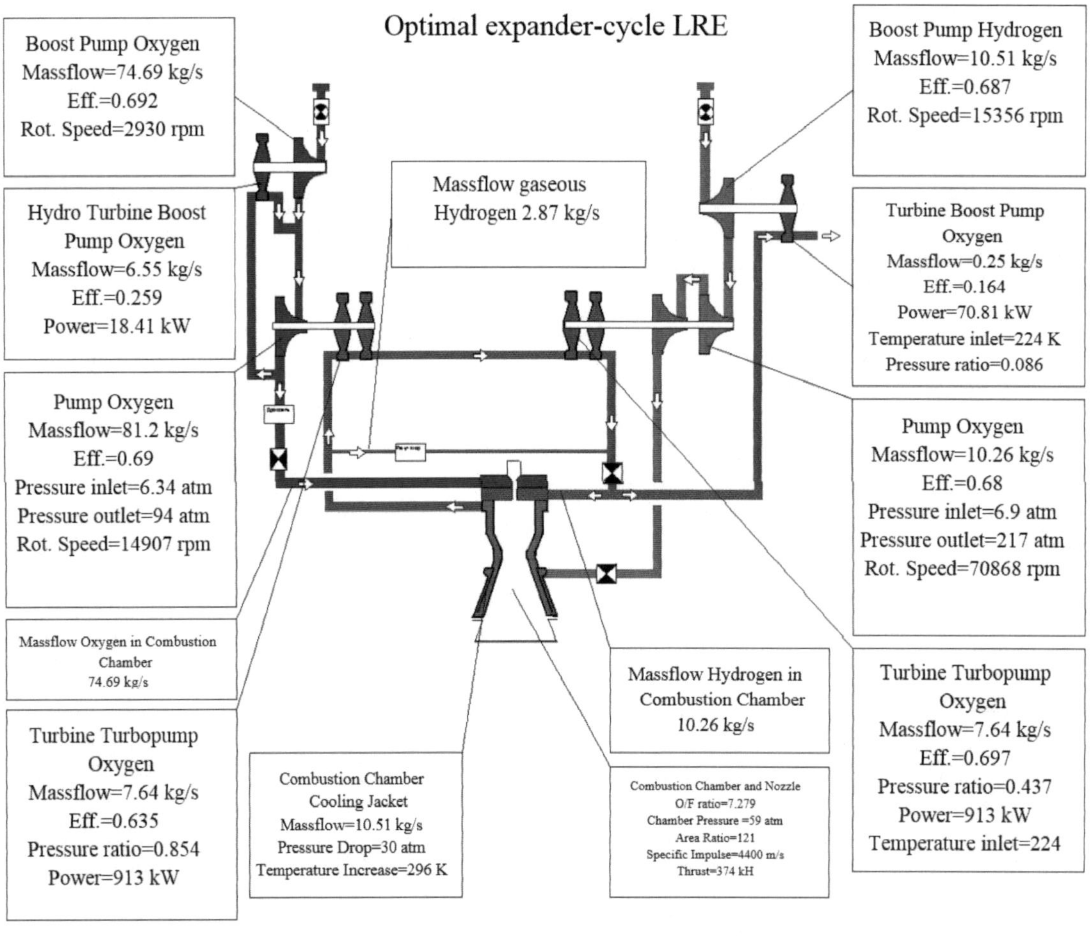

Figure 6. The technical appearance of "Federation" type R-OTV LRPS.

Figure 6 shows the technical appearance of LRPS with less heat-stressed parameters of hydraulic system power supply units. In this technical aspect, hydrogen TP turbines efficiency is 0.695, and rotational speed of its rotor is 70868 rpm. Such a decrease in required efficiency became possible due to an increase of hydrogen temperature after the cooling jacket, the hydrogen temperature is increased to 296 K, and higher value of turbine flow rate 7.64 kg / s. Combustion chamber hydrogen consumption is 10.11 kg / s, therefore, the excess hydrogen can either bypass turbines, bypass combustion chamber and nozzle cooling jacket, and immediately go to the combustion chamber from the first stage of the hydrogen pump. Of two optimized variants, less heat and vibro-stressed is more preferable; it more closely meets the reliability requirements due to the lower rotational speeds of the TP rotors. The advantages can also be attributed to the lower mass of such LRPS and lower cost of its development and production.

Disadvantages include an increased R-OTV mass on the initial orbit, up to 141,000 kg. Due to a decrease in the specific impulse from 4441 m / s to 4400 m / s, It happened due to a decreased combustion chamber pressure from $76 \cdot 10^5$ Paup to $58.8 \cdot 10^5$ Pa. The specific cost of OTV low earth orbit launch is unchanged. Cost of R-OTV and LRE design and manufacturing decreased due to a decrease in the mass of the major LRE and a decrease in the level of the required failure-free operation probability of LRPS during each on/off, and consequently, the required number of tests.

Conclusion

Optimization of OTV LRPS parameters for a specific mission allows to increase the payload mass and reduce specific cost compared to the OTV with LRPS based on an already existing engine.

For the considered mission, the use of an optimized LRPS may increase payload mass 1.8 times and at the same time reduce the specific cost of mission 1.35 times compared to the non-optimized LRPS.

The developed technical appearance of OTV LRPS provides increased reliability, resource, simplicity, and efficiency of the design.

References

[1] Pelevin, F. V., Avraamov, N. I., Ir'yanov, N. Y., Orlin, S. A., Lozovetskii, V. V., &Ponomarev, A. V. (2018). Intensification of heat exchange in the regenerative cooling system of a liquid-propellant rocket engine. Journal of Engineering Physics and Thermophysics, 91(3), 601-610. doi:10.1007/s10891-018-1781-4

[2] Pelevin, F. V., &Ponomarev, A. V. (2018). To calculation of regenerative cooling of a liquid fuel rocket engine chamber. Russian Aeronautics, 61(1), 74-80. doi:10.3103/S1068799818010117

[3] Mullin, N. N., Balugin, N. N., Morozov, K. S., Kalinkin, S. V., &Toporkov, A. G. (2018). Small meteorological rocket launch for student project payload with bio-material. Paper presented at the Proceedings of the International Astronautical Congress, IAC

[4] Aref'ev, K. Y., Voronetskii, A. V., Il'chenko, M. A., &Suchkov, S. A. (2017). Numerical and experimental study of ignition of a two-phase fuel composition (ethanol + air) in a resonance gas-dynamic system. Combustion, Explosion and Shock Waves, 53(4), 398-405. doi:10.1134/S0010508217040037

[5] Dorofeev, A. A., &Yagodnikov, D. A. (2018). Thermodynamic modeling of the composition and characteristics of combustion products of overrich liquid rocket fluids in the quenching mode. *High Temperature, 56*(2), 263-269. doi:10.1134/S0018151X18010066

[6] Serdyuk V.K., Tolyarenko N.V. Interorbital apparatus. "Rocket-making and space technology" (Results of science and technology, vol. 9, VINITI, Academy of Sciences of the USSR), M., 1985, 9, 3-288(In Russian).

[7] Huffaker F., Kelly DL Next Generation In-Space Transportation Systems. Briefing for NASA / Pennsylvania State University Transportation Propulsion Symposium, Huntsville, Alabama, June 25-28, 1990

IOP Conf. Series: Materials Science and Engineering **589** (2019) 012002 doi:10.1088/1757-899X/589/1/012002

[8] Orbital Transfer Vehicle: Concept Definition and System Analysis Study. Midterm Review (Martin Marietta Corp.) NASA-CR-183551. 112 p. 22 July 1987

[9] Project Freebird. An Orbital Transfer Vehicle. MIT. Spring 1994. NASA-CR-197201. 338 p.

[10] Shlyakhov V.I., Ovchinnikova S.V. Expander-cycle LREfor space tugs. Review of materials of the foreign press for 1980-1990. N 30. Center of scientific and technical information "Search", GONTI-8. 1991, pp. 54-56(In Russian).

[11] AIAA, International Reference Guide to Space Launch Systems, 3rd edition.

[12] BA Palaszewski, Upper Stages Using Liquid Propulsion, Metallized Propellants, Lewis Research Center, NASA Technical Paper 3191 (February 1992).

[13] Cooper LP, space-based orbital transfer vehicle. "Acta astronaut.", 1988, No. 5, 515-529 p.

[14] Galabova K. "The Scenarios of Architects of the Spaces Bounding-on-Orbital Transfer Scenarios," MS Thesis, Dept. Aeronautics and Astronautics, MIT, Cambridge, MA, 2004

[15] 2006 Commercial Space Transportation Forecasts May 2006. Federal Aviation Administration (COMSTAC). http: // ast.faa.gov.

[16] Dallas Bienhoff LEO Propellant Depot: A Commercial Opportunity?, Boeing Company, LEAG Private Sector Involvement October 1 - 5, 2007 Houston, Texas

[17] Bernard F. Kutter, Frank Zegler, Jon Barr, Mari Gravlee, Jake Szatkowski, Jeff Patton, Scott Ward. Ongoing Launch Vehicle Innovation at United Launch Alliance. IEEE 2010-1020. United Launch Alliance, 9100 E. Mineral Circle Denver, Co 80127

[18] Roscosmos. Orbital cosmodrome. [site]. URL:https: //www.roscosmos.ru/media/files/docs/2017/SpAsBus/sopov_vizionerstvo.pdf

[19] Afanasyev V. A.,Degtyarev G. L., Meshchanov A. S. The reusable space transportation system. Proceedings of higher educational institutions. Aviation technology. No. 3 2018. C 9-13.(In Russian).

[20] Y. Demyanenko, A. Dmitrenko, A. Ivanov, V. Pershin, A. Shostak, G. Zelkind, A. Minick, R. Bracken, Ground Test Demonstrator Engine Boost Turbopumps Design and Development. 41st AIAA / ASME / SAE / ASEE Joint Propulsion Conference & Exhibit 10-13 July 2005, Tucson, Arizona. AIAA 2005-3945.

[21] Borovik I.N. Development of the technical appearance of the reusable orbital transfer vehicle propulsion system: Ph.D. dis. Moscow Aviation Institute (State Technical University), Moscow, 2011. (In Russian).

[22] Computer models of liquid rocket engines / E.V. Lebedinsky, S.V. Mosolov, G.P. Kalmykov, etc .; by ed. Academician A.S.Koroteev. M .: Mashinostroenie, 2009. 376s .: ill.

[23] Rocket Propulsion Elements / by George P. Sutton, Oscar Biblarz.-7th ed. 2001. 764 p.

IOP Conf. Series: Materials Science and Engineering **589** (2019) 012003 doi:10.1088/1757-899X/589/1/012003

Finding the minimum number of optimization points for an axial aircraft pump

A Protopopov[1,2], A Mukhlaeva[1] and D Vdovin[1]

[1]Bauman Moscow State Technical University, 5 Second Baumanskaya Street, Moscow, 105005, Russian Federation

[2]E-mail: proforg6@yandex.ru

Annotation
This article discusses issues related to the calculation of aircraft axial pumps. These pumps are subject to ever-increasing demands due to high competition in the aviation industry. The highest requirements for parameters are the dimensions of the impeller and power consumption. This article discusses the method of constructing the diameter of the impeller - the expended power of an axial pump.

Introduction

Methods for calculating centrifugal pumps [1] - [4] are widely reported in the literature. Of greatest interest are publications [5] - [11]. However, the calculation of axial pumps is much harder than described in the literature. The main characteristics of hydraulic machines for aeronautical engineering include their mass and power consumption, and the mass should be minimal, and the efficiency should be maximal. As the size of hydraulic machines increases, both parameters increase, so it is necessary to find the optimal diameter of the pump, at which rather high efficiency is achieved while maintaining a low mass, a priority parameter for aviation. Additionally, using high-speed machines with relatively low pressure can reduce the pump mass, so in this study, we consider a hydraulic unit consisting of an axial pump and an axial-piston motor with an inclined disk. Since the mass is directly related to the dimensions of the hydraulic machine, and the assessment of dimensions is a simpler task, we will indirectly estimate the mass by the radial dimension of the impeller of the axial pump. At the same time, since the efficiency is directly dependent on the power expended, the change in efficiency will be estimated by the change in power consumed.

In the literature [12] - [15], hydrodynamic modeling methods are commonly used to solve the problem of calculating dynamic pumps. The disadvantages of these methods can be attributed to the high costs of human and machine time, which necessitates the creation of a calculation methodology in which it is not necessary to resort to hydrodynamic modeling. Note that the literature [1] - [4] demonstrates the excellent ability to optimize dynamic hydraulic machines using the LP-Tau-search methods. Thus, it is possible to combine known methods for calculating dynamic hydraulic machines, such as the LP-Tau-search method, with unconditional optimization by any parameter and eliminate direct hydrodynamic modeling to save the machine and human time.

Unconditional optimization is carried out according to two criteria - dimensions and power consumption. Both parameters are made dimensionless by reference to the highest value of each of the requirements, and their weights are assumed to be equal. To eliminate the influence of the sign of these parameters, both of the non-measured criteria count in the square.

IOP Conf. Series: Materials Science and Engineering **589** (2019) 012003 doi:10.1088/1757-899X/589/1/012003

Methods

To optimize the method, you need to set the number of points for the application of LP-Tau-search required for accurate calculation, which does not require the excessive expenditure of time and PC resources. Note that the accuracy of the LP-Tau sequence calculation has its limitations since it carries out the calculation with a margin, taking into account the non-rigid characteristic of the axial pump and allowing for an error.

The number of calculated points is as follows. A dimensionless generalized quality factor Φ is introduced, which is determined according to the dependence:

$$\Phi = \left(\frac{D_{\text{тр}}}{D_{\text{тр max}}}\right)^2 + \left(\frac{N_{\text{з}}}{N_{\text{з max}}}\right)^2,$$

$$F = \left(\frac{D_t}{D_{t\,max}}\right)^2 + \left(\frac{N}{N_{max}}\right)^2,$$

Here, $D_{t\ max}$ and N_{max} are the values of the maximum diameter and maximum power consumption. The smaller this ratio, the better the quality of the axial pump.

After calculating the generalized quality factor for 64 points, select its minimum values for a certain number of points: 4, 8, 16, 32, and 64. Then build the dependence where is the minimum value of the generalized quality factor for a given number of points; n — number of points.

Table 1. Calculated points of power consumption and diameter of the pipeline equal to the diameter of the pump

Point	N_{max}, Vt	D_t, m	F
1	861	0,047	0,220
2	454	0,043	0,177
3	1892	0,057	0,362
4	438	0,048	0,220
5	2795	0,072	0,610
6	948	0,053	0,279
7	756	0,036	0,131
8	278	0,042	0,167
9	1024	0,049	0,243
10	1505	0,068	0,471
11	879	0,040	0,163
12	855	0,060	0,351
13	688	0,038	0,143
14	730	0,046	0,208
15	5419	0,082	1,089
16	424	0,052	0,258
17	5217	0,099	1,346
18	1062	0,060	0,357
19	771	0,038	0,145
20	1378	0,075	0,539
21	800	0,041	0,168
22	472	0,038	0,140
23	1457	0,047	0,241

IOP Conf. Series: Materials Science and Engineering **589** (2019) 012003 doi:10.1088/1757-899X/589/1/012003

Point	N_{max}, Vt	D_t, m	F
24	528	0,053	0,269
25	545	0,036	0,127
26	532	0,043	0,179
27	2257	0,059	0,407
28	393	0,041	0,161
29	1436	0,052	0,287
30	2879	0,080	0,732
31	1118	0,041	0,178
32	291	0,045	0,192
33	1335	0,056	0,323
34	3268	0,098	1,071
35	1024	0,043	0,191
36	623	0,055	0,291
37	594	0,036	0,128
38	596	0,043	0,180
39	2791	0,063	0,495
40	896	0,068	0,448
41	677	0,040	0,158
42	400	0,038	0,139
43	1190	0,045	0,213
44	612	0,051	0,251
45	3267	0,074	0,682
46	1073	0,053	0,283
47	814	0,036	0,132
48	768	0,066	0,420
49	633	0,040	0,157
50	662	0,049	0,233
51	8031	0,103	2,000
52	350	0,041	0,160
53	1244	0,050	0,26
54	2104	0,073	0,571
55	1015	0,04	0,167
56	386	0,046	0,202
57	1957	0,063	0,433
58	797	0,050	0,246
59	696	0,035	0,123
60	2413	0,089	0,837
61	952	0,043	0,188
62	556	0,039	0,148
63	1829	0,050	0,288
64	335	0,049	0,228

This scatter of points allows us to obtain a graph of the dependence of the generalized quality factor on the number of points.

Results

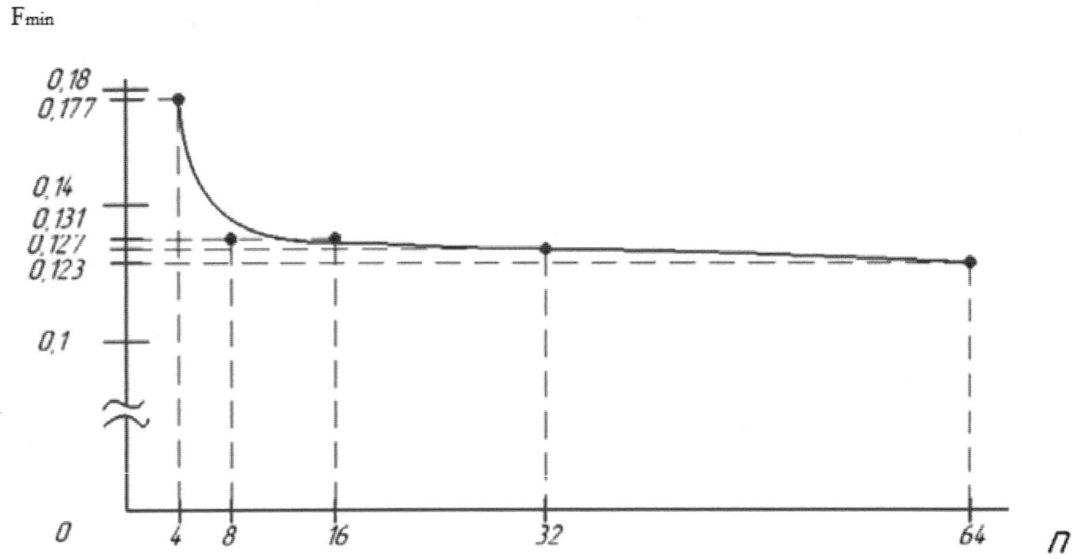

Figure 1. Scatter of the chosen points

For the successful application of LP-Tau-sequence when finding a compromise between mass and power consumption with varying head and relative time of work, at least 32 points must be used. The example of the obtained curve shows this - when considering 64 points of the LP-tau sequence, the graph slowly approaches the shape of the horizontal line for n> 16.

Conclusion

Thus, the above method can be recommended for use in calculating high-speed axial pumps of low power with a relatively large number of points and the availability of software packages that can facilitate data processing and graphical construction of a trade-off curve. It is evident that a further increase in the number of points obtained by the LP-Tau-search method will not lead to a significant improvement in the characteristics of the axial pump. At the same time, the rejection of hydrodynamic modeling in favor of simpler estimated calculations, even for the case of 64 design points, will save the machine and human time, provided that automatic methods for calculating the main design parameters of axial pumps are in use.

List of references

[1] P Chaburko and Z Kossova 2019 *IOP Conf. Ser.: Mater. Sci. Eng.* **492** 012011
[2] V Lomakin et al 2019 *IOP Conf. Ser.: Mater. Sci. Eng.* **492** 012012
[3] A Gouskov et al 2019 *IOP Conf. Ser.: Mater. Sci. Eng.* **492** 012013
[4] A Shablovskiy and E Kutovoy 2019 *IOP Conf. Ser.: Mater. Sci. Eng.* **492** 012035
[5] Zhang, S., Li, H., & Xi, D. (2019). Investigation of the integrated model of side chamber, wear-rings clearance, and balancing holes for centrifugal pumps. *Journal of Fluids Engineering, Transactions of the ASME, 141*(10) doi:10.1115/1.4043059
[6] Sengpanich, K., Bohez, E. L. J., Thongkruer, P., &Sakulphan, K. (2019). New mode to operate centrifugal pump as impulse turbine. Renewable Energy, 983-993. doi:10.1016/j.renene.2019.03.116
[7] Zhang, Z. -., Chen, H. -., Ma, Z., He, J. -., Liu, H., & Liu, C. (2019). Research on improving the dynamic performance of centrifugal pumps with twisted gap drainage blades. Journal of Fluids Engineering, Transactions of the ASME, 141(9) doi:10.1115/1.4042885

[8] Pirouzpanah, S., Patil, A., Chen, Y., & Morrison, G. (2019). Predictive erosion model for mixed flow centrifugal pump. Journal of Energy Resources Technology, Transactions of the ASME, 141(9) doi:10.1115/1.4043135

[9] Yousefi, H., Noorollahi, Y., Tahani, M., Fahimi, R., &Saremian, S. (2019). Numerical simulation for obtaining optimal impeller's blade parameters of a centrifugal pump for high-viscosity fluid pumping. Sustainable Energy Technologies and Assessments, 34, 16-26. doi:10.1016/j.seta.2019.04.011

[10] Blume, M., & Skoda, R. (2019). 3D flow simulation of a circular leading edge hydrofoil and assessment of cavitation erosion by the statistical evaluation of void collapses and cavitation structures. Wear, 428-429, 457-469. doi:10.1016/j.wear.2019.04.011

[11] Yun, R., Zuchao, Z., Denghao, W., &Xiaojun, L. (2019). Influence of guide ring on energy loss in a multistage centrifugal pump. Journal of Fluids Engineering, Transactions of the ASME, 141(6) doi:10.1115/1.4041876

[12] A Petrov et al 2019 *IOP Conf. Ser.: Mater. Sci. Eng.* **492** 012036

[13] V Lomakin and O Bibik 2019 *IOP Conf. Ser.: Mater. Sci. Eng.***492** 012037

[14] T Valiev and A Petrov 2019 *IOP Conf. Ser.: Mater. Sci. Eng.***492** 012038

[15] V Cheremushkin and V Lomakin 2019 *IOP Conf. Ser.: Mater. Sci. Eng.***492** 012039

Optimization of axial volumetric drive aviation pump

A Protopopov[1,2], A Mukhlaeva[1] and E Melnichuk[1]

[1]Bauman Moscow State Technical University, 5 Second Baumanskaya Street, Moscow, 105005, Russian Federation

[2]E-mail: proforg6@yandex.ru

Annotation
This article discusses a volumetric drive aircraft axial pump and solves the problem of its optimization by the Sobol method. The optimization criteria were the radial dimension and the power expended. In this case, the non-uniformity of the pump operation over time and the variable value of the pressure consumed by the network were taken into account. The last two parameters act as constructive variable parameters. As a result of the work, we obtained a compromise curve. The radial dimension is the power expended.

Introduction

Methods for calculating dynamic pumps [1] - [4] are widely reported in the literature. Of greatest interest are publications [5] - [12].

When developing axial pumps for aviation technology, the essential characteristics are mass and power consumption and both parameters should be as small as possible. As the size of the pump impeller increases, its efficiency increases, thereby reducing power consumption. But at the same time growing mass. As both parameters increase, it is necessary to find the optimal diameter of the pump with high efficiency, provided a relatively small mass - a priority parameter for aviation. Additionally, the mass can be reduced by using high-speed machines with relatively low pressure. Therefore, this study considers a hydraulic unit consisting of an axial pump and an axial-piston motor with an inclined disk.

The indicated problem is widely reported in the literature [13] - [16], where the main emphasis is on the use of labor-intensive methods, such as hydrodynamic modeling. In this work, a less labor-consuming LP-Tau search method is used in its application to the considered area of mechanical engineering.

According to the efficiency of a small axial-piston hydraulic motor, which rotates the pump, it can be assumed constant; the same applies to the mechanical efficiency of an axial pump. Its volumetric efficiency is approximately equal to 1, which simplifies the task, and the hydraulic efficiency can be estimated using the Lomakin formula. The required flow rate of the axial pump determines the power consumption, which is thus the focus of the paper. The mass and power consumption of the pump, among other parameters, are influenced by the pressure and the relative operating time of the hydraulic unit, the reduction of which entails an increase in the pump flow rate and, therefore, pressure losses along the length of the pipeline.

Methods

For the analysis, we use LP-Tau search, since this method does not imply a precise definition of the objective function. It generates points in a quasi-random manner in the selected interval for the two parameters we are considering — head and relative pump operation time: the relationship between pump durations and aircraft flight time. The head H can vary from 5 to 15 m; the relative operating time of the tp / in is from 0.05 to 1. Then the field of generated work points will be a rectangle (Fig. 1).

The calculated points obtained using LP-Tau search are in the table.

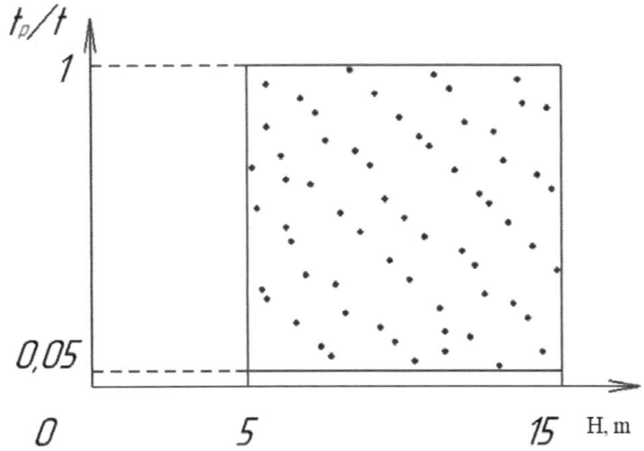

Figure 1. Point generation area.

Table 1. The calculated points of the LP-Tau search

Point #	*H*, m	t_p/t	Point #	*H*, м	t_p/t	Point #	*H*, m	t_p/t	Point #	*H*, m	t_p/t
1	10	0,5250	17	10,3125	0,0796	33	10,156	0,3320	49	10,46	0,7773
2	7,5	0,7625	18	7,8125	0,3171	34	7,6562	0,0945	50	7,9687	0,5398
3	12,5	0,2875	19	12,8125	0,7921	35	12,656	0,5695	51	12,96	0,0648
4	6,25	0,6437	20	6,5625	0,1984	36	6,4062	0,4507	52	6,7187	0,8960
5	11,25	0,1687	21	11,5625	0,6734	37	11,406	0,9257	53	11,718	0,4210
6	8,75	0,4062	22	9,0625	0,9109	38	8,9062	0,6882	54	9,2187	0,1835
7	13,75	0,8812	23	14,0625	0,4359	39	13,906	0,2132	55	14,218	0,6585
8	5,625	0,9406	24	5,9375	0,4953	40	5,7812	0,2722	56	6,0937	0,7179
9	10,625	0,4656	25	10,9375	0,9703	41	10,781	0,7476	57	11,093	0,2429
10	8,125	0,2285	26	8,4375	0,7328	42	8,2812	0,9851	58	8,593	0,4804
11	13,125	0,7031	27	13,4375	0,2578	43	13,281	0,5101	59	13,593	0,9554
12	6,875	0,3468	28	7,1875	0,8515	44	7,0312	0,6289	60	7,3437	0,1242
13	11,875	0,8218	29	12,1875	0,3765	45	12,031	0,1539	61	12,34	0,5975
14	9,375	0,5843	30	9,6875	0,1390	46	9,5312	0,3914	62	9,8437	0,8368
15	14,375	0,1093	31	14,6875	0,6140	47	14,531	0,8664	63	14,843	0,3617
16	5,3125	0,5546	32	5,15625	0,8070	48	5,4687	0,3023	64	5,0781	0,6808

Each point corresponds to values of the diameter of the pipeline, equal to the diameter of the impeller of an axial pump, and the power expended, which can be obtained by the following:

$$D_t = \frac{0{,}0827 Q^2 L_p \lambda}{H},$$

$$N = \frac{\rho g Q H}{\eta}.$$

After calculations (Table 2), it is possible to construct the distribution of all calculated points in the coordinatesD_t —N.

Table 2. Calculated points of power consumption and diameter of the pipeline equal to the diameter of the pump

Point #	*N*, Vt	D, m	Point #	*N*, Vt	D, m	Point #	*N*, Vt	D, m	Point #	*N*, Vt	D, m
1	861	0,047	17	5217	0,099	33	1335	0,056	49	634	0,04
2	454	0,043	18	1062	0,06	34	3268	0,098	50	663	0,049
3	1892	0,057	19	771	0,038	35	1024	0,043	51	8031	0,103
4	438	0,048	20	1378	0,075	36	624	0,055	52	350	0,041
5	2795	0,072	21	800	0,041	37	594	0,036	53	1244	0,05
6	949	0,053	22	473	0,038	38	597	0,043	54	2104	0,073
7	756	0,036	23	1457	0,047	39	2791	0,063	55	1015	0,04
8	278	0,042	24	528	0,053	40	897	0,068	56	386	0.046
9	1024	0,049	25	545	0,036	41	677	0,04	57	1957	0,063
10	1505	0,068	26	533	0,043	42	401	0,038	58	798	0,05
11	880	0,04	27	2257	0,059	43	1190	0,045	59	696	0,035
12	856	0,06	28	393	0,041	44	613	0,051	60	2413	0,089
13	689	0,038	29	1436	0,052	45	3267	0,074	61	952	0,043
14	731	0,046	30	2879	0,08	46	1073	0,053	62	556	0,039
15	5419	0,082	31	1118	0,041	47	814	0,036	63	1829	0,05
16	424	0,052	32	292	0,045	48	768	0,066	64	335	0,049

The obtained results allow us to construct a compromise curve.

Results

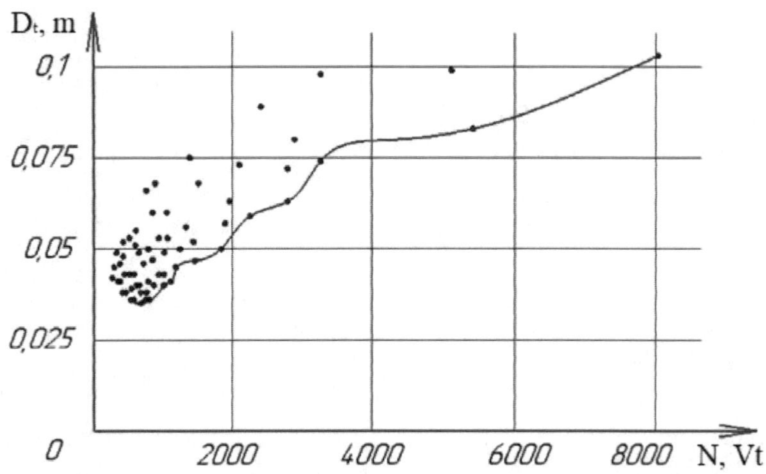

Figure 2. Distribution of calculated points and a compromise curve constructed from them.

IOP Conf. Series: Materials Science and Engineering **589** (2019) 012004 doi:10.1088/1757-899X/589/1/012004

According to this distribution, we build a compromise curve according to the following rule: with the same value of one parameter, the point with the minimum value of the second is considered optimal.

Conclusion

According to this distribution, we build a compromise curve according to the following rule: with the same value of one parameter, the point with the minimum value of the second is considered optimal. This is seen from the compromise curve obtained - when considering 64 LP-tau sequence points, 13 of them entered the Pareto set and formed a compromise curve, but with a power consumption of more than 4000 W, only two points hit the curve. Thus, the described method is recommended when calculating high-speed low-power axial pumps.

List of references

[1] P Chaburko and Z Kossova 2019 *IOP Conf. Ser.: Mater. Sci. Eng.* **492** 012011
[2] V Lomakin et al 2019 *IOP Conf. Ser.: Mater. Sci. Eng.* **492** 012012
[3] A Gouskov et al 2019 *IOP Conf. Ser.: Mater. Sci. Eng.* **492** 012013
[4] T Valiev and A Petrov 2019 *IOP Conf. Ser.: Mater. Sci. Eng.* **492** 012038
[5] V Cheremushkin and V Lomakin 2019 *IOP Conf. Ser.: Mater. Sci. Eng.* **492** 012039
[6] Zhang, S., Li, H., & Xi, D. (2019). Investigation of the integrated model of side chamber, wear-rings clearance, and balancing holes for centrifugal pumps.*Journal of Fluids Engineering, Transactions of the ASME,141*(10) doi:10.1115/1.4043059
[7] Sengpanich, K., Bohez, E. L. J., Thongkruer, P., &Sakulphan, K. (2019). New mode to operate centrifugal pump as impulse turbine. Renewable Energy, 983-993. doi:10.1016/j.renene.2019.03.116
[8] Zhang, Z. -., Chen, H. -., Ma, Z., He, J. -., Liu, H., & Liu, C. (2019). Research on improving the dynamic performance of centrifugal pumps with twisted gap drainage blades. Journal of Fluids Engineering, Transactions of the ASME, 141(9) doi:10.1115/1.4042885
[9] Pirouzpanah, S., Patil, A., Chen, Y., & Morrison, G. (2019). Predictive erosion model for mixed flow centrifugal pump. Journal of Energy Resources Technology, Transactions of the ASME, 141(9) doi:10.1115/1.4043135
[10] Yousefi, H., Noorollahi, Y., Tahani, M., Fahimi, R., &Saremian, S. (2019). Numerical simulation for obtaining optimal impeller's blade parameters of a centrifugal pump for high-viscosity fluid pumping. Sustainable Energy Technologies and Assessments, 34, 16-26. doi:10.1016/j.seta.2019.04.011
[11] Blume, M., & Skoda, R. (2019). 3D flow simulation of a circular leading edge hydrofoil and assessment of cavitation erosion by the statistical evaluation of void collapses and cavitation structures. Wear, 428-429, 457-469. doi:10.1016/j.wear.2019.04.011
[12] Yun, R., Zuchao, Z., Denghao, W., &Xiaojun, L. (2019). Influence of guide ring on energy loss in a multistage centrifugal pump. Journal of Fluids Engineering, Transactions of the ASME, 141(6) doi:10.1115/1.4041876
[13] N Egorkina and A Petrov 2019 *IOP Conf. Ser.: Mater. Sci. Eng.* **492** 012015
[14] K Dobrokhodov and A Petrov 2019 *IOP Conf. Ser.: Mater. Sci. Eng.* **492** 012016
[15] N Isaev 2019 *IOP Conf. Ser.: Mater. Sci. Eng.* **492** 012026
[16] A Shablovskiy and E Kutovoy 2019 *IOP Conf. Ser.: Mater. Sci. Eng.* **492** 012035

IOP Conf. Series: Materials Science and Engineering **589** (2019) 012005 doi:10.1088/1757-899X/589/1/012005

Resource Efficiency Compromise for Centrifugal Pump

A Protopopov[1,2], A Mukhlaeva[1] and B Tkachuk[1]

[1]Bauman Moscow State Technical University, 5 Second Baumanskaya Street, Moscow, 105005, Russian Federation

[2]E-mail: proforg6@yandex.ru

Annotation

The main characteristics of centrifugal pumps, like many other mechanisms, include resource and efficiency. Often these are two competing parameters - while increasing the first, we sacrifice the second. This is because with an increase in the rotational speed of the pump rotor, the life of the bearings drops rapidly, and the efficiency, on the contrary, increases. There is a need to find a compromise solution, when at the selected value of one parameter the second is maximum.

Introduction

The resource and efficiency of such a pump, among other parameters, are influenced by the axial clearance between the impeller and the pump casing and the frequency of rotation of the pump rotor. Modeling the front axial clearance using hydrodynamic methods [1] - [8] is extremely difficult due to the unevenness of the computational grid. Also, the use of other methods for calculating centrifugal pumps [9] - [15] for such optimization is difficult, too. Therefore, it was decided to use the LP-Tau search method, since this method does not imply an exact definition of the objective function.

Note that the currently existing methods of finding a compromise between the resource and efficiency are for small rotational speeds of the pump shaft, which creates significant errors in the calculation of the parameters of the high-speed low-flow pump.

Methods

The LP-Tau search allows generating points in a quasi-random manner in a specified interval for two parameters. For the investigated pump axial clearance, we will vary in the range of 0.5 ... 1.5 mm, the rotor speed n — in the range of 3,000 ... 8,000 rpm. Then the field of generated work points will be a rectangle (Fig. 1).

The critical effect on the radial force is the supply of pump fluid so that we will build a family of trade-off curves for different feed values. We start construction for the case when $Q = Q_o$.

For this distribution, we build a compromise curve for the following reasons: for the same value of one parameter, the point with the highest value of the second is considered to be the "winning." From this, the compromise curve must pass through these “winning” points (Fig. 2).

Similarly, we construct the trade-off curves for different flow rates Q = $0{,}5Q_o$, $0{,}75Q_o$, $1{,}25Q_o$, $1{,}5Q_o$ (Figures 3–6).

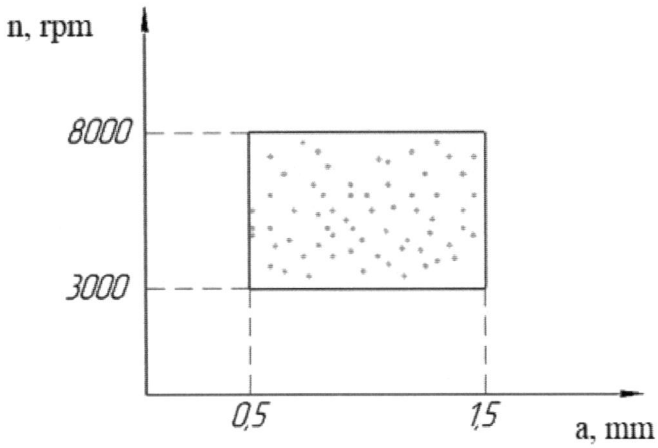

Figure 1. Points generation area.

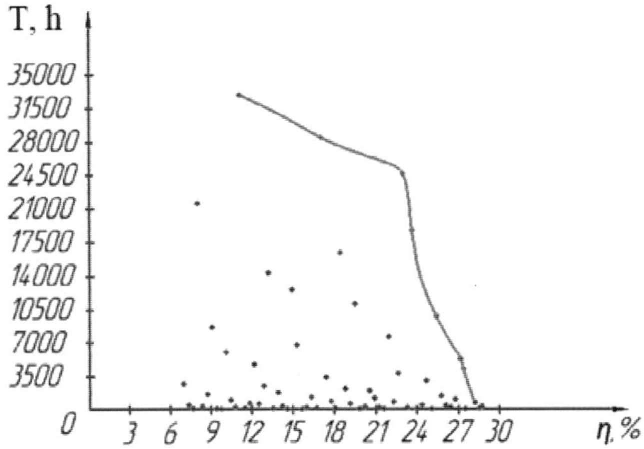

Figure 2. Compromisee curve ($Q = Q_o$)

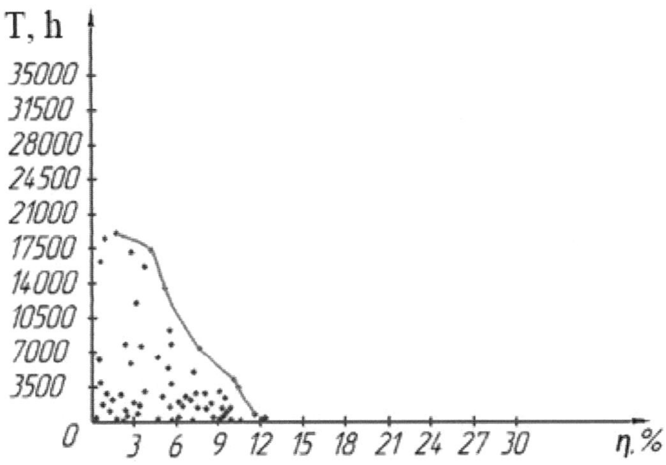

Figure 3. Compromise curve ($Q = 0{,}5Q_o$)

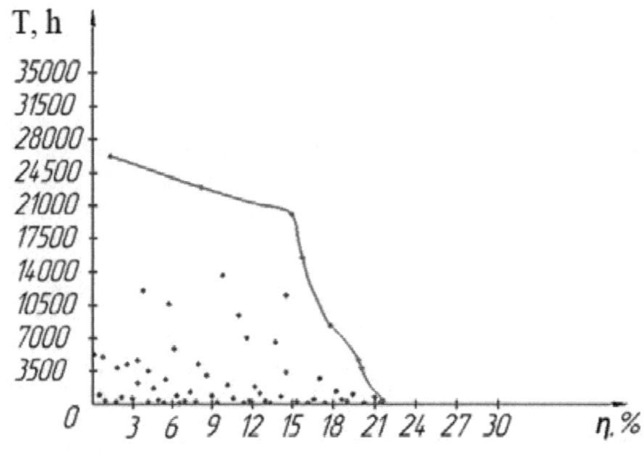

Figure 4. Compromise curve (Q =0,75Q_o)

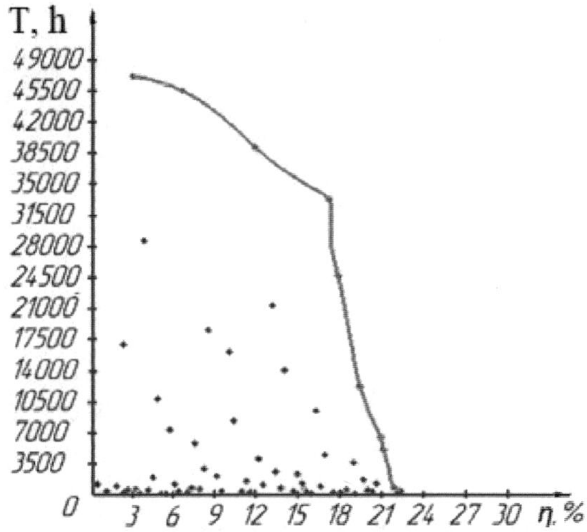

Figure 5. Compromise curve (Q =1,25Q_o)

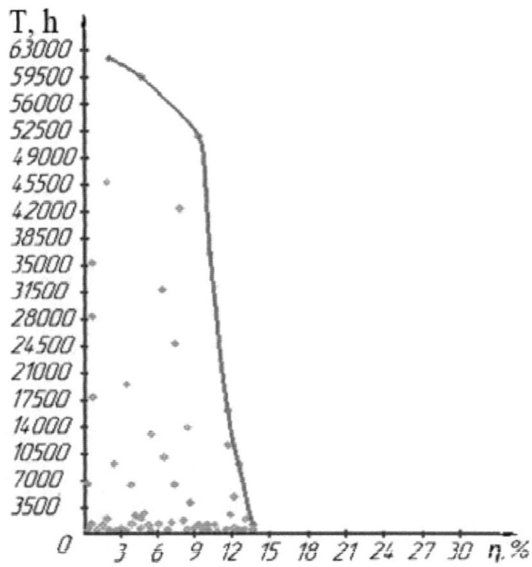

Figure 6. Compromise curve (Q =1,5Q_o)

Results

Figure 7 shows the resulting trade-off curves in one coordinate system.

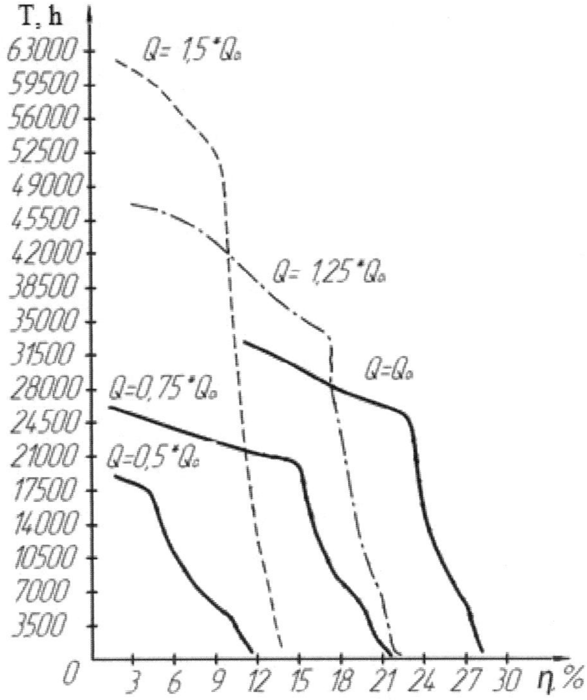

Figure 7. The family of compromise curves at different flow rates.

Conclusion

Analysis of the obtained family of curves suggests that it is undesirable to use this pump in under load conditions. This is explained quite simply. The hydrodynamic radial force with this arrangement of the diverting device is directed downwards, as a result of which a large reaction occurs in the bearing. Reducing consumption contributes to an increase in the hydrodynamic component of the radial force, which, in turn, leads to an increase in the load on the bearing, and this causes a decrease in the resource. When the pump is operating in overload conditions, the hydrodynamic radial force changes direction, as a result of which the effect on the bearing decreases. This causes a decrease in the equivalent dynamic load, which leads to an increase in resource. Efficiency increase, on the contrary, hurts efficiency. This, again, is due to the resulting dependence of efficiency on flow. Even though the increase in consumption leads to an increase in resource, the use of this pump in overload modes is unacceptable. Firstly, it does not provide the necessary pressure, as can be seen from the predictive characteristics. Secondly, the efficiency of the pump reduces.

As a result of the analysis of this family of trade-off curves, the deviation from the optimal operation mode has a substantial impact on the resource and efficiency. Therefore, it is necessary to ensure the operation of the pump in the mode as close as possible to the optimum. The obtained family of curves also allows us to estimate the effect of the rotor speed and axial clearance between the outlet body and the impeller on the resource and efficiency of the pump. Also, the constructed family of curves makes it possible to conclude that the use of frequency regulation is advisable.

References

[1] P Chaburko and Z Kossova 2019 *IOP Conf. Ser.: Mater. Sci. Eng.* **492** 012011
[2] V Lomakin et al 2019 *IOP Conf. Ser.: Mater. Sci. Eng.* **492** 012012
[3] A Gouskov et al 2019 *IOP Conf. Ser.: Mater. Sci. Eng.* **492** 012013

[4] N Egorkina and A Petrov 2019 *IOP Conf. Ser.: Mater. Sci. Eng.* **492** 012015
[5] K Dobrokhodov and A Petrov 2019 *IOP Conf. Ser.: Mater. Sci. Eng.* **492** 012016
[6] N Isaev 2019 *IOP Conf. Ser.: Mater. Sci. Eng.* **492** 012026
[7] A Shablovskiy and E Kutovoy 2019 *IOP Conf. Ser.: Mater. Sci. Eng.* **492** 012035
[8] A Petrov et al 2019 *IOP Conf. Ser.: Mater. Sci. Eng.* **492** 012036
[9] Zhang, S., Li, H., & Xi, D. (2019). Investigation of the integrated model of side chamber, wear-rings clearance, and balancing holes for centrifugal pumps. *Journal of Fluids Engineering, Transactions of the ASME, 141*(10) doi:10.1115/1.4043059
[10] Sengpanich, K., Bohez, E. L. J., Thongkruer, P., &Sakulphan, K. (2019). New mode to operate centrifugal pump as impulse turbine. Renewable Energy, 983-993. doi:10.1016/j.renene.2019.03.116
[11] Zhang, Z. -., Chen, H. -., Ma, Z., He, J. -., Liu, H., & Liu, C. (2019). Research on improving the dynamic performance of centrifugal pumps with twisted gap drainage blades. Journal of Fluids Engineering, Transactions of the ASME, 141(9) doi:10.1115/1.4042885
[12] Pirouzpanah, S., Patil, A., Chen, Y., & Morrison, G. (2019). Predictive erosion model for mixed flow centrifugal pump. Journal of Energy Resources Technology, Transactions of the ASME, 141(9) doi:10.1115/1.4043135
[13] Yousefi, H., Noorollahi, Y., Tahani, M., Fahimi, R., &Saremian, S. (2019). Numerical simulation for obtaining optimal impeller's blade parameters of a centrifugal pump for high-viscosity fluid pumping. Sustainable Energy Technologies and Assessments, 34, 16-26. doi:10.1016/j.seta.2019.04.011
[14] Blume, M., & Skoda, R. (2019). 3D flow simulation of a circular leading edge hydrofoil and assessment of cavitation erosion by the statistical evaluation of void collapses and cavitation structures. Wear, 428-429, 457-469. doi:10.1016/j.wear.2019.04.011
[15] Yun, R., Zuchao, Z., Denghao, W., & Xiaojun, L. (2019). Influence of guide ring on energy loss in a multistage centrifugal pump. Journal of Fluids Engineering, Transactions of the ASME, 141(6) doi:10.1115/1.4041876

IOP Conf. Series: Materials Science and Engineering **589** (2019) 012006 doi:10.1088/1757-899X/589/1/012006

Dynamics of the rotor of the centrifugal pump with not opposed scheme impellers

A Protopopov[1,3], E Morozova[1], A Mukhlaeva[1] and V Atamas[2]

[1]Bauman Moscow State Technical University, 5 Second Baumanskaya Street, Moscow, 105005, Russian Federation
[2]Bohdan Khmelnytsky National University of Cherkasy, 81 Shevchenko Boulevard, Cherkasy, Ukraine

[3]E-mail: proforg6@yandex.ru

Annotation
The neo-positive arrangement of the impellers is sometimes found in centrifugal pumps with hydrostatic bearings. Such a scheme is due to the high requirements for the dimensions of modern pumps. The most critical problem arising in such pumps is the problem of stable start dynamics. This article discusses the mathematical model of starting the rotor of a centrifugal pump with hydrostatic bearings. The resulting mathematical model allows us to determine the time of the output to the steady-state mode and set the angular velocity of the rotor.

Introduction

Low-consumption centrifugal pumps are used in many industries. As a result, there is a lot of literature[1]-[4] that describes the calculations (associated with finding the various parameters of these pumps) in detail. However, there is still a lot of room for research, in particular, issues related to finding the steady-state angular velocity of the pump rotor.

To solve such problems, we often use methods of hydrodynamic modeling [5]-[9]. However, the application of those methods is difficult due to a large number of calculations. Thus, we need to find other approaches.

The literature [10]-[16] reviews other methods of modeling processes occurring in centrifugal pumps, which is usually either a numerical solution of the Navier-Stokes equations or reliance on empirical methods. We need a way to solve the start problem without empirical formulas and numerical solution of the Navier-Stokes equations, which is what we attempt to do in this paper.

Methods

In this paper, we study the dynamics of the low-inertia rotor of a low-flow centrifugal cantilever pump with hydrostatic bearings at low vibrations. Figure 1 shows the scheme of the pump.

Let's examine the torques affecting the pump shaft (Figure 2).

On the figure, M_{rk} is torque from the impeller, M_{vt} is viscous friction torque, and M_{dv} is torque from the engine. The dry friction force is not considered, because there is always a thin layer of oil with thickness δ_0 between the shaft and the bearing.

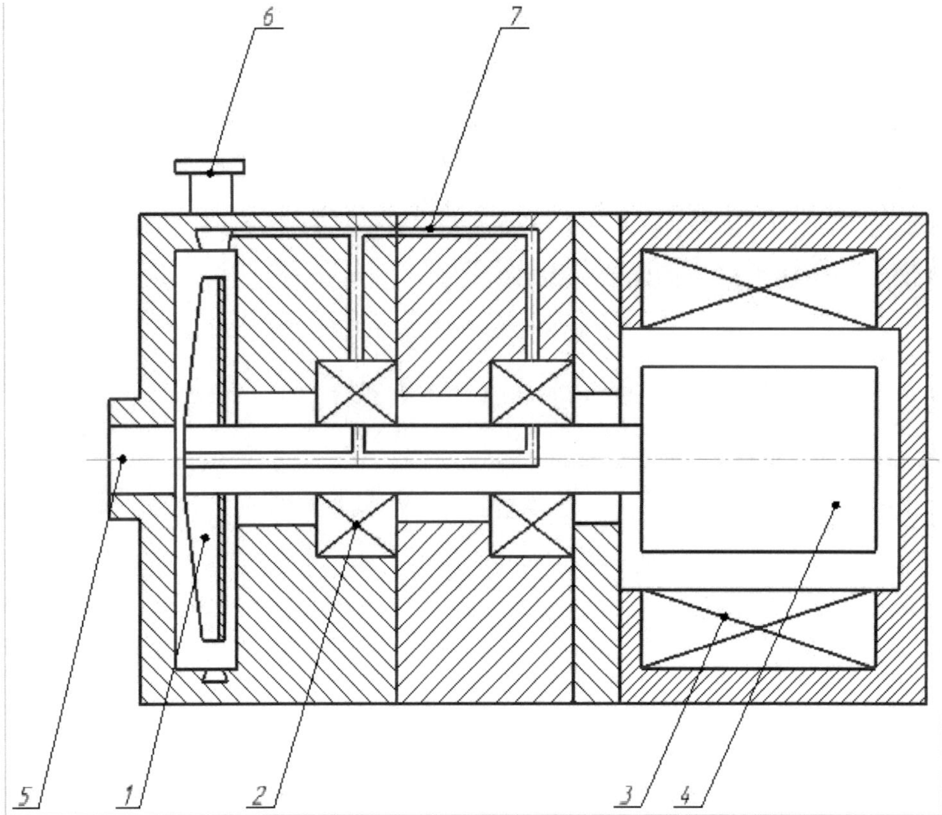

Figure 1. Pump scheme. 1 –pump impeller, 2 – hydrostatic bearings, 3 – stator, 4 – rotor, 5 – inlet pipe, 6 – outlet pipe, 7 – bearing supply channels.

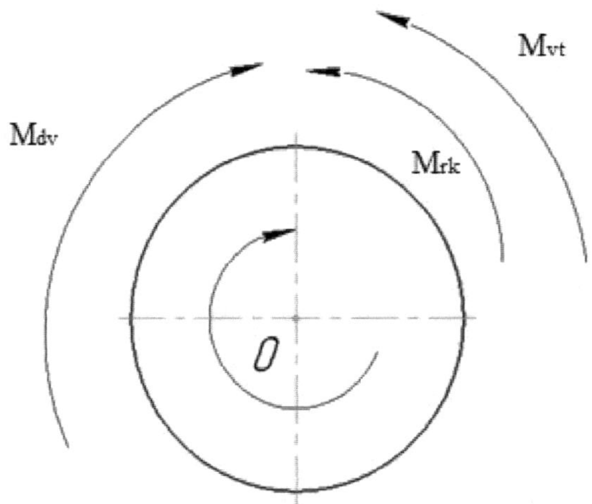

Figure 2. Torques affecting the shaft.

Let's make the equation of moments relative to the axis of the shaft:

$$J \cdot \frac{d\omega}{dt} = M_{dv}(t) - 2 \cdot M_{rk}(t) - 2 \cdot M_{vt}(t). \tag{1}$$

Here, J is the moment of inertia of the rotor relative to its axis, ω is the angular velocity of rotation of the pump shaft, t is the time from the start of the pump, g is the acceleration of gravity.

The formula below describes the moment generated by the engine.

$$M_{dv}(t) = K - K_1 \cdot \omega(t). \quad (2)$$

Here, K and K1 are the coefficients determined by the design of this particular motor.

The moment on the impeller is equal to:

$$M_{rk} = M_c + M_{df}. \quad (3)$$

Here, M_C is the centrifugal moment,M_{df} is the disk friction moment.

The centrifugal torque is calculated [1]:

$$M_c = \rho \cdot Q \cdot R_2 \cdot C_{2u}. \quad (4)$$

Here, ρ is the density of the liquid, Q is the flow rate, R_2 is the outer radius of the impeller.

Since the valve is closed, Q equals 0, respectively, $M_c = 0$.

The moment of disk friction is:

$$M_{dt}(t) = \frac{\omega(t) \cdot \pi \cdot \mu \cdot R_2{}^4}{a}. \quad (5)$$

Here, a is the axial clearance between the impeller and the pump housing, μ is the dynamic viscosity of the working fluid.

Thus, the torque of the impeller is equal to:

$$M_{rk}(t) = \frac{\omega(t) \cdot \pi \cdot \mu \cdot R_2{}^4}{a}. \quad (6)$$

Here, l is the length of the bearing pocket, d is the diameter of the pump shaft, φ is the angle, y is the size of the gap (for small oscillations we take y= δ0 - the thickness of the liquid film at the start of the start), δ is the average gap between the bearing and the shaft.

We substitute these values in equation (1) and obtain a linear inhomogeneous differential equation of the second order, the solution of which is the solution of the Cauchy problem under a known initial condition ω'+B/A·ω=K/A; ω(0)=0

The linear nonuniform differential equation corresponds to a linear homogeneous differential equation: ω'+B/A·ω=0

The solution of the equation is: ω(t)=C·e^(-B/A t)

To find the solution of the inhomogeneous equation, we use the method of variation of the constant, i.e., we consider it with the function t.

We obtain the solution in the form: ω(t)=K/B+C_1·e^(-B/A t).

From the initial conditions we find the constant C_1.

Thus, the solution of the equation looks like this: ω(t)=K/B·(1-e^(-B/A t))

Results

This technique will be tested on the pump having the parameters:

K=0.9 N*m, a=0.001 mm, μ=0.000276 PA*s, R_2=0.033 mm, J=0.0001 N*m*S2, K_1=0.0015 N*m*s , l=0.02 mm, d=0.01 mm.

We obtain a graph of the angular velocity of the pump shaft on time at these values of parameters (Figure 3).

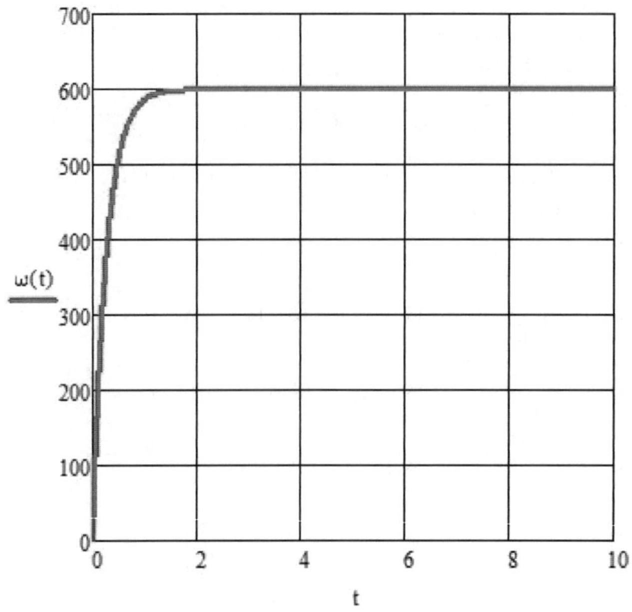

Figure 3.

The graph shows that the steady-state angular velocity of the pump shaft is ω(t)=599,515 rad/s.

Summary

Thus, the obtained technique makes it possible to find the dependence of the angular velocity on time and to predict the steady-state rotor speed, under the condition of small oscillations of the rotor in the inner ring of the bearing.

List of references

[1] N Egorkina and A Petrov 2019 *IOP Conf. Ser.: Mater. Sci. Eng.* **492** 012015

[2] Sengpanich, K., Bohez, E. L. J., Thongkruer, P., &Sakulphan, K. (2019). New mode to operate centrifugal pump as impulse turbine. Renewable Energy, 983-993. doi:10.1016/j.renene.2019.03.116

[3] K Dobrokhodov and A Petrov 2019 *IOP Conf. Ser.: Mater. Sci. Eng.* **492** 012016

[4] N Isaev 2019 *IOP Conf. Ser.: Mater. Sci. Eng.* **492** 012026

[5] V Lomakin and O Bibik 2019 *IOP Conf. Ser.: Mater. Sci. Eng.* **492** 012037

[6] A Petrov et al 2019 *IOP Conf. Ser.: Mater. Sci. Eng.* **492** 012036

[7] T Valiev and A Petrov 2019 *IOP Conf. Ser.: Mater. Sci. Eng.* **492** 012038

[8] V Cheremushkin and V Lomakin 2019 *IOP Conf. Ser.: Mater. Sci. Eng.* **492** 012039

[9] Yousefi, H., Noorollahi, Y., Tahani, M., Fahimi, R., &Saremian, S. (2019). Numerical simulation for obtaining optimal impeller's blade parameters of a centrifugal púmp for high-viscosity fluid pumping. Sustainable Energy Technologies and Assessments, 34, 16-26. doi:10.1016/j.seta.2019.04.011

[10] A Shablovskiy and E Kutovoy 2019 *IOP Conf. Ser.: Mater. Sci. Eng.* **492** 012035

[11] Blume, M., & Skoda, R. (2019). 3D flow simulation of a circular leading edge hydrofoil and assessment of cavitation erosion by the statistical evaluation of void collapses and cavitation structures. Wear, 428-429, 457-469. doi:10.1016/j.wear.2019.04.011

IOP Conf. Series: Materials Science and Engineering **589** (2019) 012006 doi:10.1088/1757-899X/589/1/012006

[12] Pirouzpanah, S., Patil, A., Chen, Y., & Morrison, G. (2019). Predictive erosion model for mixed flow centrifugal pump. Journal of Energy Resources Technology, Transactions of the ASME, 141(9) doi:10.1115/1.4043135

[13] Yun, R., Zuchao, Z., Denghao, W., &Xiaojun, L. (2019). Influence of guide ring on energy loss in a multistage centrifugal pump. Journal of Fluids Engineering, Transactions of the ASME, 141(6) doi:10.1115/1.4041876

[14] Zhang, S., Li, H., & Xi, D. (2019). Investigation of the integrated model of side chamber, wear-rings clearance, and balancing holes for centrifugal pumps. *Journal of Fluids Engineering, Transactions of the ASME, 141*(10) doi:10.1115/1.4043059

[15] Zhang, Z. -., Chen, H. -., Ma, Z., He, J. -., Liu, H., & Liu, C. (2019). Research on improving the dynamic performance of centrifugal pumps with twisted gap drainage blades. Journal of Fluids Engineering, Transactions of the ASME, 141(9) doi:10.1115/1.4042885

[16] A Shablovskiy and E Kutovoy 2019 *IOP Conf. Ser.: Mater. Sci. Eng.* **492** 012035

IOP Conf. Series: Materials Science and Engineering **589** (2019) 012007 doi:10.1088/1757-899X/589/1/012007

Hydrodynamic modelling of the impact of viscosity on the characteristics of a centrifugal pump

V Tkachuk[1], H Navas[1], A Petrov[1] and A Protopopov[1,2]

[1]Bauman Moscow State Technical University, 5 Second Baumanskaya Street, Moscow, 105005, Russian Federation

[2]E-mail: proforg6@yandex.ru

Annotation

During operation of industrial centrifugal pumps, the main factor affecting its performance is viscosity, which can change its value during operation of the pump unit. In most cases, the main cause of such changes is a change in the environmental conditions or an increase in the temperature of the unit during operation. This paper explores the possibility of using computer simulation to calculate a centrifugal pump at various viscosities of the working fluid. The results obtained are compared with data from the American Hydraulic Institute.

Introduction

During the process of operation of a pump under new conditions of use or when the performance of an existing pump is being assessed, it is often necessary to consider the influence of the viscosity of the pumped liquid. Issues related to the operation of centrifugal pumps at high viscosities are widely reported in the literature [1–16]. It is known that the pressure curves presented in the catalogs of pump suppliers are constructed using water as the pumped liquid. These curves are suitable for use when the actual fluid used for pumping has a viscosity that is less than or equal to that of water. However, in some cases, this method is not suitable, for example, for some petroleum products and other liquids at low temperatures. The performance of a viscous working fluid pump is usually estimated using corrections applied to the characteristics obtained on water, since the test facilities of the pump manufacturers provide only tests on water, and most of the accumulated data and experience relate to characteristics obtained on water.

If a highly viscous fluid circulates through the pump and the piping system during the operation, more energy is needed to pump such fluid, because the friction between the fluid layers is much greater. As a result, the following phenomena occur:

- Reduced suction lift.
- Reduced pump flow.
- Reduced efficiency.
- Increased power consumption.

The purpose of this article is to simulate the operating conditions of a type D pump with the following characteristics (for water) with different viscosity values:

Table 1. Pump Characteristics

Head pressure, м	Flow Rate, m3/h	Rotor speed, rpm	Cavitation stock, м	Efficency, %	Specific speed, n_s
80	80	2900	4,8	67	42

Using CFD STAR CCM ++ simulation software, it is necessary to compare the results obtained with the hydrodynamic simulation software and the method proposed by the American Institute of Hydraulics. [9].

Method

The process of experimental determination of the effect of fluid viscosity on a working pump has now been studied quite well. In the book Centrifugal and Axial Flow Pumps, A.J. Stepanoff [7] lists losses that affect pump performance:

- Mechanical losses;
- Losses on the impeller due to reduced rotational speed;
- Leaks;
- Losses of disk friction.

To find out what the behavior of a viscous fluid pump is, the American Hydraulic Institute (HI) has developed a graphical method based on laboratory tests. The HI method suggests taking known pump data for water and applying some correction factors for viscosity, which are obtained from Figure 1. The nomenclature is defined below:

Ha - head with water; fH is thehead correction factor for an especific viscosity;

Qa - Flow with water; fQ is the flow rate correction factor for an especific viscosity;

HPa - power with water; fη is the efficency correction factor for an especific viscosity.

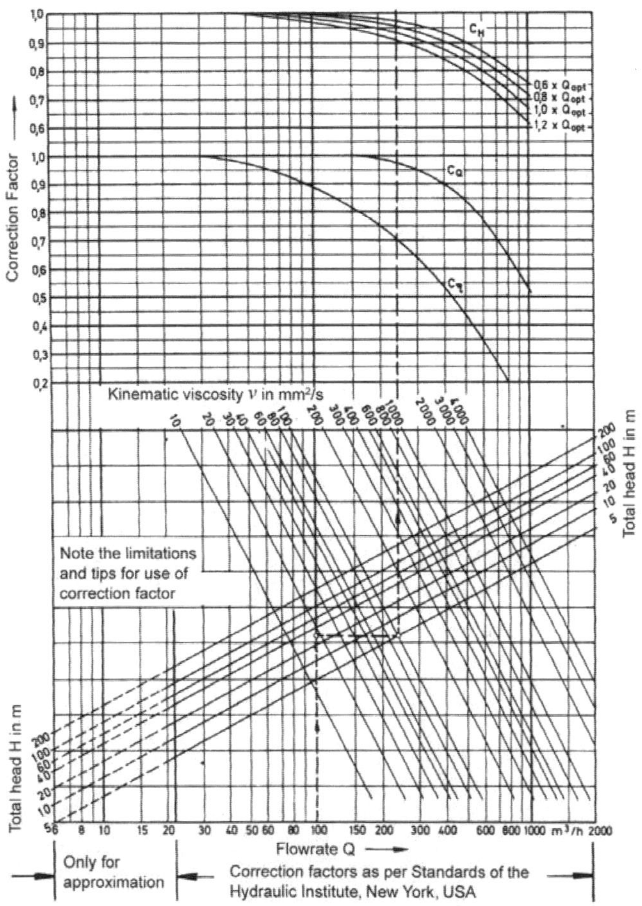

Figure 1. The graph obtained by the American Institute of hydraulics

IOP Conf. Series: Materials Science and Engineering **589** (2019) 012007 doi:10.1088/1757-899X/589/1/012007

Corrective formulas:

$$Hv = Ha \times fH. \tag{1}$$

$$Qv = Qa \times fQ. \tag{2}$$

$$HPv = HPa \times fQ\frac{fH}{f\eta}. \tag{3}$$

The following parameters are obtained through:
– Original water pump curve (Ha, Qa, HPa)
– Graph of viscosity correction from Hydraulic Institute (fH, fQ, fη)
– Application of the formula (Hv, Qv, HPv)
Limitations to consider
Before making estimates and explaining how to obtain correction factors, it is necessary to take into account the conditions of the method in order to avoid errors.

– *Flow Rate:* If the pump has a double suction impeller, consider half the flow rate of the pump, that is, the flow rate for each inlet section..

– *HeadPreassure:* If the pump has more than one stage, it is necessary to consider the pressure of one stage.

– *Fluid viscosity:* Corrections are made for Newtonian liquids, i.e. liquids whose viscosity can be considered constant in time and which does not change with stirring.

– *Curve Range:* It is not recommended to extrapolate outside the range of the curves, because it can lead to a big error.

– *Viscosity limit:* There are various criteria regarding the maximum allowable viscosity in a centrifugal pump. Some authors recommend a maximum kinematic viscosity of 1000 cSt, but there are pumps that have an operating mode with liquids up to 3000 cSt.

Mathematical model. In order to obtain the hydrodynamic modeling, the following equations were used:

Equation of continuity of the liquid medium:

$$\frac{\partial \overline{u_x}}{\partial x} + \frac{\partial \overline{u_y}}{\partial y} + \frac{\partial \overline{u_z}}{\partial z} = 0,$$

Where$\overline{u_\iota}$ – time-averaged projections of fluid velocities on the corresponding axes;

- Equation of the change in the amount of motion averaged over time:

$$\rho\left[\frac{\partial \overline{u_i}}{\partial t} + \overline{u_j}\frac{\partial \overline{u_i}}{\partial x_j}\right] = -\frac{\partial \overline{p}}{\partial x_i} + \frac{\partial}{\partial x_i}\left[T_{ij}^{(v)} - \rho u_i u_j\right],$$

Where$\overline{u_\iota}\overline{p}$ – averaged speed and pressure;

$\tilde{T}_{ij}^{(v)} = 2\mu\tilde{s}_{ij}$ – viscous stress tensor for incompressible fluid;

$\tilde{s}_{ij} = \frac{1}{2}\left[\frac{\partial \overline{u_i}}{\partial x_j} + \frac{\partial \overline{u_j}}{\partial x_i}\right]$ – strain rate tensor;

$\rho\langle u_i u_j\rangle$ – Reynolds stresses.

The Reynolds system of equations is open due to the presence of unknown Reynolds stresses. The system is closed using the k-ω SST turbulence model.

Terms of computer simulation. In the process of the computer simulation, a three-dimensional model of the flow part of the pump was created, and physical models and parameters of the

IOP Conf. Series: Materials Science and Engineering **589** (2019) 012007 doi:10.1088/1757-899X/589/1/012007

computational grid were set for it. The density value remained unchanged throughout the experiment. During the process of the hydrodynamic modeling with a change in viscosity, the flow rate, which is considered by the program, does not change, which is contrary to reality. Therefore, to obtain a valid aproximation in the computer model, the flow rate was set according to the theory described above. Also, on the basis of the limitations for considering this technique, half the flow rate from the optimal value was taken, due to this type of pump. As a result of the experiment, the parameters (head and power) found using hydrodynamic modeling were compared with the parameters found by the graphical method.

Results

As an example, it was defined the correction factors for a liquid with a kinematic viscosity of 50 cSt, according to the graphs in Figure-1. The initial data from which the recalculation will be carried out are indicated below: Qa = 40 m3 / h; Ha = 80 m; HPa = 26.4 kW

There were obtained three correction factors:fη = 0,76; fQ = 0,98; fH = 0,95

Further, analogycally, were found the correction factors and recalculate the indicators for a liquid with different kinetic viscosity values, increasing the subsequent value by 50 cSt and to 600 cSt. The results of recalculated indicators are shown in Table 2.

Table 2. Parameters obtained graphically

	Correction factors			*Head (m)*	*Flow Rate (m3 / h)*	*Power(kWt)*
	fη	**fQ**	**fH**	**Hv**	**Qv**	**HPv**
viscosity (cSt) 50	0.76	0.98	0.95	76.00	39.20	32.34
100	0.66	0.96	0.93	74.40	38.40	35.71
150	0.63	0.94	0.90	72.00	37.60	35.45
200	0.54	0.90	0.87	69.60	36.00	38.28
250	0.50	0.87	0.86	68.80	34.80	39.50
300	0.45	0.85	0.84	67.20	34.00	41.89
350	0.43	0.84	0.84	67.20	33.60	43.32
400	0.37	0.80	0.81	64.80	32.00	46.24
450	0.36	0.76	0.80	64.00	30.40	44.59
500	0.33	0.75	0.79	63.20	30.00	47.40
550	0.30	0.73	0.78	62.40	29.20	50.11
600	0.28	0.71	0.76	60.80	28.40	50.88

After the recalculation of indicators has been carried out, we compare them with the values calculated in STAR CCM + for the same kinematic viscosity values. The results of the indicators calculated in STAR CCM + are shown in Table 3.

Table 3. Parameters obtained in STAR CCM +

	Head (m)	*Flow Rate (m3 / h)*	*Efficency*
	Hv	**Qv**	
вязкость(cCm) 50	74.92	78.19	0.5789
100	67.91	76.8	0.5278
150	65.62	75.2	0.4964
200	62.92	71.74	0.4690
250	60.94	69.0	0.4402
300	58.68	68	0.4224

Окончание табл. 3

	Head (m)	*Flow Rate (m3 / h)*	*Efficency*
	Hv	**Qv**	
350	56.02	66.97	0.4014
400	55.76	64	0.3905
450	55.72	60.8	0.3750
500	53.93	59.75	0.3682
550	52.46	58.4	0.3502
600	52.16	56.59	0.3499

A comparison of all parameters is shown in Figure 2 and Figure 3.

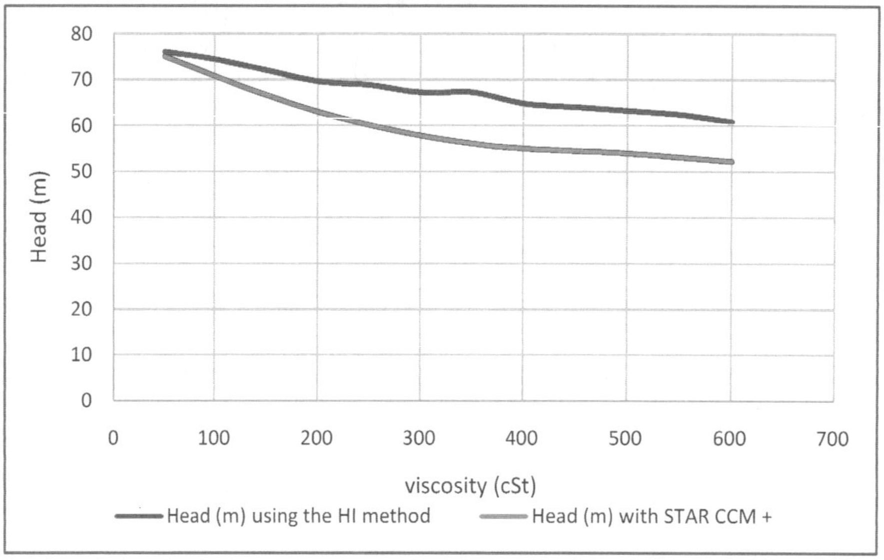

Figure 2. Head Comparison graph

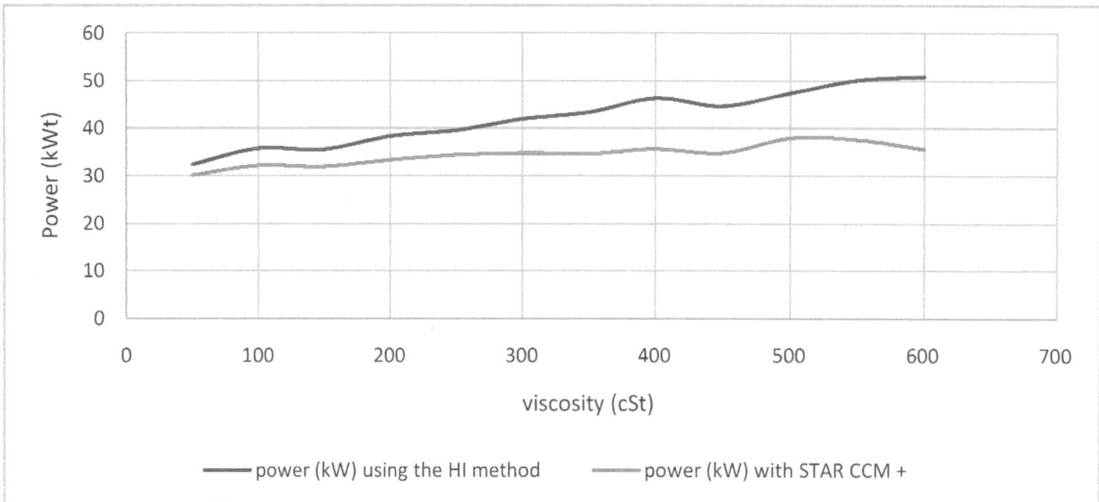

Figure 3. Power comparison graph

Conclusion

As a result of the experiment, it can be concluded that the values obtained using the bench practically correspond to the values obtained using computer simulation. However, with viscosities above 200cSt (200 mm2 / s), the indicators have significant discrepancies, which can be observed in Figure 2 and Figure 3. This phenomenon can be explained for several reasons:

- In the computer model it is not possible to specify all the conditions under which the pump will work;
- In computer simulation, a pump model with a low coefficient of speed ns was used, such pumps are not intended for pumping highly viscous liquids;
- Also in the computer model, leakages between areas and leakages were not taken into account.

Thus, in the course of the experiment, the main assumptions for viscous tests using hydrodynamic modeling were identified. Were built comparative graphs for pressure and power.

Literature

[1] DOI: 10.1016/j.cnsns.2015.10.001. Lomakin, V.O., Kuleshovav, M.S., Bozh'eva, S.M. Numerical Modeling of Liquid Flow in a Pump Station (2016) Power Technology and Engineering, 49 (5), pp. 324-327.
[2] DOI: 10.1007/s10749-016-0623-9. Lycheva, T.N., Lychev, S.A. Spectral decompositions in dynamical viscoelastic problems (2016) PNRPU Mechanics Bulletin, (4), pp. 120-150.
[3] DOI: 10.1016/S0142-727X(99)00078-8. W.-G. Li, "Effects of viscosity of fluids on centrifugal pump performance and flow pattern in the impeller," International Journal of Heat and Fluid Flow, vol. 21, no. 2, pp. 207–212, 2000.
[4] DOI: 10.1007/s10010-006-0045-1. A. Nemdili and D. H. Hellmann, "Investigations on fluid friction of rotational disks with and without modified outlet sections in real centrifugal pump casings," Forschung im Ingenieurwesen, vol. 71, no. 1, pp. 59–67, 2007.
[5] DOI: 10.1016/S0262-1762(00)87528-8. J. F. Gülich, "Pumping highly viscous fluids with centrifugal pumps—part 1," World Pumps, vol. 1999, no. 395, pp. 30–34, 1999.
[6] DOI: 10.1115/GT201542014. K. Juckelandt and F.-H. Wurm, "Applicability of wall-function approach in simulations of turbomachines," in Proceedings of ASME Turbo Expo 2015: Turbine Technical Conference and Exposition (GT '15), June 2015.
[7] Stepanoff, A. J. Centrifugal and Axial Flow Pumps. Theory, Design and Application (1957) KRIEGER PUBLISHING COMPANY, pp. 308-317
[8] Lobanoff V. S. Ross R. R. Centrifugal Pumps: Design and Application (2013) Gulf Professional Publishing, pp. 305-310
[9] Hydraulic Institute (U.S.) Hydraulic Institute standards for centrifugal, rotary & reciprocating pumps (1983) Michigan University, pp.111
[10] Dickenson T. C. Pumping Manual (1995) Oxford: Elsevier Advanced Technology, pp. 36.
[11] Turzo, Z., Takacs, G. and Zsuga, J., "Equations Correct Centrifugal Pump Curves for Viscosity" (2000), Oil & Gas Journal, pp. 57-61
[12] Mathematical modeling of the mechanisms of volumetric hydraulic machines.S. Semenov and D. Kulakov 2019 IOP Conf. Ser.: Mater. Sci. Eng. 492 012042
[13] Methodology of dynamic pumps testing on high-viscosity liquids. K. Dobrokhodov and A. Petrov 2019 IOP Conf. Ser.: Mater. Sci. Eng. 492 012016
[14] Karassik, I. y R. Carter. Bombas centrífugas. Selección, operación y mantenimiento (1985). Compañía Editorial Continental, México, pp. 143
[15] Turiño, I. M. "Determinación aproximada de las características de funcionamiento de una bomba centrífuga" (1994), Centro Azúcar, Cuba, pp.34
[16] A. T. Ippen, "The influence of viscosity on centrifugal pump performance," Transactions of the American Society of Mechanical Engineers, vol. 68, no. 8, pp. 823–848, 1946

[17] W. G. Li and Z. M. Hu, "An experimental study on performance of centrifugal oil pump," Fluids Machinery, vol. 25, no. 2, pp. 3–7, 1997.
[18] A Protopopov and D Bondareva 2019 IOP Conf. Ser.: Mater. Sci. Eng. 492 012002
[19] A Protopopov and V Vigovskij 2019 IOP Conf. Ser.: Mater. Sci. Eng. 492 012003
[20] A Petrov et al 2019 IOP Conf. Ser.: Mater. Sci. Eng. 492 012040
[21] The effect of design parameters of the closed type regenerative pump the energy characteristics. N Isaev 2019 IOP Conf. Ser.: Mater. Sci. Eng. 492 012026

IOP Conf. Series: Materials Science and Engineering **589** (2019) 012008 doi:10.1088/1757-899X/589/1/012008

Optimization of the radial channel guide vane of a centrifugal pump

E Morozova[1,2], N Belov[1] and V Cheremushkin[1]

[1]Bauman Moscow State Technical University, 5 Second Baumanskaya Street, Moscow, 105005, Russian Federation

[2]Email: zakharovaelenaeng@mail.ru

Abstract
The guide vane of a centrifugal multistage pump mostly determines the hydraulic efficiency of the pump and as a consequence the optimal mode of its operation. In order to increase the pump efficiency, the radial channel guide vane has been optimized. Three geometrical parameters and the number of channels of the guide vane were selected as optimization parameters. Hydraulic efficiency was chosen as a criterion. 128 computational models were obtained by the method of automated parametrized construction. The efficiency parameters were obtained via hydrodynamic modeling. As a result, a significant increase in the pump efficiency was achieved.

Introduction
The majority efforts to increase pump efficiency are aimed at upgrading the pump impeller, while guide vane has asecondary role [1] - [11]. However, on the basis of VNIIGidromash, Professor S. S. Rudnev proved that the output device is not less important than the impeller. hydrauliclossesare largely determined by output device as well as the hydraulic efficiency of the pump and the optimal mode of its operation [12] - [14].

One of the varieties of output devices is a guiding apparatus.The primary direction of increasing the multistage centrifugal pump efficiency is an optimization of the guide vane.

Method
The considered pump CNS 300-500 has 5stages and nominal parameters: $Q = 300\ m^3 / h$ –volume flow, $H = 500$ m - pump head, $n = 2900$ rpm - pump shaft rotational speed, $ns = 97$ - pump specific speed. A radial channel transfer guide was selectedfor the study (Fig. 1).

Figure1. Radial channel vane guide.

As a result of preliminary calculations of a guide vane, 3 geometrical parameters which have the greatest influence on the selected optimization criterion (hydraulic efficiency)were obtained. These are the entrance diameter of the guide vane (D_3), the ratio of the input and output diameters (D_4 / D_3) and the radial direction diffuser angle of the channel (Θ). In addition, z, the number of channels of the guide vane, was takenasan optimization parameter since it has a significant impact on the efficiency. The values of the optimization parameters are presented in Table 1.

Table 1. Vane guide optimization parameters.

Parameter	Min	Max
D_3, мм	292	300
D_4/D_3	1,3	1,5
Θ, °	8	12
z	6	9

It was decided to use LP-tau search method as a quasi-random sequence generator to create an initial population. 128 models with different values of geometrical parameters for different z numbers were generated.

Table 2. Vane guide optimization parameter values.

Modelno	D_3	D_4/D_3	Θ
0	296	1,4	10
1	294	1,45	9
2	298	1,35	11
3	293	1,425	11,5
4	297	1,325	9,5
5	295	1,375	10,5
6	299	1,475	8,5
7	292,5	1,488	10,75
8	296,5	1,387	8,75
9	294,5	1,338	11,75
10	298,5	1,438	9,75
11	293,5	1,363	9,25
12	297,5	1,462	11,25
13	295,5	1,413	8,25
14	299,5	1,313	10,25
15	292,25	1,406	9,625
16	296,25	1,306	11,625
17	294,25	1,356	8,625
18	298,25	1,456	10,625
19	293,25	1,331	10,125
20	297,25	1,431	8,125
21	295,25	1,481	11,125
22	299,25	1,381	9,125
23	292,75	1,394	11,375
24	296,75	1,494	9,375

Окончание табл. 2

Modelno	D3	D_4/D_3	Θ
25	294,75	1,444	10,375
26	298,75	1,344	8,375
27	293,75	1,469	8,875
28	297,75	1,369	10,875
29	295,75	1,319	9,875
30	299,75	1,419	11,875
31	292,125	1,459	11,813

3d models were built for all 128 combinations of parameters. An example of a solid model is presentedon Fig. 2.

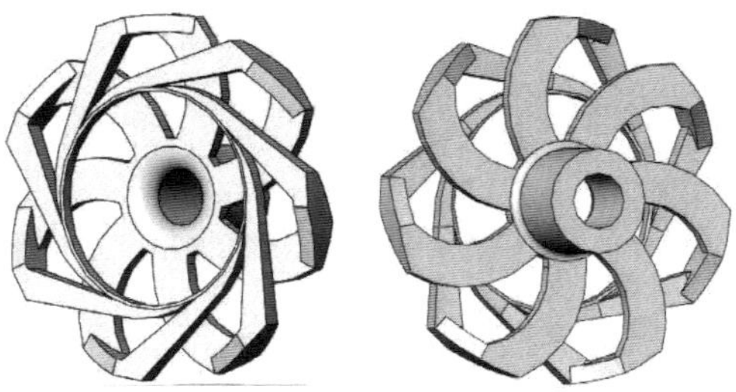

Figure2. Vaneguidefluidvolume 3dmodel.

The calculation was carried out by the methods of hydrodynamic modeling. The following equations were used: Reynolds averaged Navier-Stokes, mass conservation and two additional differential equations responsible for turbulence modeling. For the task the k-ω SST turbulence model was used [15, 16].

Results

According to the results of the calculation, the pump efficiency values at the nominal flow rate for 128 modelswere obtained and are summarized in Table 3.

Table3. Efficiency values for the models.

№	z=6	z=7	z=8	z=9
0	76,54	78,97	78,61	79,46
1	78,55	75,79	77,20	78,07
2	74,51	80,95	81,23	81,17
3	77,74	77,32	77,57	79,32
4	76,23	76,10	76,48	78,60
5	76,73	75,35	76,20	76,36
6	73,78	82,14	81,77	81,83
7	80,55	79,06	78,83	79,88
8	76,69	80,36	80,69	81,03
9	78,11	76,82	76,95	77,97
10	75,13	81,25	81,23	81,17

Окончание табл. 3

№	z=6	z=7	z=8	z=9
11	79,44	80,49	80,73	81,04
12	75,39	77,11	77,77	78,29
13	75,54	79,68	79,44	80,82
14	77,39	74,52	74,94	77,19
15	73,19	81,90	81,72	81,56
16	79,92	77,71	73,62	79,33
17	75,73	79,88	80,00	80,33
18	78,36	76,16	76,46	77,20
19	74,41	80,72	80,60	75,32
20	78,76	77,29	77,49	79,00
21	75,65	79,00	79,18	80,04
22	77,23	77,65	76,89	78,62
23	79,91	75,38	75,95	76,80
24	73,86	81,95	81,80	82,29
25	81,36	81,62	81,44	82,00
26	75,36	77,73	78,06	79,40
27	78,11	79,97	79,62	80,47
28	73,28	75,76	76,63	77,11
29	80,74	82,18	81,54	78,09
30	75,63	78,28	78,55	79,43
31	78,87	80,44	79,61	80,89

For the obtained resultsthe dependence of the efficiency on the optimization parameters should be considered in more detail. fig. 3 shows the dependence of the efficiency on a value equal to the ratio of the diameters of the entrance to the guide vane and the impeller output D_3 / D_2, where D_2 is the impeller diameter, for different z numbers.

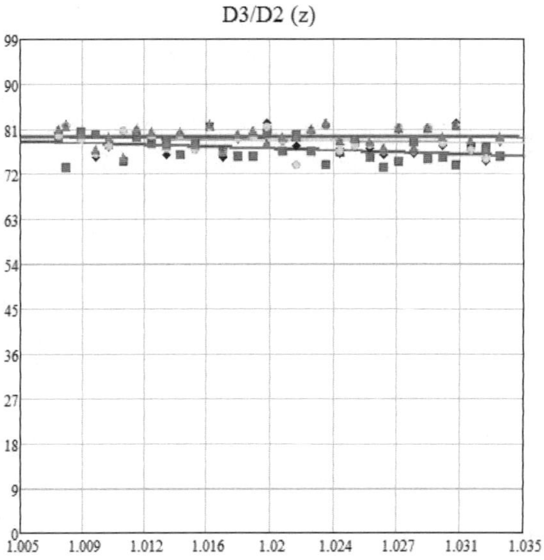

Figure 3. Dependenceoftheefficiencyon D_3/D_2, where z=6 with red color, z=7 with blue, z=8 with green, z=9 with violet.

Fig. 4 show sefficiency dependence on the ratio of input and output diameters of the guide vane D_4/D_3, for different z.

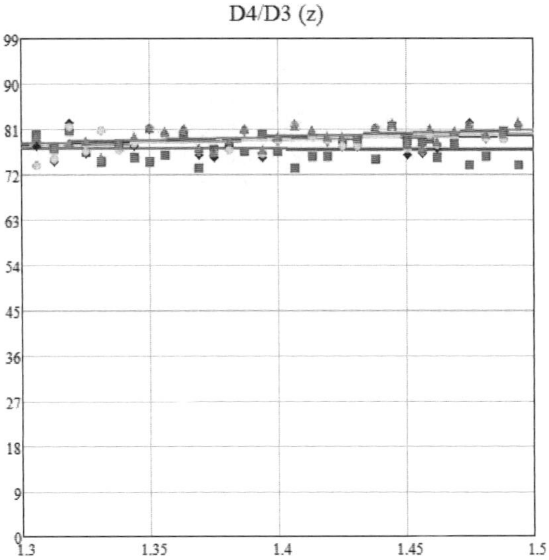

Figure 4. Efficiencydependenceon D_4/D_3, where
fed dots markz=6, z=7 - blue, z=8 – green andz=9 - violet.

Figure 5 shows efficiency dependence on the channel diffusivity Θfor different z numbers.

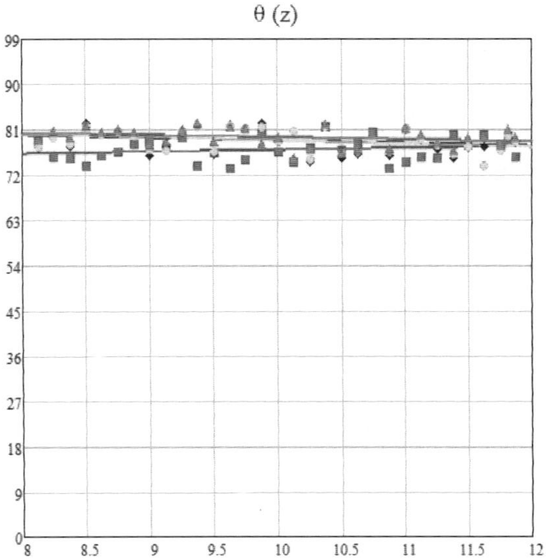

Fig. 5. Efficiencydependenceon Θ, where
fed dot smark z=6, z=7 - blue, z=8 – green and z=9 - violet.

Diagramon figure 6 illustrates zinfluence on efficiency. Maximum efficiency value is achieve dot z equal to 9 in 50% of all cases.

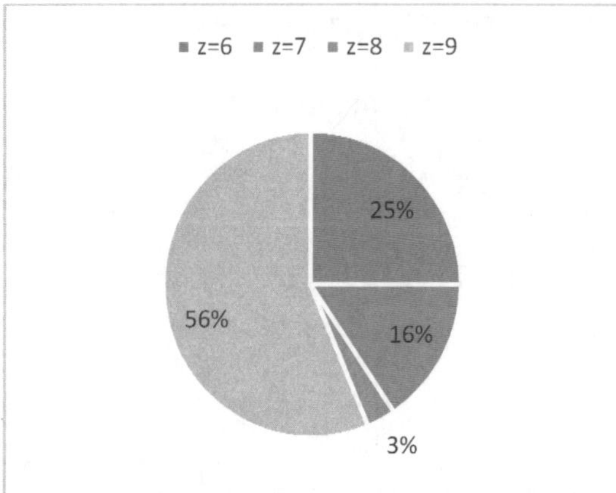

Fig. 6. Maximum efficiency value distributiondiagram for different z numbers.

The best combination of geometrical parameters for the pump is model No. 24 with z = 9, the efficiency value of the model is 82.29%, which is almost 10% more than that of the model with the minimum efficiency value. Parameters value and KPD of the best model are presented in table 4.

Table4. Optimal model.

№	D_3, mm	D_4/D_3	Θ, °	z	η, %
24	296,75	1,494	9,375	9	82,29

Discussion

In this work, the influence of the main geometrical parameters of the guide vane of a multistage centrifugal pump on its efficiency was determined. After the results of hydrodynamic modeling were analyzed, the dependence of the parameter values on the optimization criterion was formulated. These regularities can serve as a recommendation when designing a pump guide vane. A significant increase of the efficiency was achieved. In the future, it is planned to expand the parameter space and conduct a similar study of other types of guide vanes, which will make further pump efficiency increase possible.

Reference

[1] A Protopopov and D Bondareva 2019 *IOP Conf. Ser.: Mater. Sci. Eng.***492** 012002
[2] A Protopopov and V Vigovskij 2019 *IOP Conf. Ser.: Mater. Sci. Eng.***492** 012003
[3] N Egorkina and A Petrov 2019 *IOP Conf. Ser.: Mater. Sci. Eng.***492** 012015
[4] K Dobrokhodov and A Petrov 2019 *IOP Conf. Ser.: Mater. Sci. Eng.***492** 012016
[5] N Isaev 2019 *IOP Conf. Ser.: Mater. Sci. Eng.***492** 012026
[6] S Korsakova and A Protopopov 2019 *IOP Conf. Ser.: Mater. Sci. Eng.***492** 012032
[7] A Protopopov and C Jakovich 2019 *IOP Conf. Ser.: Mater. Sci. Eng.***492** 012034
[8] A Shablovskiy and E Kutovoy 2019 *IOP Conf. Ser.: Mater. Sci. Eng.***492** 012035
[9] A Petrov et al 2019 *IOP Conf. Ser.: Mater. Sci. Eng.* **492** 012036
[10] V Lomakin and O Bibik 2019 *IOP Conf. Ser.: Mater. Sci. Eng.***492** 012037
[11] T Valiev and A Petrov 2019 *IOP Conf. Ser.: Mater. Sci. Eng.***492** 012038
[12] Zhang, S., Li, H., & Xi, D. (2019). Investigation of the integrated model of side chamber, wear-rings clearance, and balancing holes for centrifugal pumps. *Journal of Fluids Engineering, Transactions of the ASME, 141*(10) doi:10.1115/1.4043059
[13] Sengpanich, K., Bohez, E. L. J., Thongkruer, P., &Sakulphan, K. (2019). New mode to operate centrifugal pump as impulse turbine. Renewable Energy, 983-993.

doi:10.1016/j.renene.2019.03.116

[14] Zhang, Z. -., Chen, H. -., Ma, Z., He, J. -., Liu, H., & Liu, C. (2019). Research on improving the dynamic performance of centrifugal pumps with twisted gap drainage blades. Journal of Fluids Engineering, Transactions of the ASME, 141(9) doi:10.1115/1.4042885

[15] V Lomakin et al 2019 *IOP Conf. Ser.: Mater. Sci. Eng.* **492** 012012

[16] V Cheremushkin and V Lomakin 2019 *IOP Conf. Ser.: Mater. Sci. Eng.* **492** 012039

IOP Conf. Series: Materials Science and Engineering **589** (2019) 012009 doi:10.1088/1757-899X/589/1/012009

Optimization of a radial guide device with a no-vane transfer channel

N Isaev[1], T Valiev[1], E Morozova[1,2] and D Bondareva[1]

[1]Bauman Moscow State Technical University, 5 Second Baumanskaya Street, Moscow, 105005, Russian Federation

[2]Email: zakharovaelenaeng@mail.ru

Abstract
Various studies show that the centrifugal multistage pump guide vanegreatly determines the hydraulic efficiency of the pump and, as a result, the optimal mode of its operation. In order to increase the pump efficiency, the radial guide vane with a no-blade transfer channelwas optimized. Three geometrical parameters and the number of channels of the guide vanewere chosen as optimization parameters; the hydraulic efficiency was the optimization criterion. 128 computational models were made, and the efficiency values at the operating point were obtained by the method of hydrodynamic modeling. As a result, significant increase in the pump efficiency was achieved.

Introduction

Nowadays research of centrifugal pumps in our time are rising in number [1] - [10]. In these studies, various methods of mathematical modeling are becoming increasingly important. Publications [11] - [16] are of the greatest interest. It should be noted that for a long time the impeller was considered to be the main part of the pump, as it almost completely determines all its energy qualities. In the VNIIGidromash, Professor S. S. Rudnev together with the staff conducted a research, the results of which radically changed the status of the diverting device as a secondary pump unit. They have shown that the output device is no less important than the impeller. While the impeller determines the energy transfer to the fluid from the drive, the output device largely determines the hydraulic losses and, therefore, the hydraulic efficiency of the pump and the optimal mode of its operation.

One of the varieties of output devices is the guiding apparatus. Thus, the primary direction of a multistage centrifugal pump efficiency increase is the optimization of the guide vane.

TheconsideredpumpCNS 300-500has 5 stages with nominal parameters: $Q = 300$ m^3/h -volumetric flow,H=500 m–head, n = 2900 rpm - rotor rotation speed, n_s=97 –specific speed coefficient.

Methods and results

The guide vane consisting of direct, transfer and reverse channels can be of several main types: radial with a no-vane transfer channel, radial channel and radial with a no-vane diffuser. Alsoguide vanes can be divided into blade and channel (tubular). For the study, the most common type of guide vane was chosen: a radial with a no-vanetransfer channel (Fig. 1).

Fig. 1. Radial channel guide vane with a no-vanetransfer channel.

As a result of preliminary calculations of a guide vane, 3 geometrical parameters which have the greatest influence on the selected optimization criterion (hydraulic efficiency) were obtained. These are the entrance diameter of the guide vane (D3), the ratio of the input and output diameters (D_4 / D_3) and the radial direction diffuser angle of the channel (Θ). In addition, z, the number of channels of the guide vane, was taken as an optimization parameter since it has a significant impact on the efficiency. The values of the optimization parameters are presented in Table 1.

Table 1. Vane guide optimization parameters.

Parameter	Min	Max
D_3, мм	292	300
D_4/D_3	1,3	1,5
Θ, ◦	5	15
z	6	9

It was decided to use LP-tau search method as a quasi-random sequence generator to create an initial population. 128 models with different values of geometrical parameters for different z numbers were generated.

Table 2. Vane guide optimization parameter values.

Model no	D_3, мм	D_4/D_3	Θ, ◦
0	296	1,4	10
1	294	1,45	7,5
2	298	1,35	12,5
3	293	1,425	13,75
4	297	1,325	8,75
5	295	1,375	11,25
6	299	1,475	6,25
7	292,5	1,488	11,875
8	296,5	1,387	6,875
9	294,5	1,338	14,375
10	298,5	1,438	9,375

IOP Conf. Series: Materials Science and Engineering **589** (2019) 012009 doi:10.1088/1757-899X/589/1/012009

Окончание табл. 2

Modelno	D_3, мм	D_4/D_3	Θ, °
11	293,5	1,363	8,125
12	297,5	1,462	13,125
13	295,5	1,413	5,625
14	299,5	1,313	10,625
15	292,25	1,406	9,063
16	296,25	1,306	14,063
17	294,25	1,356	6,563
18	298,25	1,456	11,563
19	293,25	1,331	10,313
20	297,25	1,431	5,313
21	295,25	1,481	12,813
22	299,25	1,381	7,813
23	292,75	1,394	13,438
24	296,75	1,494	8,438
25	294,75	1,444	10,938
26	298,75	1,344	5,938
27	293,75	1,469	7,188
28	297,75	1,369	12,188
29	295,75	1,319	9,688
30	299,75	1,419	14,688
31	292,125	1,459	14,531

The calculation of the characteristics of the guide vane was carried outvia the hydrodynamic modeling methods. To carry out the calculations, it is necessary to have a solid model of the liquid in the guide vane (wet part) [17, 18]. These models were built for all 128 combinations of parameters. An example of a 3d model is presented on Fig. 2

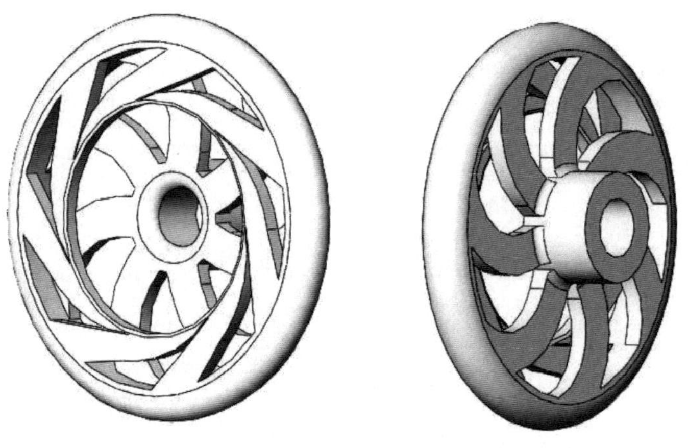

Fig. 2. 3dmodelofthewetpartofaguide.

According to the results of the calculation, the pump efficiency values at the nominal flow rate for 128 models were obtained and are summarized in Table 3.

Table 3. Efficiency values for the models.

№	z=6	z=7	z=8	z=9
0	71,84615	74,63696	76,16436	77,1819
1	72,13242	76,23065	76,64311	77,45546
2	71,62466	72,60727	74,96188	76,17113
3	72,87196	78,00551	77,58177	78,30279
4	70,52284	72,61815	74,64398	75,71789
5	71,34676	75,12726	76,28521	77,16787
6	71,84011	72,46152	75,10506	76,2552
7	73,24223	78,91635	77,83937	78,81076
8	71,18466	73,72123	75,46981	76,33718
9	71,31495	74,97471	75,99549	76,92168
10	71,84262	73,04025	75,51601	76,71559
11	72,04613	75,78705	75,8797	76,66931
12	72,26775	74,44898	76,72787	77,88402
13	71,34509	74,24394	75,44609	76,60036
14	72,29774	70,85749	73,75359	74,99172
15	72,90423	77,96368	76,84712	77,7681
16	69,43522	72,98789	75,01044	75,90298
17	71,04041	74,76312	75,45487	76,1858
18	71,41798	73,67538	76,19555	77,38556
19	71,32955	76,05073	75,82998	76,78656
20	70,75869	73,10948	75,03615	76,04021
21	72,88367	76,25514	77,46307	78,55237
22	69,79679	71,75854	74,40408	75,585
23	72,67881	77,68059	77,02212	77,94046
24	71,88522	74,79831	76,82994	77,92008
25	72,46815	76,24634	77,14632	77,99227
26	69,34801	71,35854	73,70893	74,95876
27	72,66588	76,41285	76,60459	77,66013
28	70,46876	73,04595	75,29463	76,46306
29	70,02588	73,53144	75,06103	76,10218
30	70,35392	72,18364	75,38302	76,25135
31	73,93425	79,14716	78,13022	78,68652

Forgreaterclarityfigure 3 shows a histogram of the pump efficiency distribution on the number of the computational model with $z = 9$.

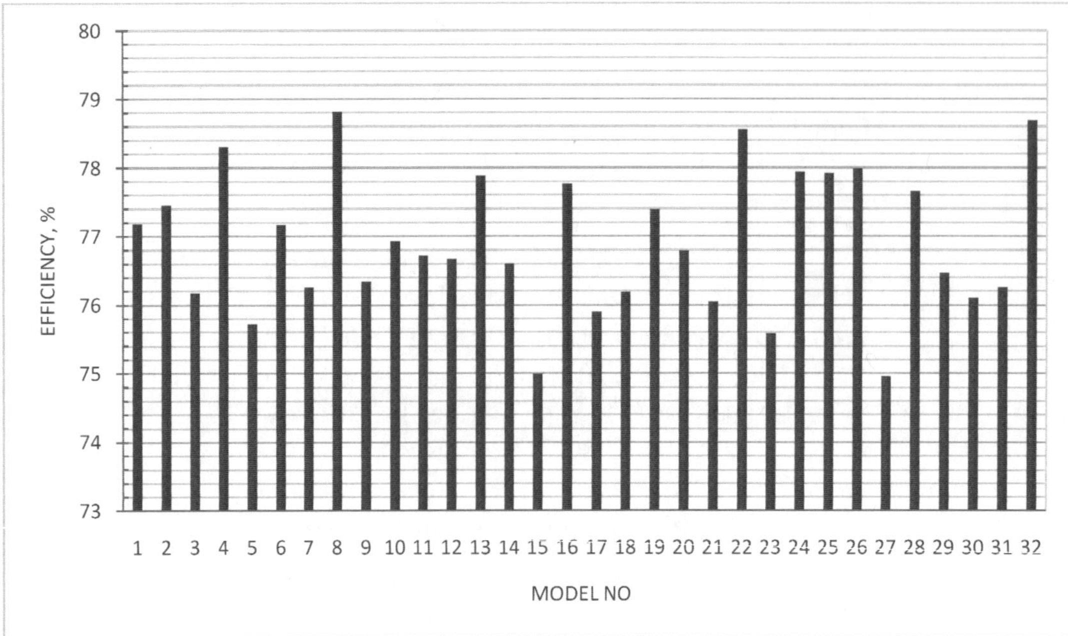

Fig. 3.histogramofefficiencydistributionat z=9.

Fig. 4 shows the dependence of the efficiency on a value equal to the ratio of the diameters of the entrance to the guide vane and the outputof the impeller, D_3 / D_2, where D_2 is the impeller diameter with z = 6.

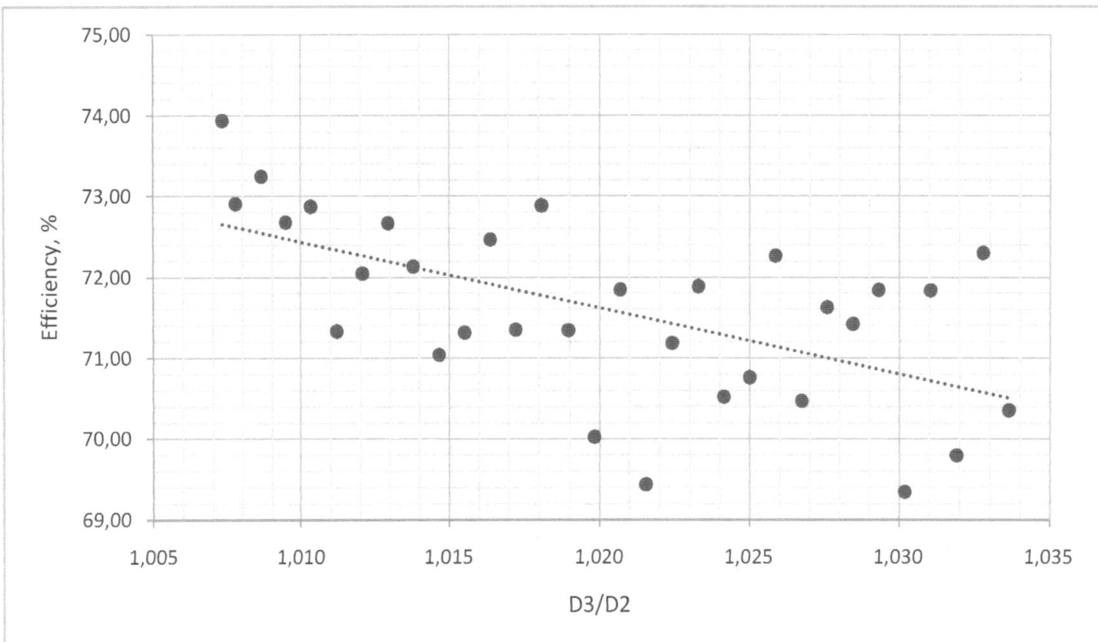

Fig. 4. Efficiency dependence on D_3/D_2at z=6.

Figure5showsefficiencydependenceon D_4/D_3, forz=6.

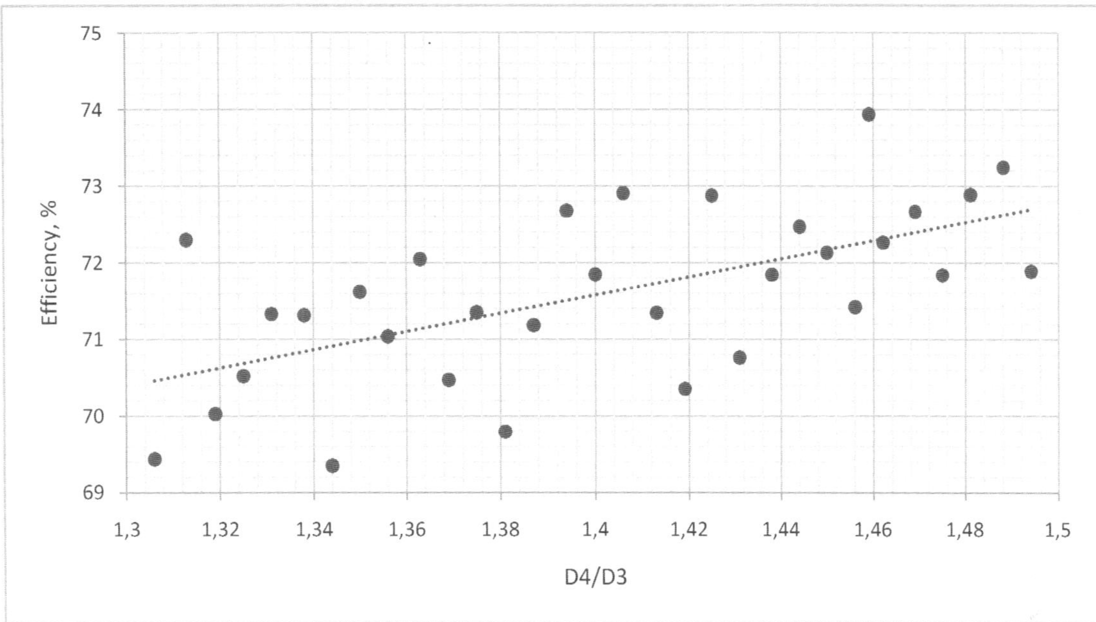

Fig. 5. Efficiency dependence on D_4/D_3at z=6.

Figure5showsefficiencydependenceon the channel diffuser angle Θ, forz=6.

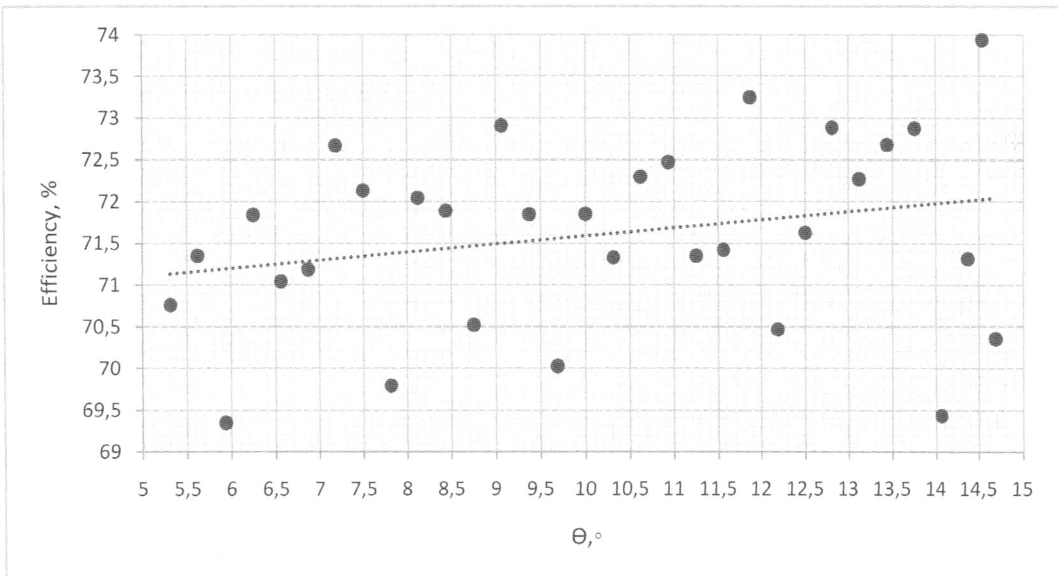

Fig. 6. Efficiency dependence onΘatz=6.

Diagram on figure 7 illustrates z influence on efficiency. Maximum efficiency value is achieved at z equal to 9 in 90,2 % of all cases which proves advisability of 9 channels.

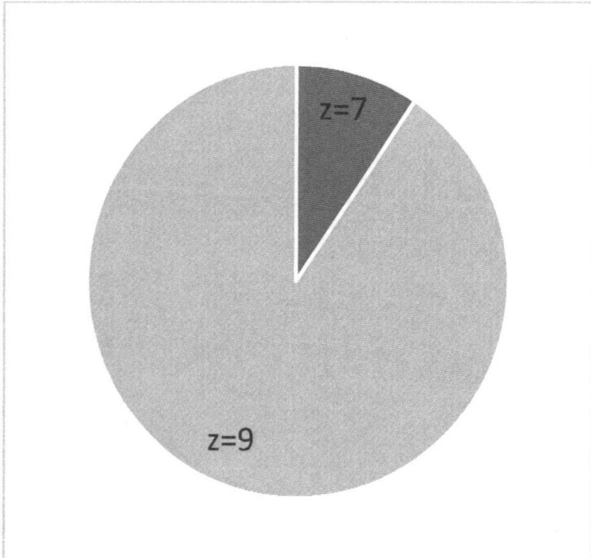

Fig. 7. Maximum efficiency value distribution diagram for different z numbers.

The best option for combining geometrical parameters for the pump is model No. 7 with z = 9, the efficiency value of this model is 78.8%, which is almost 10% more than that of the model with the minimum efficiency value. Parameter values of the best model are presented in Table 4.

Table4. Optimal model.

№ модели	D_3, мм	D_4/D_3	Θ, °	z	η, %
7	292,5	1,488	11,875	9	78,81076

Conclusion.
In this work, the influence of the main geometrical parameters of a multistage centrifugal pump guide vane on its efficiency was determined. After analyzing the results of hydrodynamic modeling, the laws of the parameter valuesinfluence on the optimization criterion were formulated. These patterns can serve as a recommendation when designing a pump guide vane.

In the future it is planned to expand the parameter space and conduct a similar study of the radial channel translational and radial transfer guide vanes with a no - vane diffuser, which will make further pump efficiency increase possible.

References
[1] A Protopopov and D Bondareva 2019 *IOP Conf. Ser.: Mater. Sci. Eng.***492** 012002
[2] A Protopopov and V Vigovskij 2019 *IOP Conf. Ser.: Mater. Sci. Eng.***492** 012003
[3] P Chaburko and Z Kossova 2019 *IOP Conf. Ser.: Mater. Sci. Eng.***492** 012011
[4] A Gouskov et al 2019 *IOP Conf. Ser.: Mater. Sci. Eng.***492** 012013
[5] N Egorkina and APetrov 2019 *IOP Conf. Ser.: Mater. Sci. Eng.***492** 012015
[6] K Dobrokhodov and APetrov 2019 *IOP Conf. Ser.: Mater. Sci. Eng.***492** 012016
[7] N Isaev 2019 *IOP Conf. Ser.: Mater. Sci. Eng.***492** 012026
[8] S Korsakova and AProtopopov 2019 *IOP Conf. Ser.: Mater. Sci. Eng.***492** 012032
[9] A Protopopov and C Jakovich 2019 *IOP Conf. Ser.: Mater. Sci. Eng.***492** 012034
[10] A Shablovskiy and E Kutovoy 2019 IOP Conf. Ser.: Mater. Sci. Eng.492 012035
[11] A Petrov et al 2019 IOP Conf. Ser.: Mater. Sci. Eng. 492 012036

[12] V Lomakin and O Bibik 2019 IOP Conf. Ser.: Mater. Sci. Eng.492 012037
[13] T Valiev and APetrov 2019 IOP Conf. Ser.: Mater. Sci. Eng.492 012038
[14] Blume, M., & Skoda, R. (2019). 3D flow simulation of a circular leading edge hydrofoil and assessment of cavitation erosion by the statistical evaluation of void collapses and cavitation structures. Wear, 428-429, 457-469. doi:10.1016/j.wear.2019.04.011
[15] Yun, R., Zuchao, Z., Denghao, W., &Xiaojun, L. (2019). Influence of guide ring on energy loss in a multistage centrifugal pump. Journal of Fluids Engineering, Transactions of the ASME, 141(6) doi:10.1115/1.4041876
[16] V Lomakin et al 2019 *IOP Conf. Ser.: Mater. Sci. Eng.* **492** 012012
[17] V Cheremushkin and V Lomakin 2019 *IOP Conf. Ser.: Mater. Sci. Eng.* **492** 012039

IOP Conf. Series: Materials Science and Engineering **589** (2019) 012010 doi:10.1088/1757-899X/589/1/012010

The starting-up dynamics of the pumping station

V Tkachyk[1], S Yakovich[1] and A Shablovskiy[1,2]

[1]Bauman Moscow State Technical University, 5 Second Baumanskaya Street, Moscow, 105005, Russian Federation

[2]Email: shabl@bmstu.ru

Abstract
A mathematical model of the starting-up of the centrifugal pump rotor with an asynchronous electric motor has been built as part of a pumping station operating on the grid. The described method takes into account such factors as the pressure loss in the pipeline, inertial pressure occurring during the operation of the pump, the torque of the impeller and the torque of the engine.
The obtained method of dynamic analysis makes it possible to estimate the overheating of the electric motor of a centrifugal pump depending on various factors and, as a consequence, to predict its possible failure.

Introduction

Centrifugal pumps are widely used in industrial enterprises, in agriculture and for urban water supply, as a transport system for various working fluids.

The issues related to the centrifugal pumps analysis are widely reported in the literature [1–15]. The works [16]-[22] are of particular interest. However, the issues related to the temperature of the motor during the pump starting-up require separate consideration.

Let's consider a pump set, a schematic diagram of which is shown in figure 1.

A method for analyzing starting phenomena in the pumps has been developed in the course of work on the mathematical model of the considered pump. The correctness of this technique would be checked on an experienced pump with the following parameters: working fluid density ρ = = 1000 kg/m^3; total efficiency of the pump $\eta_\Sigma = 0.7$; radius of the impeller at the output $R_2 = 0.24$ m; pipe diameter $d = 0.142$ m; pipeline length $l = 20$ m; flow coefficient $\mu = 0{,}5 \cdot 10^{-4}$; front axial clearance between the impeller and the housing $a = 0.001$ m; the coefficient of torque-mechanical characteristics of the electric drive K_1=35 N · m · s; the coefficient of the torque-mechanical characteristics of the motor K = 5000 N · m; winding mass $m = 2$ kg; motor operating voltage U = 220 V; starting factor $K_p = 20$; specific heat capacity $C_{\text{medi}} = 500$; friction resistance coefficient $l = 0.04$; coefficients $K_2 = 200$ and $K_3 = 0$; rotor inertia moment $J = 132 \cdot 10^{-4}$kg · m^2.

All graphic dependencies are for the certain pump.

IOP Conf. Series: Materials Science and Engineering **589** (2019) 012010 doi:10.1088/1757-899X/589/1/012010

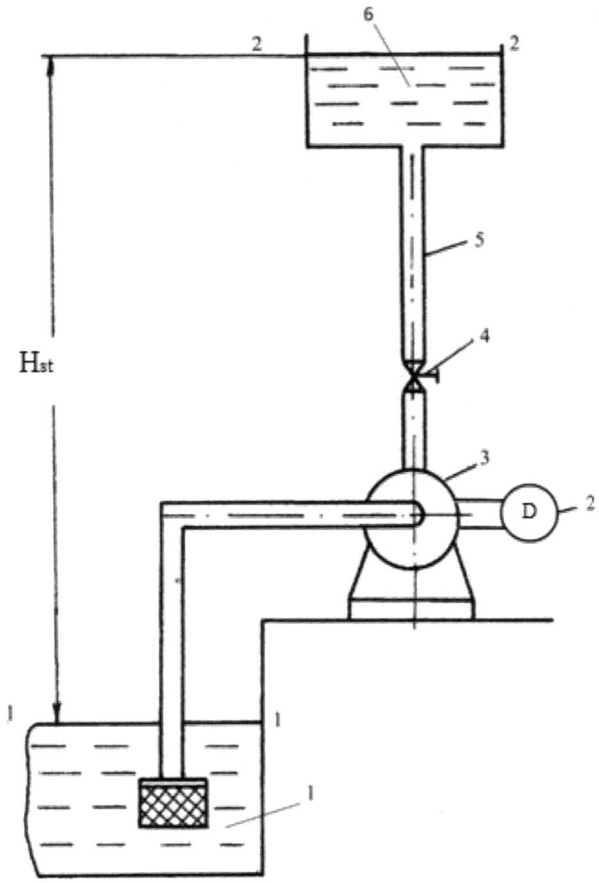

Fig. 1. The schematic diagram of a centrifugal pump with asynchronous electric motor: *1*—feeder tank; *2*—asynchronous electric drive; *3*— centrifugal pump; *4*—gate valve; *5*—pipeline; *6*—receiver tank

Mathematical model

Based on the momentum theorem, we'll compose the moment equation:

$$J\frac{d\omega}{dt} = M_{dv}(t) - \alpha M_{rk}(t). \tag{1}$$

Where J — the rotor moment of inertia about the axis; ω is the angular speed of rotation of the pump shaft; M_{dv}— the moment of the drive without load; M_{rk}— the moment of the electrodrive at the time of launch; α — losses on the coupling, bearings, pump seals during torque transmission,$\alpha > 1$.

Let's consider the terms of the equation (1).

The torque on the impeller can be calculated like this

$$M_{rk} = M_c + M_{dt}, \tag{2}$$

whereM_c— centrifugal moment, M_{dt} — frictional disk torque.

Centrifugal moment:

$$M_c(t) = \rho Q R_2^2 \omega(t). \tag{3}$$

$$M_{dt}(t)=\frac{\omega(t)\mu\pi R_2^2}{a}. \quad (4)$$

where a — the axial clearance between the impeller and the pump housing.

Thus, the torque of the impeller

$$M_{rk}(t)=\frac{\omega(t)\mu\pi R_2^2}{a}+\rho Q R_2^2\omega(t). \quad (5)$$

The torque, generated by drive, is described by formula [8]:

$$M_d(t)=K-K_1\omega(t), \quad (6)$$

where K and K_1 — coefficients of torque-mechanical characteristics of the electric drive.

Let's derive the balance equation for the required head:

$$H_H=H_{st}+H_{tr}+h_{in}, \quad (7)$$

whereH_H — pump head required to overcome losses; H_{st}— static pressure between the tanks of the feeder and receiver; H_{tr} — pressure loss in the pipeline; h_{in} - inertial head.

Let's consider the terms of the equation (7).

We'll use the dynamic similarity formulas for a pump with a constant linear size for recalculation from different frequencies:

$$H_{\text{н}}(Q;\omega)=H_0\left(\frac{\omega_0}{\omega}\right)^2, \quad (8)$$

0 — the initial values ω — head and angular velocity of the pump at starting-up; H0, ω where Hн, of the pump head and angular velocity.

The inertial head is caused by the acceleration or deceleration of the fluid flow, therefore, to find it, we'll use the second Newton law for the element of the stream of an ideal incompressible fluid, and as a result the equation would be obtained:

$$h_{in}=\frac{1}{g}\int_{l_1}^{l_2}\frac{\partial V}{\partial t}dl=\frac{j}{g}l, \quad (9)$$

where j— the acceleration of the fluid flow; l— length of the pipeline.

The acceleration of fluid flow j will be obtained after differentiating the flow formulas bytime t:

$$\frac{dV}{dt}=j=\frac{1}{F}Q'. \quad (10)$$

The pressure losses in the pipeline are the sum of friction losses along the length and losses in local resistance, expressed by flow rate. With this in mind, we'll write down the formula of general loss:

$$H_{tr}=\left(\frac{\lambda\frac{l}{d}+\zeta(t)}{2gF^2}\right)Q^2(t), \quad (11)$$

where*l* is the friction resistance coefficient; *z*(*t*) — the total coefficient of local resistance; *l*, *d* — the length and diameter of the pipeline; $F=\frac{\pi d^2}{4}$ — the cross-section area of the pipeline; *Q*(*t*) — the flow rate through the section.

IOP Conf. Series: Materials Science and Engineering **589** (2019) 012010 doi:10.1088/1757-899X/589/1/012010

The initial conditions for the task:

$$\omega(0)=0; \quad (12)$$

$$Q(0)=0. \quad (13)$$

So, the mathematical model of the starting-up process of the considered pump is:

$$\begin{cases} H_0\left(\dfrac{\omega_0}{\omega}\right)^2 = H_{st} + \left(\dfrac{\lambda \dfrac{l}{d} + \zeta(t)}{2gF^2}\right) Q^2(t) + \dfrac{Q`(t)l}{Fg} \\ J\dfrac{d\omega}{dt} = K - K_1\omega(t) - \alpha\left(\dfrac{\omega(t)\mu\pi R_2^2}{a} + \rho Q R_2^2 \omega(t)\right) \\ \omega(0)=0 \\ Q(0)=0 \end{cases} \quad (14)$$

Let's find the solution of the system of equations (14). To do this, we'll rewrite it as follows:

$$\begin{cases} Q`(t) = \dfrac{Fg}{l}\left(H_0\left(\dfrac{\omega_0}{\omega}\right)^2 - H_{st} - \left(\dfrac{\lambda \dfrac{l}{d} + \zeta(t)}{2gF^2}\right) Q^2(t)\right) \\ \dfrac{d\omega}{dt} = \dfrac{K - K_1\omega(t) - \alpha\left(\dfrac{\omega(t)\mu\pi R_2^2}{a} + \rho Q R_2^2 \omega(t)\right)}{J} \\ \omega(0)=0 \\ Q(0)=0 \end{cases} \quad (15)$$

Results

The system of equations (15) will be solved by the Mathcad soft ware package using the Runge-Kutt method of the 4-th order. The graphical dependence of the angular velocity on time will have the form shown in Fig. 2, and the dependence of the flow on time is shown in Fig. 3

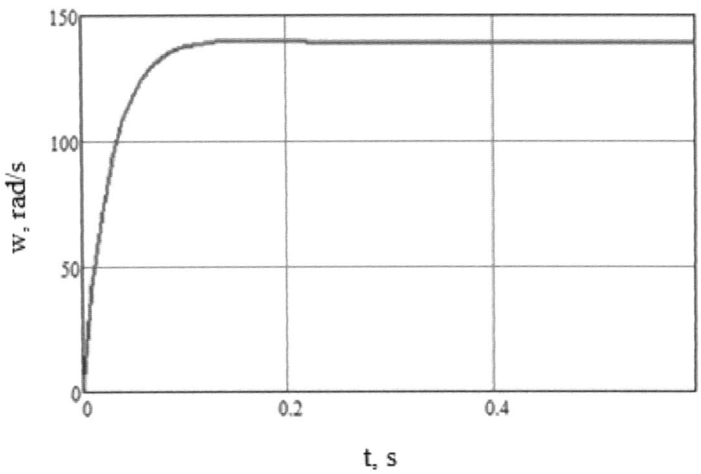

Fig. 2. The dependence of the angular velocity of the rotor on time

The dependence of the flow on time:

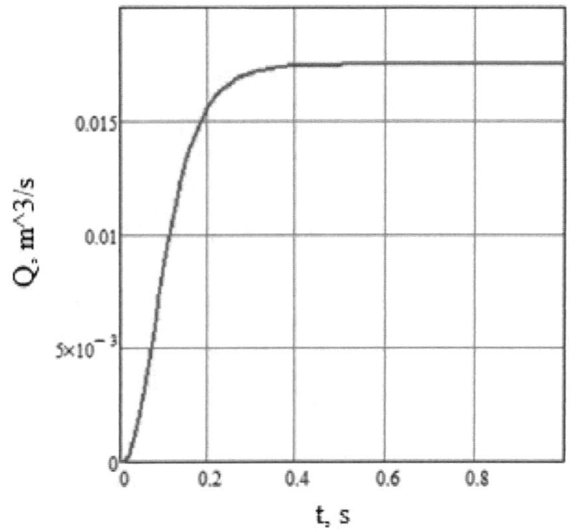

Fig. 3. The flow versus time graph

The Joule - Lenz law for the amount of heat:

$$I_{nom}K_pUt_{pp} = C_{medi}m_{prov}\Delta T, \tag{16}$$

whereI_{nom}— current intensity; K_p— start factor; t_{pp}— transient time; U — voltage; C_{medi}— copper specific heat capacity; m_{prov}— the mass of the conductor; ΔT —drive over heating temperature.

Power of electric current in the drive winding:

$$P = I_{nom}U = \frac{\rho gHQ}{\eta}, \tag{17}$$

whereη is the pump efficiency.

Let's express the value of the current $I_{nom}$from the equation (17)

$$I_{nom} = \frac{\rho gQH}{\eta U}. \tag{18}$$

A general equation for temperature based on the equations (16) and (18):

$$\Delta T = \frac{t_{pp}I_{nom}K_pU}{C_{medi}m_{prov}} = \frac{t_{pp}\rho gQHK_p}{C_{medi}m_{prov}\eta}. \tag{19}$$

So, we obtain the calculation of the temperature of the drive overheating during starting-up.

Conclusion

A mathematical model of starting-up of a centrifugal pump rotor with an asynchronous electric drive has been constructed. The described method takes in to account such factors as the pressure loss in the pipeline H_{tr}, inertial pressure h_{in}arising during pump operation, the moment of the impeller $M_{rk}$and the moment of the engine M_d.

The obtained method of dynamic calculation makes it possible to estimate the over heating of the electric motor of a centrifugal pump depending on various factors and, as a result, to predict its possible failure.

A number of assumptions has been made during the construction of the model, in particular, it has been assumed that during the transition process the starting current is considered to be constant and a multiply greater than the nominal with the same parameters; also, the cooling process was not taken in to account during starting up because of the smallness of the drive starting time.

References

[1] A Protopopov and D Bondareva 2019 *IOP Conf. Ser.: Mater. Sci. Eng.***492** 012002
[2] A Protopopov and V Vigovskij 2019 *IOP Conf. Ser.: Mater. Sci. Eng.***492** 012003
[3] P Chaburko and Z Kossova 2019 *IOP Conf. Ser.: Mater. Sci. Eng.***492** 012011
[4] V Lomakin et al 2019 *IOP Conf. Ser.: Mater. Sci. Eng.***492** 012012
[5] A Gouskov et al 2019 *IOP Conf. Ser.: Mater. Sci. Eng.***492** 012013
[6] N Egorkina and APetrov 2019 *IOP Conf. Ser.: Mater. Sci. Eng.***492** 012015
[7] K Dobrokhodov and APetrov 2019 *IOP Conf. Ser.: Mater. Sci. Eng.***492** 012016
[8] N Isaev 2019 *IOP Conf. Ser.: Mater. Sci. Eng.***492** 012026
[9] S Korsakova and AProtopopov 2019 *IOP Conf. Ser.: Mater. Sci. Eng.***492** 012032
[10] A Protopopov and C Jakovich 2019 *IOP Conf. Ser.: Mater. Sci. Eng.***492** 012034
[11] A Petrov et al 2019 *IOP Conf. Ser.: Mater. Sci. Eng.* **492** 012036
[12] V Lomakin and O Bibik 2019 *IOP Conf. Ser.: Mater. Sci. Eng.***492** 012037
[13] T Valiev and APetrov 2019 *IOP Conf. Ser.: Mater. Sci. Eng.***492** 012038
[14] V Cheremushkin and V Lomakin 2019 *IOP Conf. Ser.: Mater. Sci. Eng.***492** 012039
[15] Zhang, S., Li, H., & Xi, D. (2019). Investigation of the integrated model of side chamber, wear-rings clearance, and balancing holes for centrifugal pumps. *Journal of Fluids Engineering, Transactions of the ASME, 141*(10) doi:10.1115/1.4043059
[16] Sengpanich, K., Bohez, E. L. J., Thongkruer, P., &Sakulphan, K. (2019). New mode to operate centrifugal pump as impulse turbine. Renewable Energy, 983-993. doi:10.1016/j.renene.2019.03.116
[17] Yousefi, H., Noorollahi, Y., Tahani, M., Fahimi, R., &Saremian, S. (2019). Numerical simulation for obtaining optimal impeller's blade parameters of a centrifugal pump for high-viscosity fluid pumping. Sustainable Energy Technologies and Assessments, 34, 16-26. doi:10.1016/j.seta.2019.04.011
[18] Blume, M., & Skoda, R. (2019). 3D flow simulation of a circular leading edge hydrofoil and assessment of cavitation erosion by the statistical evaluation of void collapses and cavitation structures. Wear, 428-429, 457-469. doi:10.1016/j.wear.2019.04.011
[19] Yun, R., Zuchao, Z., Denghao, W., &Xiaojun, L. (2019). Influence of guide ring on energy loss in a multistage centrifugal pump. Journal of Fluids Engineering, Transactions of the ASME, 141(6) doi:10.1115/1.4041876

IOP Conf. Series: Materials Science and Engineering **589** (2019) 012011 doi:10.1088/1757-899X/589/1/012011

On the centrifugal pump optimization question

V Tkachyk[1], V Atamas[2] and A Shablovskiy[1,3]

[1]Bauman Moscow State Technical University, 5 Second Baumanskaya Street, Moscow, 105005, Russian Federation
[2]Bohdan Khmelnytsky National University of Cherkasy, 81 Shevchenko Boulevard, Cherkasy, Ukraine

[3] Email: shabl@bmstu.ru

Abstract
There is a wide range of rotor rotational frequencies and cavitation cavitatin margins used in the industrial centrifugal pumps. The effect of these parameters on the dimensions criteria and cavitation characteristics of a pump is studied in this paper, the Sobol method is proposed for it and the task is to find the minimum number of point sat which the best ratio of the parameters will be achieved.

Introduction

There is a huge abundance of researches [1] - [11] in the area of centrifugal pump analysis. The studies [12] - [22] are of particular interest. A common feature of these studies is the use of complex computational methods involving various empirical formulas and hydrodynamic modeling. These methods are quite difficult to implement and therefore there is a need to find less time-consuming methods for centrifugal pumps optimization.

The main characteristics of centrifugal pumps for general industrial use are considered to be overall dimensions and cavitation qualities, characterized by the value of the suction specific speed, which determines the risk of cavitation occurrence in the pump. Cavitational destruction in the pump, the value of efficiency, pump head depend on how good the cavitational properties of the hydraulic machine are, and the compactness and amount of consumed material for manufacturing depend on the dimensions of the impeller.

The solving of the optimizing task for these characteristics is possible using the LP-tau search method, which generates points in a quasi-random manner in a selected interval of two parameters. For the considered centrifugal pumps these parameters are the cavitation margin (the overpressure of the working fluid at the pump inlet above the pressure of its saturated vapor) Δh will be varyied from 1 to 9 m of height of water and the rotor rotation speed n – from 500 to 12000 rpm. Then the generated work points field will be a rectangle.

Methods

Thus, in order to find the variables we need, we will use the formulas known from the literature that define the radial dimension of the impeller D_2 and the suction specific speed C. Rudnev's formula for calculating C:

$$C = 5.62 \cdot n_p \cdot \frac{\sqrt{Q}}{\Delta h^{\frac{3}{4}}}, (1)$$

where Q – pump flow, m^3/s. The formula for calculating D_2, obtained from Euler's formula, taking into account the efficiency of the pump η:

$$D_2 = 2\sqrt{\frac{H \cdot g}{\eta \cdot \omega^2}}, \quad (2)$$

where H – pump head, m, ω – angular velocity, rad/s. The number of calculation points will be determined as follows. A dimensionless generalized quality factor Φ is introduced. It is determined by the dependence

$$F = \left(\frac{D_2}{D_{2(1)}}\right)^2 + \left(\frac{C}{C_{(1)}}\right)^2,$$

where $D_2(1)$ and C(1) – values of diameter (in m) and suction specific speed (dimensionless) for the first design point. The smaller the ratio, the better the quality of the centrifugal pump is. After calculating the generalized quality factor for 64 points, its minimum values for a certain number of points will be chosen: for 4, 8, 16, 32 and 64 points. Then the dependence F_{min} (n) will be formulated, where F_{min} is the minimum the value of the generalized quality factor for a given number of points, n – the number of points. To formulate this dependence, we'll calculate the values of the generalized quality factor depending on the number of the test point and put it in the table.

Table 1. The calculated points of the suction specific speed and wheel diameter

Point №	n_p, rpm	Δh, m	C	D_2, m	F	Point №	n_p, rpm	Δh, m	C	D_2, m	F
1	6250	5.50	1458	0.111	1.991	33	6429	3.67	2031	0.107	2.877
2	3375	7.75	609	0.205	3.574	34	3554	1.42	2287	0.194	5.525
3	9125	3.25	3158	0.076	5.157	35	9304	5.92	2053	0.074	2.43
4	1937	6.62	393	0.357	10.388	36	2117	4.79	547	0.326	8.78
5	7687	2.12	3659	0.090	6.953	37	7867	9.29	1238	0.088	1.347
6	4812	4.37	1333	0.144	2.508	38	4992	7.04	966	0.138	1.994
7	10562	8.87	1721	0.065	1.74	39	10742	2.54	4464	0.064	9.71
8	1218	9.43	190	0.567	26.087	40	1398	3.10	500	0.494	19.918
9	6968	4.93	1763	0.099	2.26	41	7148	7.60	1307	0.097	1.561
10	4093	2.68	1634	0.169	3.567	42	4273	9.85	643	0.162	2.315
11	9843	7.18	1879	0.070	2.06	43	10023	5.35	2384	0.069	3.059
12	2656	3.81	816	0.260	5.801	44	2835	6.48	584	0.244	4.976
13	8406	8.31	1439	0.082	1.522	45	8585	1.98	4302	0.080	9.231
14	5531	6.06	1199	0.125	1.942	46	5710	4.23	1621	0.121	2.423
15	11281	1.56	6763	0.061	21.82	47	11460	8.73	1890	0.060	1.975
16	859	5.78	193	0.804	52.45	48	1039	3.39	348	0.665	35.923
17	6609	1.28	4598	0.105	10.832	49	6789	7.89	1208	0.102	1.527
18	3734	3.53	1215	0.185	3.471	50	3914	5.64	895	0.176	2.905
19	9484	8.03	1666	0.073	1.736	51	9664	1.14	7336	0.071	25.731
20	2296	2.40	996	0.301	7.807	52	2476	9.01	398	0.279	6.388

Point №	n_p, rpm	Δh, m	C	D_2, m	F	Point №	n_p, rpm	Δh, m	C	D_2, m	F
21	8046	6.90	1582	0.086	1.775	53	8226	4.51	2225	0.084	2.901
22	5171	9.15	823	0.134	1.766	54	5351	2.26	2428	0.129	4.125
23	10921	4.65	2887	0.063	4.245	55	11101	6.76	2217	0.062	2.626
24	1578	5.21	382	0.438	15.617	56	1757	7.32	330	0.393	12.583
25	7328	9.71	1115	0.094	1.306	57	7507	2.82	2884	0.092	4.6
26	4453	7.46	825	0.155	2.273	58	4632	5.07	1147	0.149	2.423
27	10203	2.96	3779	0.068	7.09	59	10382	9.57	1598	0.067	1.56
28	3015	8.59	503	0.229	4.377	60	3195	1.70	1796	0.216	5.31
29	8765	4.09	2552	0.079	3.568	61	8945	6.20	1907	0.077	2.195
30	5890	1.84	3119	0.117	5.692	62	6070	8.45	1026	0.114	1.546
31	11640	6.34	2440	0.059	3.086	63	11820	3.95	3532	0.058	6.146
32	679	8.17	118	1.016	83.826	64	589	6.97	115	1.171	111.305

Thus, we have the opportunity to construct the dependence of the minimum value of the generalized quality criterion F_{min} on the number of calculation points n (see figure).

Results

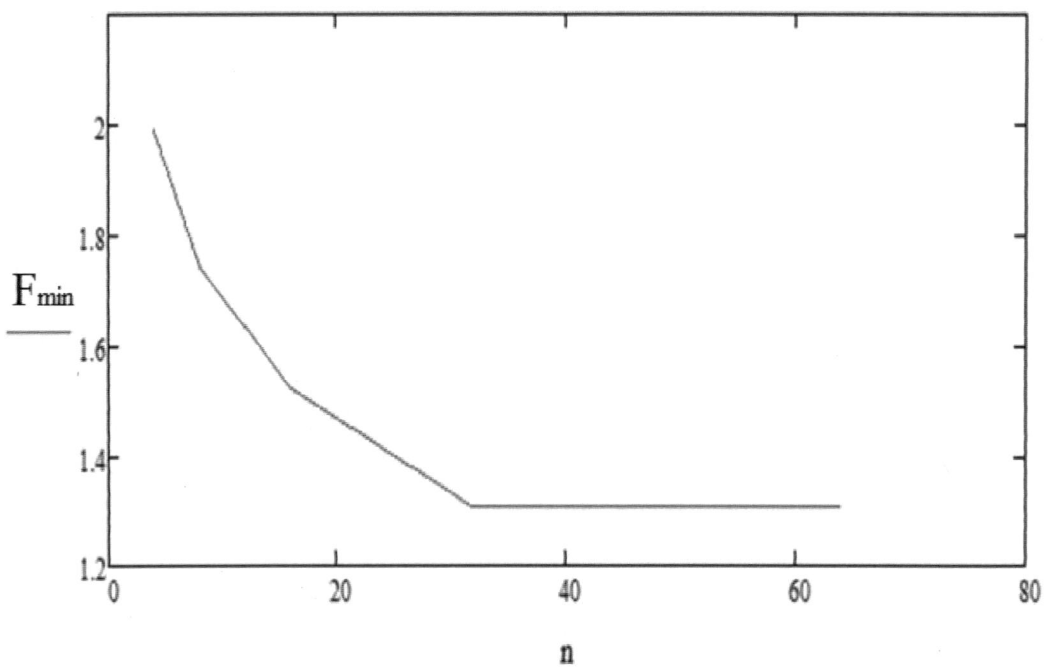

Figure 1. Dependence of the minimum value of the generalized quality criterion on the number of points

Conclusion

For successful application of the LP-tau sequence, at least 32 points have to be used to find acompromise between the dimensions and the suction specific speed. This can be clearly seenfrom the resulting curve - the graph $F_{min}(n)$ approached the shape of the horizontal line at n> 32 when64 points of the LP-tau sequence are considered. Thus, the above method can be recommended forcavitation analysis of centrifugal pumps in order to achieve a high degree of protection againstcavitation.

References

[1] A Protopopov and D Bondareva 2019 *IOP Conf. Ser.: Mater. Sci. Eng.***492** 012002
[2] A Protopopov and V Vigovskij 2019 *IOP Conf. Ser.: Mater. Sci. Eng.***492** 012003
[3] P Chaburko and Z Kossova 2019 *IOP Conf. Ser.: Mater. Sci. Eng.***492** 012011
[4] V Lomakin et al 2019 *IOP Conf. Ser.: Mater. Sci. Eng.***492** 012012
[5] A Gouskov et al 2019 *IOP Conf. Ser.: Mater. Sci. Eng.***492** 012013
[6] N Egorkina and A Petrov 2019 *IOP Conf. Ser.: Mater. Sci. Eng.***492** 012015
[7] K Dobrokhodov and A Petrov 2019 *IOP Conf. Ser.: Mater. Sci. Eng.***492** 012016
[8] N Isaev 2019 *IOP Conf. Ser.: Mater. Sci. Eng.***492** 012026
[9] S Korsakova and A Protopopov 2019 *IOP Conf. Ser.: Mater. Sci. Eng.***492** 012032
[10] A Protopopov and C Jakovich 2019 *IOP Conf. Ser.: Mater. Sci. Eng.***492** 012034
[11] B Ksenofontov et al 2019 *IOP Conf. Ser.: Mater. Sci. Eng.* **492** 012001
[12] A Petrov et al 2019 *IOP Conf. Ser.: Mater. Sci. Eng.* **492** 012036
[13] V Lomakin and O Bibik 2019 *IOP Conf. Ser.: Mater. Sci. Eng.***492** 012037
[14] T Valiev and A Petrov 2019 *IOP Conf. Ser.: Mater. Sci. Eng.***492** 012038
[15] V Cheremushkin and V Lomakin 2019 *IOP Conf. Ser.: Mater. Sci. Eng.***492** 012039
[16] Zhang, S., Li, H., & Xi, D. (2019).Investigation of the integrated model of side chamber, wear-rings clearance, and balancing holes for centrifugal pumps. *Journal of Fluids Engineering, Transactions of the ASME, 141*(10) doi:10.1115/1.4043059
[17] Seng panich, K., Bohez, E. L. J., Thongkruer, P., &Sakulphan, K. (2019). New mode to operate centrifugal pump as impulse turbine. Renewable Energy, 983-993. doi:10.1016/j.renene.2019.03.116
[18] Zhang, Z. -., Chen, H. -., Ma, Z., He, J. -., Liu, H., & Liu, C. (2019). Research on improving the dynamic performance of centrifugal pumps with twisted gap drainage blades. Journal of Fluids Engineering, Transactions of the ASME, 141(9) doi:10.1115/1.4042885
[19] Blume, M., & Skoda, R. (2019).3D flow simulation of a circular leading edge hydrofoil and assessment of cavitation erosion by the statistical evaluation of void collapses and cavitation structures. Wear, 428-429, 457-469. doi:10.1016/j.wear.2019.04.011
[20] Yun, R., Zuchao, Z., Denghao, W., &Xiaojun, L. (2019).Influence of guide ring on energy loss in a multistage centrifugal pump. Journal of Fluids Engineering, Trans actions of the ASME, 141(6) doi:10.1115/1.4041876

IOP Conf. Series: Materials Science and Engineering **589** (2019) 012012 doi:10.1088/1757-899X/589/1/012012

Optimization of centrifugal pump radial guide vanes

E Morozova[1,2], D Bondareva[1] and A Shablovsky[1]

[1]Bauman Moscow State Technical University, 5 Second Baumanskaya Street, Moscow, 105005, Russian Federation

[2]Email: zakharovaelenaeng@mail.ru

Abstract
Centrifugal multi-stage pump hydraulic efficiency and, as a result, its optimal mode of operation depend primarily on guide vanes. Radial crossover guide vanes were optimized in order to increase pump hydraulic efficiency. Three geometric parameters and the number of guide vane channels were chosen as optimization parameters, and pump hydraulic efficiency was chosen as the criterion. Automatic parameterized design method produced 128 design models; hydraulic efficiency in work mode was calculated with the help of hydrodynamic simulation. Optimization brought about significant increase of pump hydraulic efficiency.

Introduction

More and more research papers on centrifugal pumps are published today [1]-[15]. Various methods of mathematical modeling gain increased importance in those research papers. Most interesting are publications [16]-[20] and in most cases attempts to increase pump efficiency are focused on the impeller with guide vanes relegated to a secondary role. However professor S.S. Rudnev from NPO«GIDROMASH» and his team showed that a diverting device was no less important than the impeller. Hydraulic losses – and consequently pump hydraulic efficiency and optimal work mode – depend to a large extenton the diverting device.

Guide vanes are a certain type of diverting devices. As a result, optimizing guide vanes should be the primary task when increasing centrifugal multi-stage pump hydraulic efficiency. Guide vanes are used primarily in multi-stage pumps with n_s<120 and insingle-stage pumps with n_s= = 250-500.Guidevanes collects liquid and supplies it to the next stage of the pump, converting kinetic energy to potential energy and decreasing moment of velocities.

The CPS 300-500 pump has 5 stages and the following rated parameters: Q = 300 m^3/h -flow rate, H=500 m - head, n = 2900 r/min - rotational speed, n_s=97 - specific speed.

Guide vanes consist of direct, crossover and inverse channels. There are several types of guide vanes, including radial crossover with vane less crossover channel, radial crossover and radial crossover with vane less space. Also guide vanes can be divided into blading and channel categories. The research was carried out with the use of radial crossover guide vanes (Fig. 1).

As a result of preliminary calculations, which most heavily influence the chosen criterion of optimization (hydraulic efficiency), 3 geometric parameters of guide vanes were obtained: inlet diameter of the guide vanes (D_3),inlet to outlet diameters ratio (D_4/D_3) and diffusion airfoil angle of a channel in radial direction (Θ).z – number of guide vane channels–was also used as an optimization parameter because it seriously influences pump efficiency. Optimization parameters values are shown in Table 1.

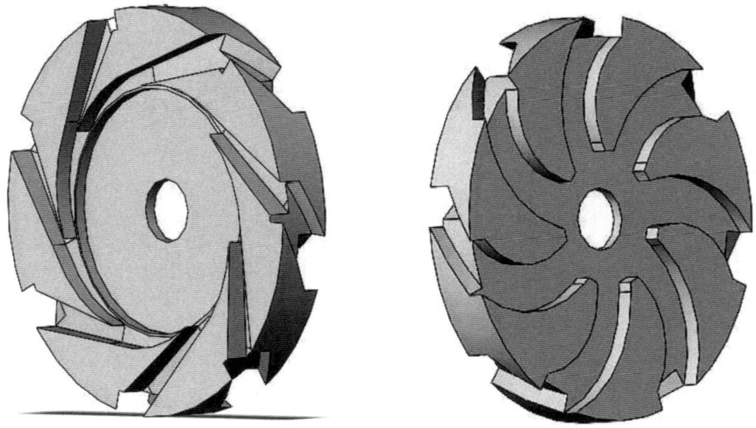

Fig. 1. Radial crossover guide vanes with vane less crossover channel.

Table1. Guide vanes optimization parameters.

Parameter	Min	Max
D_3, mm	292	300
D_4/D_3	1,3	1,5
Θ, °	8	12
z	6	9

Sobol sequences (LP-tausequences) were used as a pseudo-random sequence generator to produce initial values, because this method delivers more uniform population of parameter space compared with other generators. 128 models were generated with different values of geometric parameters for different values of z.

Table2. Combinations of guide vanes geometric parameters.

№ of model	D_3	D_4/D_3	Θ
0	296	1,4	10
1	294	1,45	9
2	298	1,35	11
3	293	1,425	11,5
4	297	1,325	9,5
5	295	1,375	10,5
6	299	1,475	8,5
7	292,5	1,488	10,75
8	296,5	1,387	8,75
9	294,5	1,338	11,75
10	298,5	1,438	9,75
11	293,5	1,363	9,25
12	297,5	1,462	11,25
13	295,5	1,413	8,25

IOP Conf. Series: Materials Science and Engineering **589** (2019) 012012 doi:10.1088/1757-899X/589/1/012012

14	299,5	1,313	10,25
15	292,25	1,406	9,625

Окончание табл. 2

№ ofmodel	D_3	D_4/D_3	Θ
16	296,25	1,306	11,625
17	294,25	1,356	8,625
18	298,25	1,456	10,625
19	293,25	1,331	10,125
20	297,25	1,431	8,125
21	295,25	1,481	11,125
22	299,25	1,381	9,125
23	292,75	1,394	11,375
24	296,75	1,494	9,375
25	294,75	1,444	10,375
26	298,75	1,344	8,375
27	293,75	1,469	8,875
28	297,75	1,369	10,875
29	295,75	1,319	9,875
30	299,75	1,419	11,875
31	292,125	1,459	11,813

Calculation of guide vanes characteristics was carried out using methods of hydrodynamic simulation in STARCCM+ software package. The model data was built for all 128 parameter combinations by automatically creating parameterized models in SolidWorks. The example of a solid state model is shown in Fig. 2.

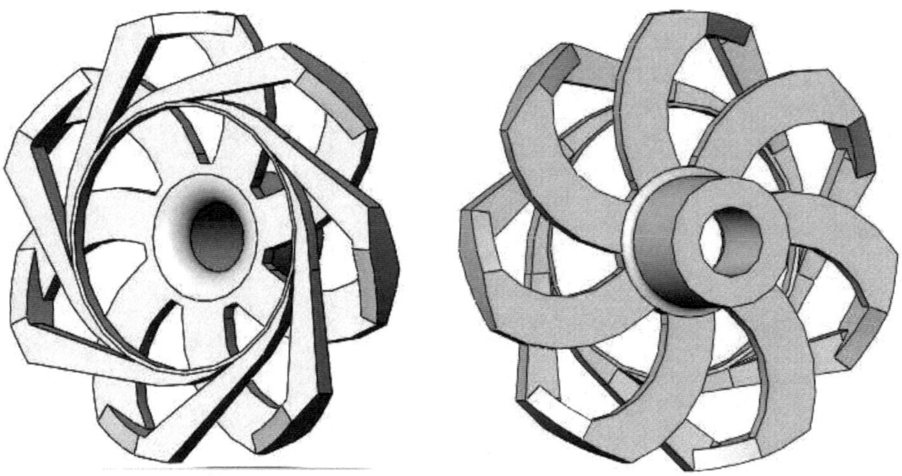

Fig. 2. Solidstate model of the inner liquid volume in guide vanes.

IOP Conf. Series: Materials Science and Engineering **589** (2019) 012012 doi:10.1088/1757-899X/589/1/012012

Pump efficiency values were obtained undernominal value of pump flow rate for all 128 models. They are shown in Table 3.

Table 3. Pump efficiency valuesforconcernedmodels.

№	z=6	z=7	z=8	z=9
0	76,54	78,97	78,61	79,46
1	78,55	75,79	77,20	78,07
2	74,51	80,95	81,23	81,17
3	77,74	77,32	77,57	79,32
4	76,23	76,10	76,48	78,60
5	76,73	75,35	76,20	76,36
6	73,78	82,14	81,77	81,83
7	80,55	79,06	78,83	79,88
8	76,69	80,36	80,69	81,03
9	78,11	76,82	76,95	77,97
10	75,13	81,25	81,23	81,17
11	79,44	80,49	80,73	81,04
12	75,39	77,11	77,77	78,29
13	75,54	79,68	79,44	80,82
14	77,39	74,52	74,94	77,19
15	73,19	81,90	81,72	81,56
16	79,92	77,71	73,62	79,33
17	75,73	79,88	80,00	80,33
18	78,36	76,16	76,46	77,20
19	74,41	80,72	80,60	75,32
20	78,76	77,29	77,49	79,00
21	75,65	79,00	79,18	80,04
22	77,23	77,65	76,89	78,62
23	79,91	75,38	75,95	76,80
24	73,86	81,95	81,80	82,29
25	81,36	81,62	81,44	82,00
26	75,36	77,73	78,06	79,40
27	78,11	79,97	79,62	80,47
28	73,28	75,76	76,63	77,11
29	80,74	82,18	81,54	78,09
30	75,63	78,28	78,55	79,43
31	78,87	80,44	79,61	80,89

For better understanding Fig. 3 shows pump efficiency distribution bar chart depending on the number of design model for z=9.

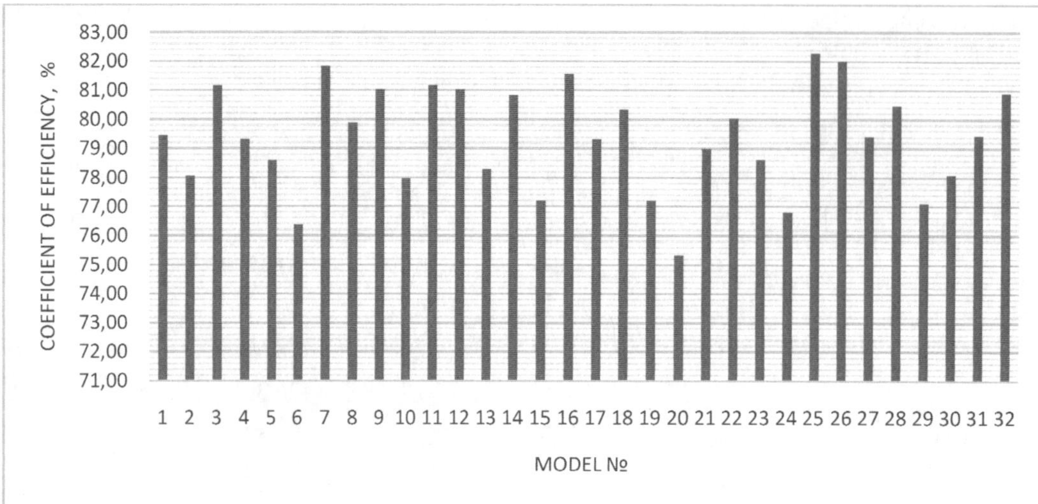

Fig. 3. Pump efficiency distribution bar chart depending on the number of design model for z=9.

In order to analyze the obtained results let's look at dependence of efficiency on optimization parameters in more detail. Fig. 4 shows dependency of efficiency on value calculated as ratio of diameters of guide vanes inlet and wheel outlet D_3/D_2, where $D_2$is a diameter ofwheel outlet for different values of z.

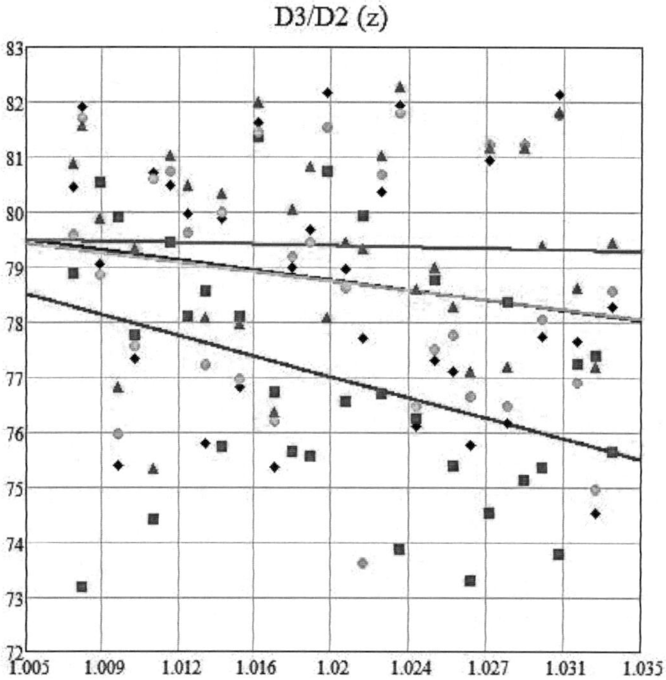

Fig. 4. Dependencyofefficiencyonvalue D_3/D_2, where red represents dependency at z=6, blue – at z=7, green – at z=8, pink – at z=9.

Fig. 5shows dependency of efficiency on ratio of outlet and inlet diameters of guide vanes D_4/D_3 for different values of z.

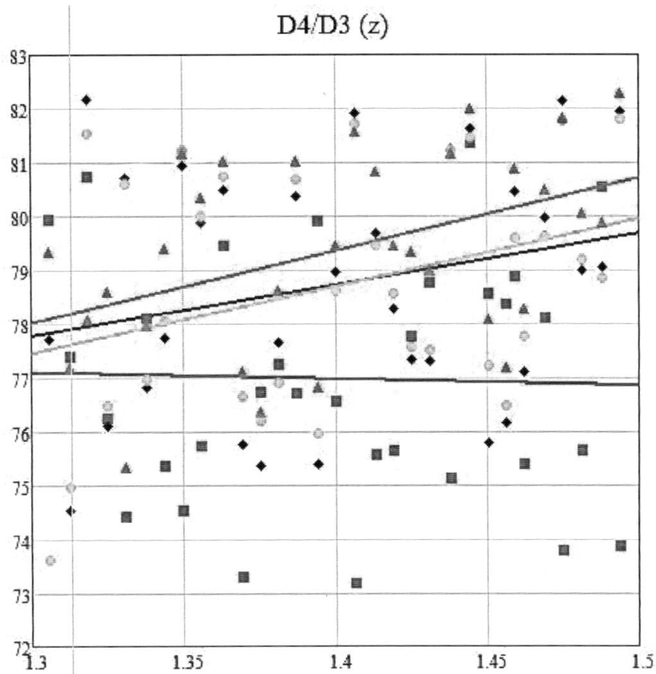

Fig. 5. Dependency of efficiency on D_4/D_3 value, where red represents dependency at Z=6, blue – at z=7, green – at z=8, pink – at z=9.

Fig. 6 shows dependency of efficiency on diffusion airfoil angle of a channel, Θ, for different z values.

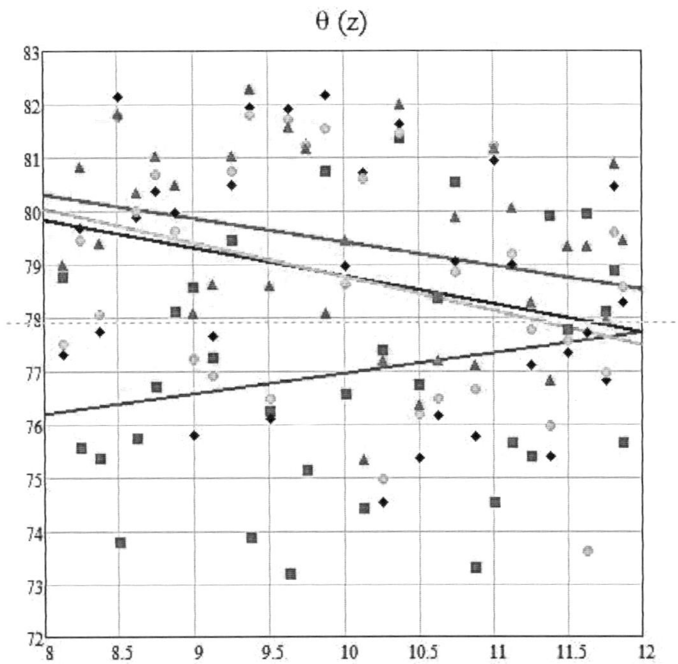

Fig. 6. Dependency of efficiency on Θ, where red represents dependency at z=6, blue – at z=7, green – at z=8, pink – at z=9.

Let's analyze dependency of the number of channels z on efficiency and build a corresponding pie chart (Fig 7). The maximum efficiency is achieved at z=9 in 56% of cases.

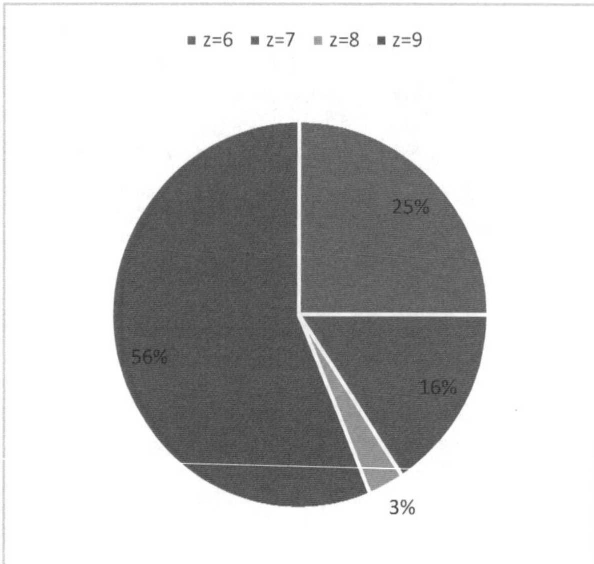

Fig. 7. Pie chart of maximum efficiency values at different z.

The best combination of geometric parameters for the pump is achieved with model №24 at z=9. The efficiency of this model is 82,29 %, almost 10% more than that of the model with the smallest efficiency. Parameter values and efficiency of that model are represented in Table 4.

Table 4. Optimal model.

№ of model	D_3, mm	D_4/D_3	Θ, °	z	η, %
24	296,75	1,494	9,375	9	82,29

Conclusion. This paper determines how major geometric parameters of centrifugal multistage pump guide vanes influence pump efficiency. Patterns of influence of parameter values on optimization criterion were formulated after analyzing results of hydrodynamic simulation. These patterns can serve as recommendations when designing pump guide vanes.

In the future it is planned to expand parameter space and conduct a similar research for other types of guide vanes. This will allow to increase pump efficiency even further.

References

[1] A Protopopov and D Bondareva 2019 *IOP Conf. Ser.: Mater. Sci. Eng.***492** 012002
[2] A Protopopov and V Vigovskij 2019 *IOP Conf. Ser.: Mater. Sci. Eng.***492** 012003
[3] P Chaburko and Z Kossova 2019 *IOP Conf. Ser.: Mater. Sci. Eng.***492** 012011
[4] V Lomakin et al 2019 *IOP Conf. Ser.: Mater. Sci. Eng.***492** 012012
[5] A Gouskov et al 2019 *IOP Conf. Ser.: Mater. Sci. Eng.***492** 012013
[6] N Egorkina and A Petrov 2019 *IOP Conf. Ser.: Mater. Sci. Eng.***492** 012015
[7] K Dobrokhodov and A Petrov 2019 *IOP Conf. Ser.: Mater. Sci. Eng.***492** 012016
[8] N Isaev 2019 *IOP Conf. Ser.: Mater. Sci. Eng.***492** 012026
[9] S Korsakova and A Protopopov 2019 *IOP Conf. Ser.: Mater. Sci. Eng.***492** 012032
[10] A Protopopov and C Jakovich 2019 *IOP Conf. Ser.: Mater. Sci. Eng.***492** 012034
[11] A Petrov et al 2019 *IOP Conf. Ser.: Mater. Sci. Eng.* **492** 012036

[12] V Lomakin and O Bibik 2019 *IOP Conf. Ser.: Mater. Sci. Eng.* **492** 012037
[13] T Valiev and A Petrov 2019 *IOP Conf. Ser.: Mater. Sci. Eng.* **492** 012038
[14] V Cheremushkin and V Lomakin 2019 *IOP Conf. Ser.: Mater. Sci. Eng.* **492** 012039
[15] Zhang, S., Li, H., & Xi, D. (2019). Investigation of the integrated model of side chamber, wear-rings clearance, and balancing holes for centrifugal pumps. *Journal of Fluids Engineering, Transactions of the ASME, 141*(10) doi:10.1115/1.4043059
[16] Sengpanich, K., Bohez, E. L. J., Thongkruer, P., &Sakulphan, K. (2019). New mode to operate centrifugal pump as impulse turbine. Renewable Energy, 983-993. doi:10.1016/j.renene.2019.03.116
[17] Zhang, Z. -., Chen, H. -., Ma, Z., He, J. -., Liu, H., & Liu, C. (2019). Research on improving the dynamic performance of centrifugal pumps with twisted gap drainage blades. Journal of Fluids Engineering, Transactions of the ASME, 141(9) doi:10.1115/1.4042885
[18] Pirouzpanah, S., Patil, A., Chen, Y., & Morrison, G. (2019). Predictive erosion model for mixed flow centrifugal pump. Journal of Energy Resources Technology, Transactions of the ASME, 141(9) doi:10.1115/1.4043135
[19] Yousefi, H., Noorollahi, Y., Tahani, M., Fahimi, R., &Saremian, S. (2019). Numerical simulation for obtaining optimal impeller's blade parameters of a centrifugal pump for high-viscosity fluid pumping. Sustainable Energy Technologies and Assessments, 34, 16-26. doi:10.1016/j.seta.2019.04.011

IOP Conf. Series: Materials Science and Engineering **589** (2019) 012013 doi:10.1088/1757-899X/589/1/012013

Study of multi-stage centrifugal pump guide vanes in a package of hydrodynamic simulating STAR CCM+

K Abramov

Bauman Moscow State Technical University, 5 Second Baumanskaya Street, Moscow, 105005, Russian Federation

Email: abramovki96@mail.ru

Abstract
For multistage centrifugal pumps, one of the main components is the guide vane. A rather important task for the guide vane is compliance with the requirements of the technical specification. It is equally important technologically correct to produce a guide vane. The manufacture of such parts is a time consuming process. To avoid a number of difficulties, this article explores the two most frequently used guide vanes of hydrodynamic simulating tools.

Introduction

Multistage pumps are one of the most popular among those used in industry. The main advantage of this type of pump is the ability to change the head. This energy characteristic is directly related to the number of stages in the pump. The pump head is equal to the sum of the head generated by each stage. The pump flow rate remains constant at a given rotor speed.

In multistage pumps, the flow of the pumped liquid moves sequentially between several impellers mounted on the same shaft. Fluid is delivered from one stage to another through a guide vane. One of the main tasks of the guide vane is to supply the liquid to the next stage with the least hydraulic losses. Therefore, to perform this task requires high manufacturability of the design of the guide vane. [1, 2]

There are many kinds of guide vanes. In this article, the following types will be considered: a radial channel translational guide vane and a guide vane with a non-blade translation channel.

In channel-type guide vane of a multistage pump, conditionally there are three sections: spiral, transfer blade and diffuser (in which the main conversion of kinetic energy takes place). The transfer channel and the blade inlet area provide a uniform liquid supply with a given circulation into the suction area of the next-stage impeller.

A guide vane with a no-blade transfer channel is much easier to manufacture than the channel type. This guide vane consists of a diffuser blade channel, a transfer blade channel and a blade return (supply) channel to the next stage of the centrifugal pump. The circular no-blade channel begins and ends at the same diameter, while the flow rate in the circular space does not change.

This article provides further research and comparison of the above described guide vanes.

The initial data for the calculation of the guide vane are:

Q – flow rate through the vane without counting leakage in the seals of the impeller;

H – pump head;

n – rotation speed.

The calculation of the guide vanes is made for a multistage centrifugal pump, which is shown in Figure 1 with the following initial parameters:

Q = 150 m^3/h;

H_p= 1000 m;

n = 2900 r/min.

By design, it was determined that the number of stages is equal to 9. The head per each stage of the pump is 111.1 m.

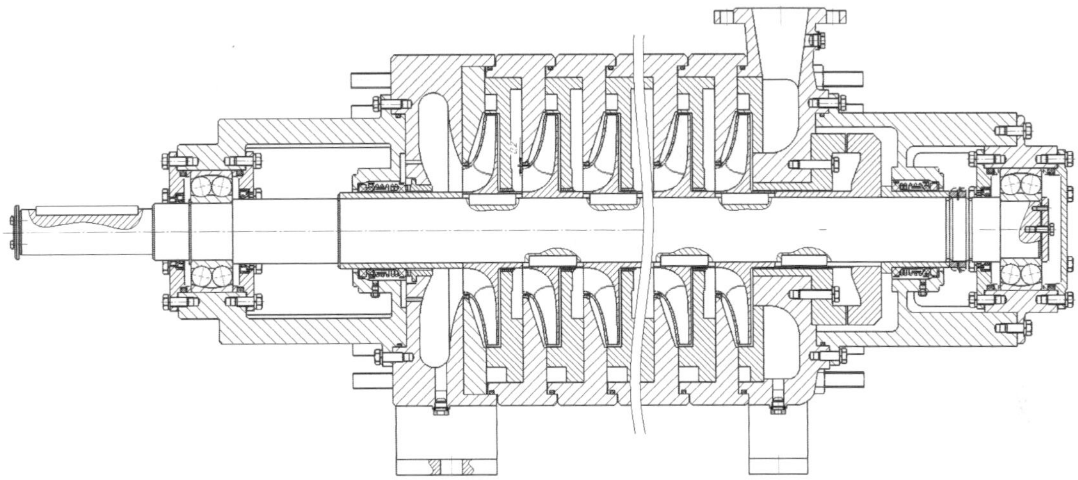

Figure 1. Multistage centrifugal pump

Outlet diameter of impeller D_2 = 300 mm.

The study of guide vanes is carried out in the STAR CCM + hydrodynamic simulation package. To do this, it is necessary to create a solid-state three-dimensional model of each guide vane. [3, 4]

3D–models of the channel-type guide vane and the guide vane with no-blade transfer channel are shown in Figures 2 and 3, respectively.

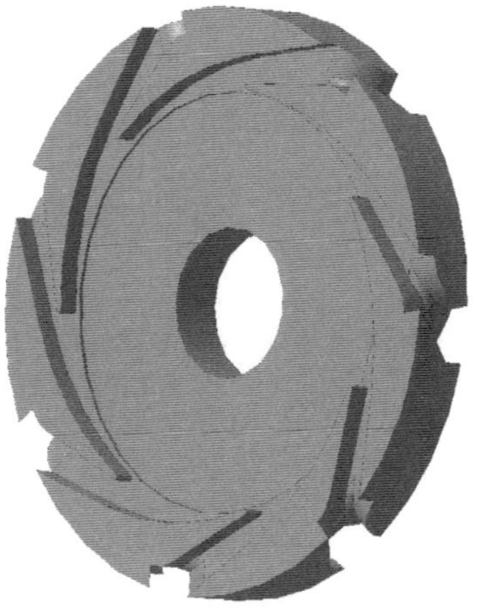

Figure 2. Channel-type guide vane

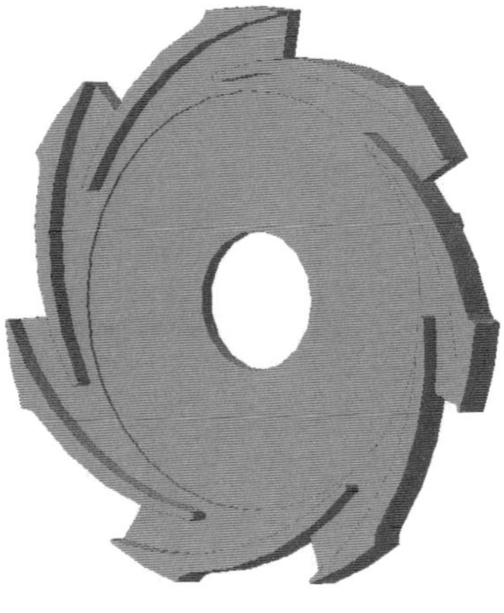

Figure 3. Guide vane with no-blade transfer channel

The criteria for comparing the work of guide vanes are the energy characteristics such as the head and hydraulic efficiency.

The study is conducted under the same conditions for both guide vanes.

The mathematical model of the processes, on the basis of which the hydrodynamic modeling is carried out, is presented below.

Mathematical model

Pump stage head:

$$H = \frac{p_2 - p_1}{\rho g} + \frac{V_2^2 - V_1^2}{2g},$$

wherep_2– the pressure at the outlet of the pump stage;

p_1– the pressure at the inlet of the pump stage;

V_2– the average speed at the outlet from the pump stage;

V_1– the average speed at the inlet from the pump stage.

Rotor speed and pump flow rate are parameters which are set in the numerical model, except for some special cases.

In the overwhelming majority of cases, the flow of fluid in the flow path of the pump is turbulent.

In a general form, the equations of dynamics of an incompressible viscous medium for a turbulent flow mode are as follows:

- fluidcontinuityequation:

$$\frac{\partial \overline{u_x}}{\partial x} + \frac{\partial \overline{u_y}}{\partial y} + \frac{\partial \overline{u_z}}{\partial z} = 0,$$

where $\overline{u_i}$ – time-averaged projections of fluid velocities on the corresponding axes.

- the mass conservation equation is written taking into account the availability of two phases and their concentrations:

$$\frac{\partial u_x}{\partial t} + \frac{\partial u_y}{\partial t} + \frac{\partial u_z}{\partial t} = 0,$$

where $\overline{u_i}$ — time-averaged projections of fluid velocities on the corresponding axes.

- equation of the change in the amount of motion averaged over time:

$$\rho\left[\frac{\partial \overline{u_i}}{\partial t} + \overline{u_j}\frac{\partial \overline{u_i}}{\partial x_j}\right] = -\frac{\partial \overline{p}}{\partial x_i} + \frac{\partial}{\partial x_i}\left[T_{ij}^{(v)} - \rho\langle u_i u_j\rangle\right];$$

where$\overline{u_i}\,\overline{p}$ — averaged velocity and pressure;

$\tilde{T}_{ij}^{(v)} = 2\mu\tilde{s}_{ij}$ — viscous stress tensor for incompressible fluid;

$\tilde{s}_{ij} = \frac{1}{2}\left[\frac{\partial \overline{u_i}}{\partial x_j} + \frac{\partial \overline{u_j}}{\partial x_i}\right]$ — strain rate tensor;

$\rho\langle u_i u_j\rangle$ — Reynolds stresses.

The Reynolds system of equations is open due to the presence of unknown Reynolds stresses. The system is closed by using the k-ω SST turbulence model. This model combines the advantages of both the k-ω and k-ε models: in the near-wall region the k-ω model is used, and in the core of the flow is the k-ε model. [5 – 9]

This model includes two equations of turbulence parameters transfer:

- turbulence kinetic energy transfer equation:

$$\frac{\partial k}{\partial t}+\overline{u}_j\frac{\partial k}{\partial x_j}=P_k-\beta\cdot k\omega+\frac{\partial}{\partial x_i}\left[\left(\nu+\sigma_k\nu_T\right)\frac{\partial k}{\partial x_j}\right],$$

where $k=\dfrac{u'^2_x+u'^2_y+u'^2_z}{2}$ — kinetic energy of turbulence;

u'_i — velocity pulsations;

P_k — turbulence power generation member;

ω — relative rate of turbulence dissipation;

ν_T — turbulent viscosity.

- transfer equation for the relative rate of energy dissipation of turbulence:

$$\frac{\partial\omega}{\partial t}+\overline{u}_j\frac{\partial\omega}{\partial x_j}=\alpha S^2-\beta\omega^2+\frac{\partial}{\partial x_i}\left[\left(\nu+\sigma_\omega\nu_T\right)\frac{\partial\omega}{\partial x_j}\right]+2\left(1-F_j\right)\sigma_\omega\frac{1}{\omega}\frac{\partial k}{\partial x_j}\frac{\partial\omega}{\partial x_j}.$$

Reynolds stresses in the equations of dynamics are based on the Boussinesq hypothesis:

$$\rho\left\langle u_i u_j\right\rangle=2\mu_T\left[\frac{1}{2}\left(\frac{\partial\overline{u_i}}{\partial x_j}+\frac{\partial\overline{u_j}}{\partial x_i}\right)-\frac{1}{3}\frac{\partial\overline{u_k}}{\partial x_k}\delta_{ij}\right]-\frac{2}{3}\rho k\delta_{ij},$$

where δ_{ij} — Kronecker symbol.

The calculation can be carried out both in steady mode and in non-steady.[10 – 13]

When conducting a stationary calculation in the hydrodynamic equations, there are no terms with time differentiation. This calculation allows you to quite accurately calculate the parameters of the pump at the optimum operating point and near it, and also allows you to significantly reduce the calculation time. However, far from the nominal point, the error of the stationary calculation significantly affects the calculation due to the intensive vortex formation, since the vortex formation processes are non-stationary processes in their essence.[14 – 21]

Also, with a stationary calculation, it is impossible to obtain by calculation the oscillations of hydrodynamic parameters caused, for example, by rotation of the impeller, therefore, if necessary, take into account all the abovementioned factors, as well as evaluate the pump behavior not only at the optimum point, but throughout the entire flow rate range. The calculation is carried out in non-stationary mode.

To carry out the calculation in STAR CCM +, it is necessary to import into the hydrodynamic package a solid-state three-dimensional model of the flow part of the pump stage.

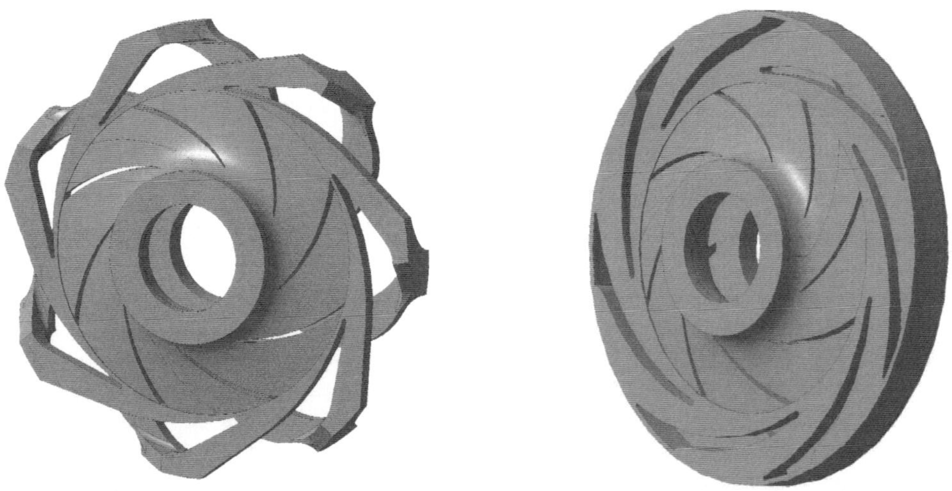

IOP Conf. Series: Materials Science and Engineering **589** (2019) 012013 doi:10.1088/1757-899X/589/1/012013

Figure 4. 3D model of the flow part of the pump stage with a channel-type guide vane

Figure 5. 3D model of the flow part of the pump stage with a guide vane with a no-blade transfer channel

Calculated mesh

In the study, in order to set the parameters of the computational grid, the surface partitioning and the generator of polyhedral cells are used together with the generator of the wall layers. For clarity, the calculated grids in sections perpendicular to the impeller rotation are presented below. The calculated grid for a model with a channel-type guide vane is shown in Figure 6, for a model with a guide vane with a no-blade transfer channel in Figure 7.

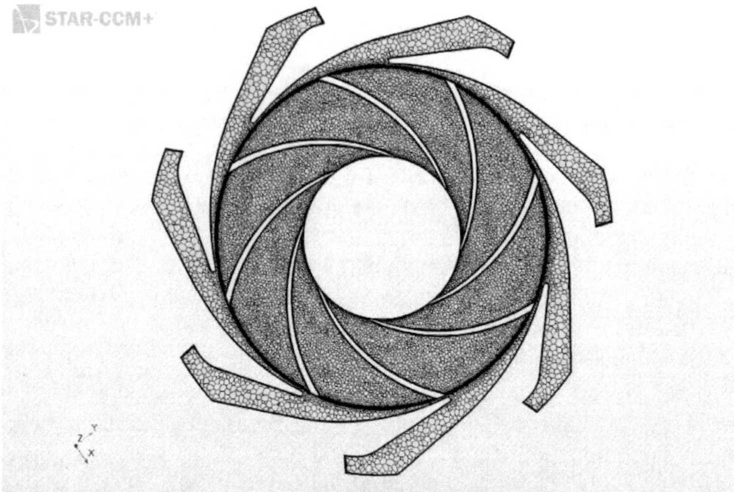

Figure 6

Figure 7

After generating the volume mesh and further the computational mesh, it is necessary to set a physical model. The following models are selected:

- usteady task;
- working medium - liquid;
- equation of state - constant density;

- solver - divided;
- flow mode - turbulent;
- k-omega SST model of turbulence.

The results of hydrodynamic calculations are scalar scenes of the distribution of velocity and pressure in the flow part of the pump stage. The energy characteristics of the pump stage were calculated as criteria for comparison: pressure and hydraulic efficiency. In the hydrodynamic calculation were not incorporated volumetric and mechanical losses.

Scalar scenes of speed and pressure distribution for a stage with a channel-type guide vane are shown in Figures 8 and 9.

Figure 8. Scalar scenes of speed distribution

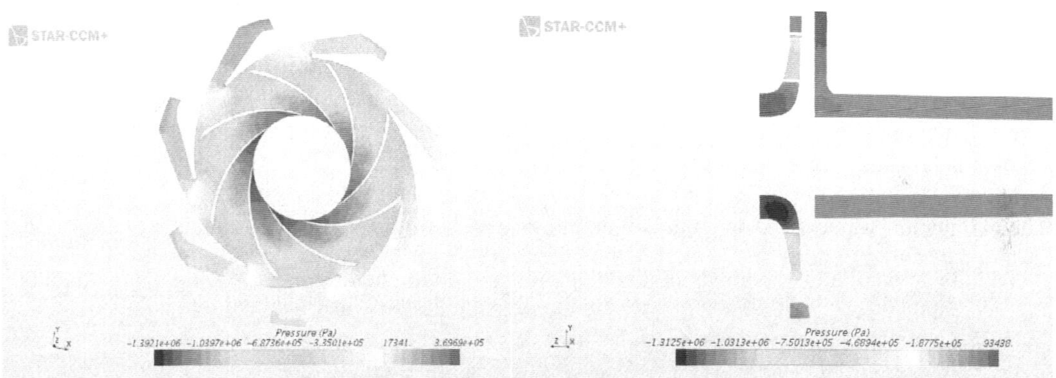

Figure 9. Scalar scenes of pressure distribution

For a stage with a guide vane with a no-blade transfer channel, scenes of speed and pressure distribution are also similarly represented in Figures 10 and 11.

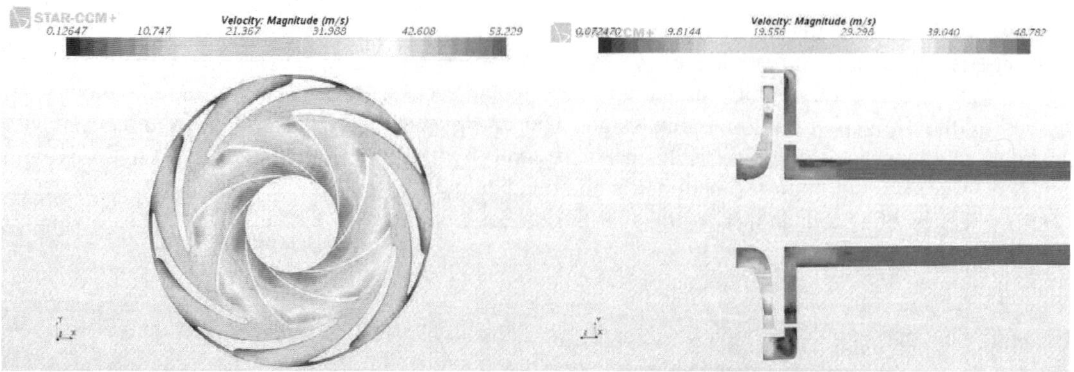

Figure 10. Scalar scenes of speed distribution

Figure 11. Scalar scenes of pressure distribution

The calculation of stages for both types of guide vanes was carried out under the same conditions, physical models, and reference values of the grid builder. The boundary conditions remained unchanged.

The calculation was carried out at 2100 iterations. Such a number of iterations is sufficient for the full rotation of the impeller.

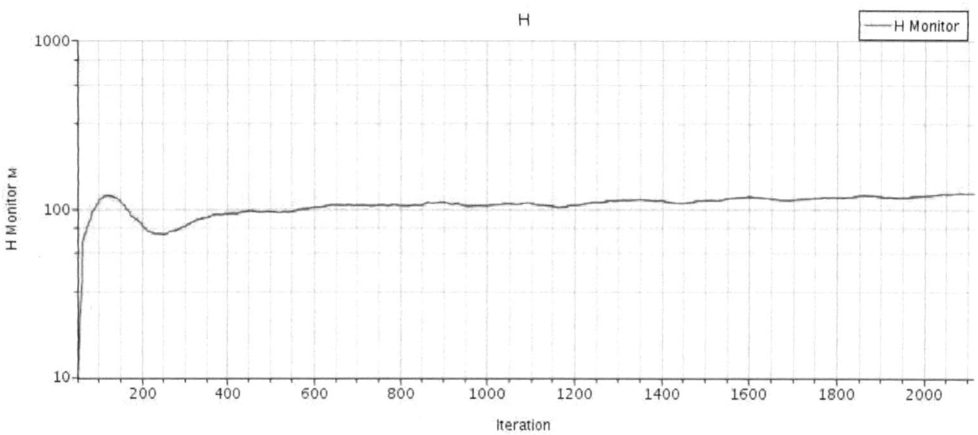

Figure 12. Head graph

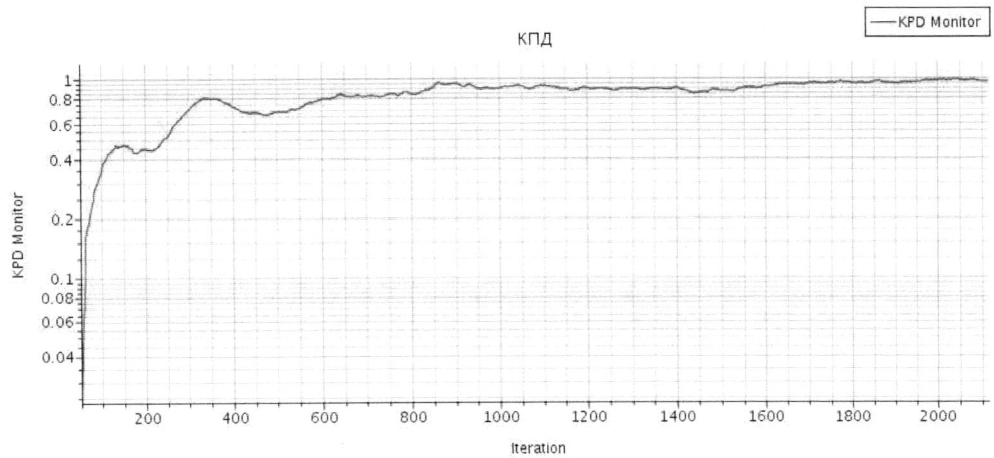

Figure 13. Hydraulic efficiency graph

Energy characteristics are obtained in the form of iterative graphs. To calculate the final value of head and efficiency, it is enough to consider the last 25% of iterations. The pressure and hydraulic efficiency graphs for a stage with a channel-type guide vane are shown in Figures 12 and 13, respectively.

The average value of the hydraulic efficiency is 93%. The average head is 120 meters. The obtained energy characteristics of the stage correspond to the required values of the hydraulic efficiency - 88% and a head -111.1 m., obtained by mathematical calculations.

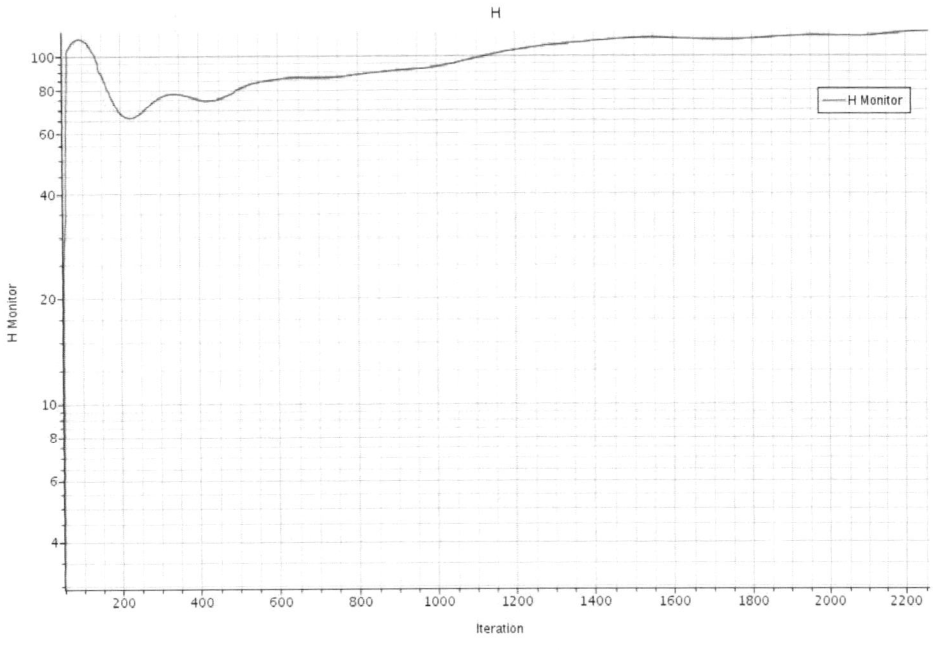

Figure 14. Head graph

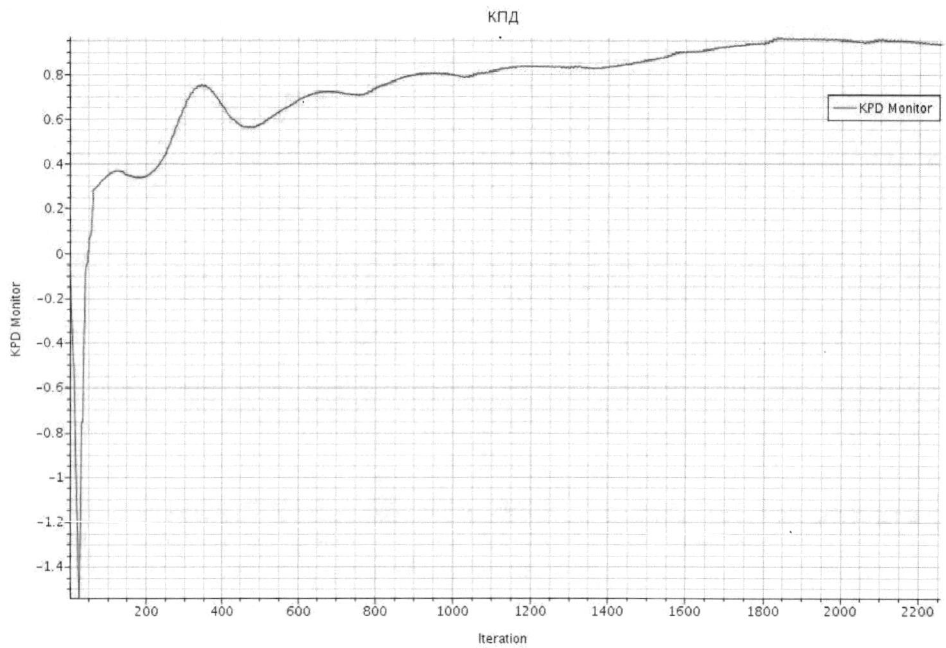

Figure 15. Hydraulic efficiency graph

Similarly, the graphs of head (Figure 14) and hydraulic efficiency (Figure 15) were calculated for a stage with a guide vane with a no-blade transfer channel.

The average value of hydraulic efficiency - 91%, head - 112 meters.

The characteristics obtained in this case also satisfy the required. However, stabilization of processes in the second case took longer.

Conclusion

For a visual demonstration of the processes occurring in the stages of a multistage centrifugal pump, scalar scenes of the distribution of speed and pressure were presented.

The calculation was carried out in a non-steady mode, despite the fact that the hydrodynamic modeling was carried out near the optimal point.

Thus, although the research took place under equal conditions and the results met the requirements, they differ. The model of a stage with a channel-type guide vane proved to be better from an energetic point of view. In spite of the fact that the guide vane with the no-blade transfer channel is easier to manufacture, its test model turned out to be less technologically than the model of the channel-type guide vane, what could have caused a number of hydraulic losses. In the future, it is necessary to consider in more detail the processes in the transfer channel, since the rotation of the fluid from the forward channel to the transfer channel and from transfer to the return channel (leading to the next stage) turned out to be quite steep, which could have been the cause of unnecessary losses.

References

[1] Lomakin, V.O., Kuleshovav, M.S., Bozh'eva, S.M. Numerical Modeling of Liquid Flow in a Pump Station (2016) Power Technology and Engineering, 49 (5), pp. 324-327.

[2] Ivanov, A.S., Murkin, S.V. Threaded joints in drives (2016) Russian Engineering Research, 36 (8), pp. 611-619.

[3] Romanov, A.S. Method of hydrodynamic description of the spreading of a partially wetting liquid over a flat solid surface (1990) Colloid journal of the USSR, 52 (1), pp. 72-78.

[4] Perevezentsev, V.V. Experimental study of the hydrodynamics of gas-solids flows in transient (accelerating) motion of the solid particles (1988) Fluid mechanics. Soviet research, 17 (3), pp. 1-6.

[5] Krapivtsev, V.G. Model Studies of Coolant Flow Hydrodynamics in VVER-1000 In-Reactor Pressure Channel (2017) Atomic Energy, 122 (5), pp. 304-310.

[6] Arefyev, K.Y., Voronetsky, A.V., Suchkov, S.A., Ilchenko, M.A. Computational and experimental study of the two-phase mixing in gas-dynamic ignition system (2017) Thermophysics and Aeromechanics, 24 (2), pp. 225-237.

[7] Leont'ev, A.I., Kuzma-Kichta, Y.A., Popov, I.A. Heat and mass transfer and hydrodynamics in swirling flows (review) (2017) Thermal Engineering, 64 (2), pp. 111-126.

[8] Marchevsky, I.K., Puzikova, V.V. Numerical Simulation of Wind Turbine Rotors Autorotation by Using the Modified LS-STAG Immersed Boundary Method (2017) International Journal of Rotating Machinery, 2017, article № 6418108.

[9] Shevchenko, S.Y., Melnik, Y.A., Smirnov, A.E., Htet, W.Y.M. Comparative evaluation of methods for the determination of heat transfer coefficients of liquid and gaseous quenching media (2017) Mechanics and Industry, 18 (7), article № 703.

[10] Devisilov, V.A., Sharai, E.Y. Numerical study of the flow structure in a hydrodynamic filter (2016) Theoretical Foundations of Chemical Engineering, 50 (2), pp. 209-216.

[11] Isaev, S.A., Schelchkov, A.V., Leontiev, A.I., Baranov, P.A., Gulcova, M.E. Numerical simulation of the turbulent air flow in the narrow channel with a heated wall and a spherical dimple placed on it for vortex heat transfer enhancement depending on the dimple depth (2016) International Journal of Heat and Mass Transfer, 94, pp. 426-448.

[12] Grishin, Y.A., Bakulin, V.N. Numerical Investigation of Flow in a Centrifugal Compressor (2015) Journal of Engineering Physics and Thermophysics, 88 (5), pp. 1274-1279.

[13] Pozhalostin, A.A., Goncharov, D.A. Free axisymmetric oscillations of a two-layer liquid with an elastic separator between layers (2015) Russian Aeronautics, 58 (1), pp. 37-41.

[14] Belov, P.A., Kobets, L.P., Borodulin, A.S. Impregnation kinetics of fibers with liquids: Simulation within the generalization of Navier-Stokes equations (2014) Inorganic Materials: Applied Research, 5 (4), pp. 403-406.

[15] Derevich, I.V., Ermolaev, V.S., Mordkovich, V.Z., Galdina, D.D. Simulation of fluid dynamics in a microchannel Fischer-Tropsch reactor (2012) Theoretical Foundations of Chemical Engineering, 46 (1), pp. 8-19.

[16] Derevich, I.V., Ermolaev, V.S., Mordkovich, V.Z. Modeling of hydrodynamics in microchannel reactor for Fischer-Tropsch synthesis (2012) International Journal of Heat and Mass Transfer, 55 (5-6), pp. 1695-1708.

[17] Lozovetskii, V.V., Pelevin, F.V. Resistance of spherical fills in the case of flow of single- and two-phase media (2009) Journal of Engineering Physics and Thermophysics, 82 (2), pp. 277-282.

[18] Mikulin, E.I., Shevich, Yu.A., Lysyi, O.A. Boiling heat transfer and hydrodynamics in matrix-type channels (2001) Experimental Thermal and Fluid Science, 25 (1-2), pp. 65-72.

[19] Krut'ko, P.D. Analytical solution of the Voznesenskiy problem for stationary and nonstationary linear systems (1996) Journal of Computer and Systems Sciences International, 34 (5), pp. 47-59.

[20] Bystrov, P.I., Ivlyutin, A.I., Kharchenko, V.N., Shul'ts, A.N. Physical mechanisms for transfer of heat, mass, and momentum in a short low-temperature heat pipe. I. Hydrodynamics of vapor flows (1991) Journal of Engineering Physics, 60 (1), pp. 1-7.

[21] Romanov, A.S. Spreading of small liquid drops along a flat surface (1990) Journal of Engineering Physics (English Translation of Inzhenerno-FizicheskiiZhurnal), 57 (2), pp. 884-887.

IOP Conf. Series: Materials Science and Engineering **589** (2019) 012014 doi:10.1088/1757-899X/589/1/012014

Comparison of various simulation methods of a two-phase flow in a multiphase pump

A Boyarshinova[1,2], V Lomakin[1] and A Petrov[1]

[1]Bauman Moscow State Technical University, 5 Second Baumanskaya Street, Moscow, 105005, Russian Federation

[2]E-mail address: a.boyarsh@yandex.ru

Annotation. The article presents a mathematical model of a two-phase flow in a multiphase helico-axial pump. In this paper, the model of a multiphase incompressible fluid flow (ρ = const) were used. Numerical simulation is based on solving discrete anagoges of the basic hydrodynamic equations. The hydrodynamic simulation of the described mathematical model was performed in two ways: in a simplified form, using a multiphase volumeoffluid model (VOF), and the segregated flow method. The results of the calculation of two-phase flows for various values of the gas contaent are given. The influence of gas content on the pump performance and its pressure is considered. A comparison of calculation results of two methods of hydrodynamic simulation of two-phase flow is shown.

Introduction

Multiphase flow is a flow of a mixture of substances in various aggregative states - phases. A continuos or discrete phase can be a solid, liquid or gaseous state of matter. In terms of simulation, a phase can be defined as the amount of a substance in a system that has its own physical properties in order to distinguish it from other phases in the system[1]-[2]. Two-phase flow is a special case of multiphase flow, containing only two different components. These are liquids with solid or gas inclusions, gases with liquid drops or solid particles. The structure of two-phase flows can be very diverse. It is determined by the size, distribution of the elements of the dispersed phase in a continuous medium and their direct interaction. Multiphase media in nature are smoke, smog, fog, rain, in turn, in the technique typical examples of such media are emulsion, sludge, boiling liquid, etc.[3]-[4].

The problem of numerical simulation of the hydrodynamics of two-phase flows is relevant, since for many years one of the main methods for studying such media has been experiment. However, the development of numerical methods and computing hardware made it possible to calculate the fields of velocities, pressures and other parameters of two-phase flows in different engineering tasks[5]-[7].

The aim of the work is to describe a mathematical model of a two-phase flow and develop methods for hydrodynamic simulation of water flow with different gas content in a multiphase pump, as well as to determine the best way for calculating two-phase flow.

One of the possible solutions to the task is to use a helico-axial multiphase pump for the calculation [8]-[9]. It was designed for pumping a media with gas content up to 99%, during continuous operation[10]. Such pumps were designed primarily for oil production with associated gas without the use of additional units[11]-[12].This two-phase pumping technology reduces pollutants and CO_2 emissions, minimizes environmental impact, avoids gas or crude oil leaks, and is also more cost-effective than oil extraction using separation plants [13].The design of one stage of the pump is shown in Fig. one.

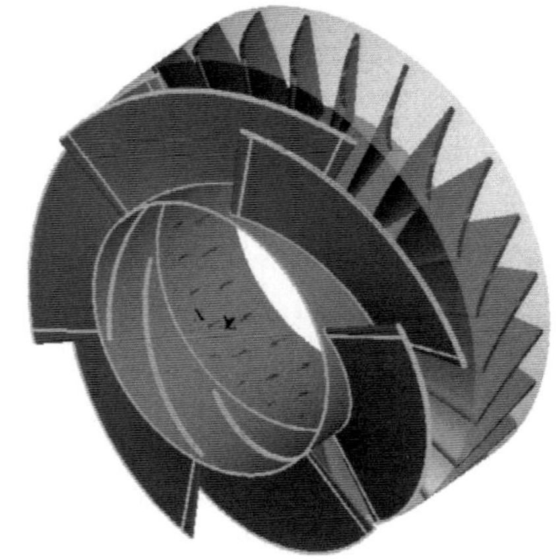

Fig. 1. The design of the stage of the helico-axial multiphase pump

The stage of the helico-axial multiphase pump is an axial impeller followed by a guide vanes[14].The idea is that the impeller disperses the multiphase flow, and due to expansion of cross section of guide vane, the volume of gas bubbles decreases [15]-[16].Thus, the gas is not shut off the flow.

In this paper hydrodynamic simulation was carried out using several methods of two-phase flow. The simulation was performed in the STAR CCM + software package.

Methodology

Hydrodynamic simulation of a two-phase flow can be calculated by following methods:

1) homogeneous multiphase volumeof fluid model (VOF), is a simplified model that simulates a multiphase flow in the form of a suspension. It is used when two phases of the same substance (for example, water and steam) are involved, which are in thermodynamic equilibrium and have common fields of speed, pressure and temperature[17]-[18].

Basic assumptions of the two-phase mixture model:

- The mixture of phases behaves like a single fluid. A set of following equations: transport equations for mass, momentum, and energy, is solved for the mixture as a whole.
- Phases are in thermal equilibrium. The two phases have the same temperature, and the phase distribution is calculated on the basis of the heat balance.
- There is no relative (sliding) speed between phases.

2) segregatedmultiphaseflow model uses the same set of equations as the two-phase thermodynamic equilibrium model (mass and momentum transfer equations), but solves them separately for each phase (except for the general pressure field).

The mathematical model of segregated multiphase flow consists of a set of differential and algebraic equations[19]-[20]:

1. The volume of the i-th phase in each cell is calculated as:

$$V_i = \int_V \alpha_i dV,$$

where α_i – is the concentration of the i-th phase in the cell.

The sum of the concentrations of all phases in the cell is equal the one.

$$\sum_{i6}^{n}\alpha_i = 1.$$

2. The equation of mass conservation (continuity equation):

$$\frac{\partial}{\partial t}\int_V \alpha_i\rho_i dV + \int_A \alpha_i\rho_i \vec{V_i} d\vec{a} = 0,$$

where ρ_i – i-th phase density;

$\vec{V_i}$ – i-th phase velocity (in the case of turbulent flow simulation by a RANS-type model, velocity averaged over time[21]-[23]).

3. The quantity of motion changing:

$$\frac{\partial}{\partial t}\int_V \alpha_i\rho_i \vec{V_i} dV + \int_A \alpha_i\rho_i\left(\vec{V_i}\vec{V_i}\right)d\vec{a} = -\int_V \alpha_i \nabla p dV +$$

$$+\int_V \alpha_i\rho_i \vec{g} dV + \int_A \left[\alpha_i\left(T_i + T_i^t\right)\right]d\vec{a} + \int_V \vec{M_i} dV,$$

where $\left(\vec{V_i}\vec{V_i}\right)$ – tensormultiplication of the velocity vector of the i-th phase;

p — pressure;

$\vec{g}$ – mass intensity vector (in this case, the gravity force is 9.81 m/s^2 and the inertial force due to the rotation of the computational domain);

T_i —molecular viscosity stress tensor;

T_i^t – turbulent stress tensor;

$\vec{M}$ – the vector of the total intensity of phase interaction forces per unit volume, for the vector $\vec{M_i}$ the equality is true:

$$\sum_i \vec{M_i} = 0.$$

The vector $\vec{M_i}$ characterizes all the forces that separate phases interact with each other.

$$\vec{M_i} = \sum_{i\neq j}\left(\vec{F_{ij}^D} + \vec{F_{ij}^{VM}} + \vec{F_{ij}^L} + \vec{F_{ij}^{TD}} + \vec{F_{ij}^{WL}}\right),$$

где $\vec{F_{ij}^D}$ – resistance force;

$\vec{F_{ij}^{VM}}$ – virtual mass force;

$\vec{F_i j^L}$ – lifting force;

$\vec{F_{ij}^{TD}}$ – force caused by turbulent dispersion;

$\vec{F_{ij}^{WL}}$ – force caused by wall effect.

4. Resistance force

In flow simulation of continuous and dispersed media, the resistance force acting on the dispersed medium I from the side of the continuous medium j is equal to:

$$\vec{F_{ij}^D} = A_D \vec{V_r}.$$

WhereA_D – linearized resistant coefficient;

$\vec{V_r} = \vec{V_j} - \vec{V_i}$ – relative velocity of mediums;

$$A_D = C_D \frac{1}{2} \rho_C \left|\vec{V_r}\right| \left(\frac{a_{cd}}{4}\right) C$$

WhereC_D – resistant coefficient;

ρ_C– continuous phase density;

α_{cd}– interaction area of the phases (in this case, the term $\frac{a_{cd}}{4}$ is the area of the projection of the spherical particle on the plane).

Resistant coefficient is based on the relation:

$$C_D = f_D C_{D\infty}$$

Where $C_{D\infty}$ - coefficient of resistance of a single spherical particle moving in an infinite flow;

f_D– coefficient taking into account the concentration of particles.

$C_{D\infty}$ can be found from relation:

$$C_{D\infty} = \frac{24}{Re_d}\left(1 + 0.15 Re_d^{0.687}\right) \text{ when } Re_d < 1000$$

or

$$C_{D\infty} = 0,44 \text{ when } Re_d > 1000$$

$$Re_d = \frac{\rho_C \left|V_r\right| l}{\mu_C}$$

l– characteristic interaction length or bubble size;

μC– dynamic viscosity coefficient of continuous medium;

Coefficient$f_D$can be found from relation:

$$f_D = \alpha_C^{\ nD}$$

α_C– continuous phase concentration;

$n_D = -8.3$ for spherical particle.

5. Virtual mass

The inertia of the surrounding fluid affects the acceleration of the particle immersed in the fluid. This effect is simulated by adding mass to the dispersed particle.

The force from the virtual mass acting on phase i moving accelerated relative to phase j is found as:

$$\overrightarrow{F_{ij}^{VM}} = C_{VM} \rho_C \alpha_C \left(\vec{a_j} - \vec{a_i}\right)$$

$\overrightarrow{a_{j,i}}$ – acceleration of j-th and i-th phase;

$C_{VM} = 0,5$– virtual mass coefficient for a spherical particle.

6. Lifting force

In the case when the flow surrounding the dispersed particle is inhomogeneous or the particle is swirling, the force is perpendicular to the relative velocity..

$$\overrightarrow{F_{ij}^{L}} = C_{Leff} \rho_C \alpha_d \left(\vec{V_r} \times \left(\nabla \times \vec{V_c}\right)\right)$$

$\vec{V_r}$ – relative phase velocity;

$\overrightarrow{V_c}$ – continuous phase speed;

α_d– dispersed phase concentration;

C_L = 0,25–lifting force coefficient.

7. Turbulentdispersionforce

Additional change in phase concentrations caused by flow turbulence is modeled as the force of turbulent dispersion

$$\overrightarrow{F_{ij}^{TD}} = A_D \overrightarrow{V_{TD}}$$

A_D– resistant force coefficient;

$\overrightarrow{V_{TD}}$ – relative slip speed;

$$\overrightarrow{V_{TD}} = D_{TD}\left(\frac{\nabla\alpha_d}{\alpha_d} - \frac{\nabla\alpha_c}{\alpha_c}\right)$$

$D_{TD} = C_0 \frac{\nu_c^t}{\sigma_\alpha} I$ – turbulent diffusion tensor;C_0 = 1;

ν_c^t –kinematiccoefficient of turbulent viscosity;

σ_α – turbulentPrandtl number;

I– unit tensor;

$$\sigma_\alpha = \sigma_0 \sqrt{1 + C_\beta \xi^2} \frac{1+\eta}{b+\eta}$$

$\sigma_0 = 1$ – unmodified turbulent Prandtl number;

$C_\beta = 1{,}8$ – correction factor;

ξ – article sliding velocity related to the speed of turbulent fluctuations;

η– particle interaction time related to relaxation time;

b– the ratio of the accelerations of the continuous / dispersed phase;

$$b = \frac{1 + C_{VM}}{\frac{\rho_d}{\rho_c} + C_{VM}}$$

$$\eta = \frac{\tau_I}{\tau_R}$$

τ_I –characteristic time of interaction of a particle and a turbulent vortex;

τ_R – particle relaxation time;

$$\tau_I = \frac{\tau_T}{\sigma_0} \frac{1}{\sqrt{1 + C_\beta \xi^2}}$$

$$\tau_T = \frac{2}{3} C_\mu \frac{k_c}{\varepsilon_c}$$

ε_c - continuous phase turbulence energy dissipation rate;

$$\tau_R = \tau_D \left(1 + \frac{\rho_c}{\rho_d} C_{VM} \right)$$

$\tau_D = \dfrac{\rho_d d^2}{18\mu_c}$ – characteristic time scale for a dispersed particle;

$$\xi = \frac{\left|\vec{V_r}\right|}{\sqrt{\frac{2}{3}k_c}}$$

k_c – kinetic turbulence energy of the continuous phase.

8. Wall influence

Whengasbubble located near the solid wall, it undergoes an asymmetrical action from the liquid. The force per unit volume that a gas bubble experiences is equal to:

$$\overrightarrow{F_{ij}^{WL}} = -C_{WL} y_w \alpha_D \rho_C \frac{\left|\overrightarrow{V_{r,\tau}}\right|^2}{d_p} \vec{n}$$

где y_w – distance from the wall;

α_D – dispersed phase concentration;

ρ_C – dispersed phase density;

d_p – bubble diameter;

C_{WL} – coefficient, which is the function of the distance from the wall and decreasing with increasing distance;

$\vec{n}$ – normally oriented to the wall unit vector at the point closest to the bubble;

$\overrightarrow{V_{r,\tau}}$ – tangent to the wall component of the relative velocity;

Coefficient C_{WL} is as:

$$C_{WL} = max\left(C_{W1} + \left(\frac{C_{W2}}{y_w} \right) d_p, 0 \right)$$

Coefficientsequal to: $C_{W1} = -0,01$, C_{W2}=0,05. Thus the force caused by the influence of the wall disappears at a distance from the wall equal to the five diameters of the bubble.

The 3D model of the helico-axial multiphase pump for the calculation is shown in Fig. 2.

IOP Conf. Series: Materials Science and Engineering **589** (2019) 012014 doi:10.1088/1757-899X/589/1/012014

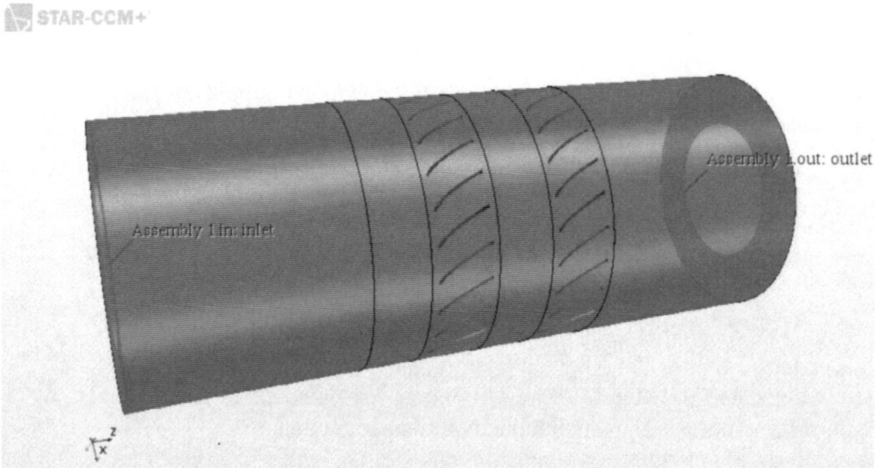

Fig. 2. 3D model of the flow part of the pump

For the calculation two pump stages were chosen for a more visual representation of the two-phase flow.

Boundary conditions: inlet velocity and outlet pressure, and for a split flow model, it is necessary to indicate the velocity of both the liquid and the gas. The calculation was carried out by gradually increasing the gas content value at the inlet boundary of the pump.

The grid in longitudinal and cross-section, presented in Fig. 3, consists of 508,000 cells. Near the solid walls, the cells elongated in the direction perpendicular to the wall (boundary layer), in the core of the flow the grid is not structured, consists of polyhedra of various shapes and sizes.

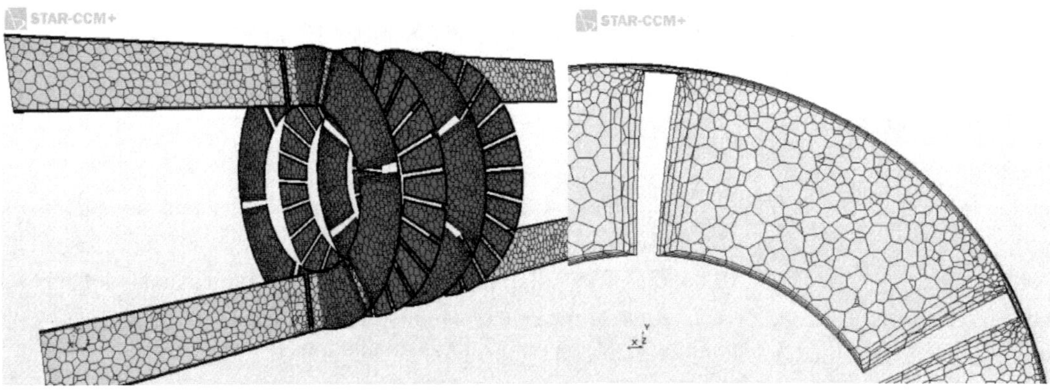

Fig. 3. The grid in the cross section of the pump

Although the simulation was performed in different ways, the grid was constructed identically.

Results

The flow of the two-phase medium in the flow part of the pump, obtained as a result of the simulation, is shown in the illustrations below.

For a visual representation of the two-phase flow, the scalar distribution of the volume fraction of water is showed on Fig.4 (the liquid phase is red, the gas phase is blue).

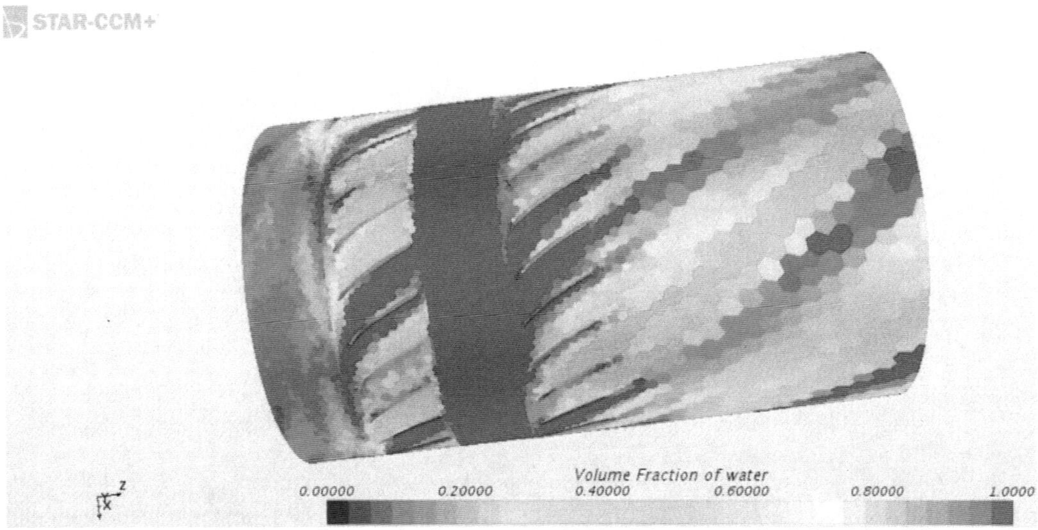

Fig. 4. The volume fraction of water in the pump at 50% of the gas at the inlet (simplified model)

AtFig. 4 it can be seen that the pump push through the liquid-gas mixture in a 50/50 ratio, and after the first stage the amount of gas has decreased.

AtFig. 5 showed the current flow lines in first stage of a multiphase pump.

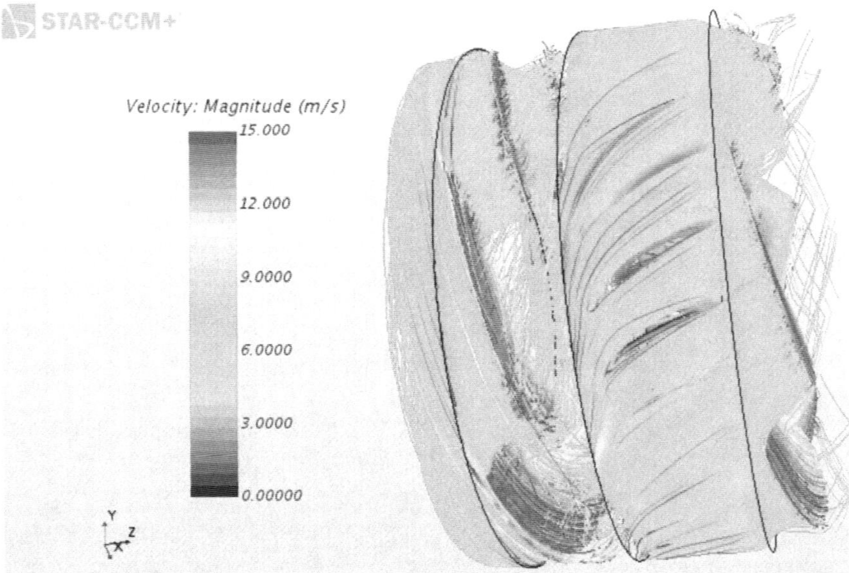

Fig. 5. Current flow lines in the pump stage

The following are the scalar scenes of the distribution of the volume fraction of water in the cross section of the pump so as to get a visual representation of what is happening in impeller and guide vanes. In order to compare the two calculation methods, all the results are presented for the simplified model, then for the segregate flow model. In Fig. 6-11 are presented from left to right: impeller of the first stage,guide vanes of first stage, then impeller of second stage and the guide vanes of second stage (the liquid phase is red, the gas phase is blue).

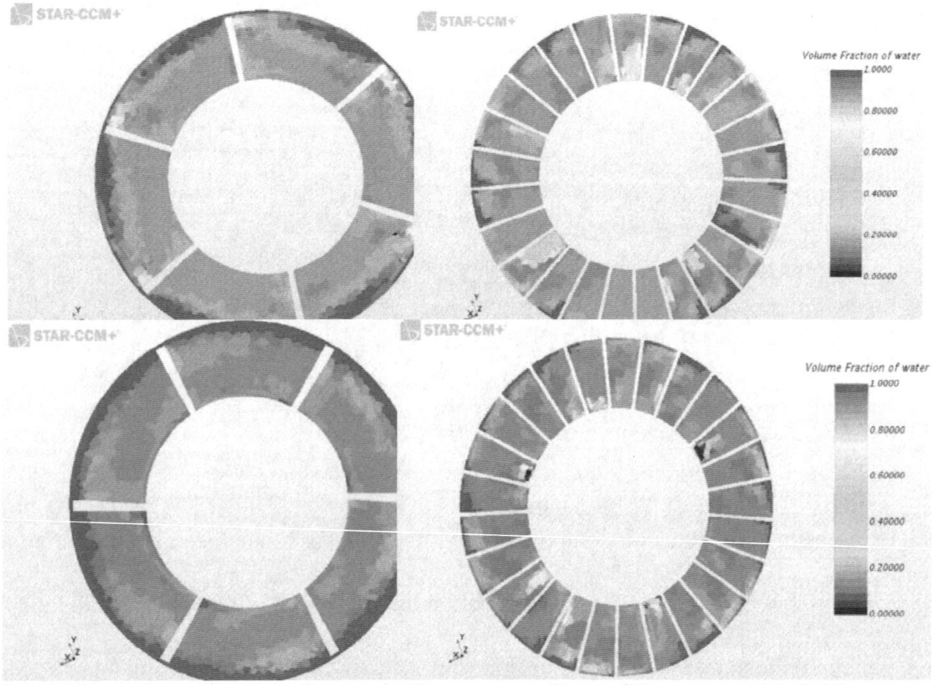

Fig. 6. The volume fraction of water in the cross sections of the pump with 20% of the gas at the inlet (simplified model)

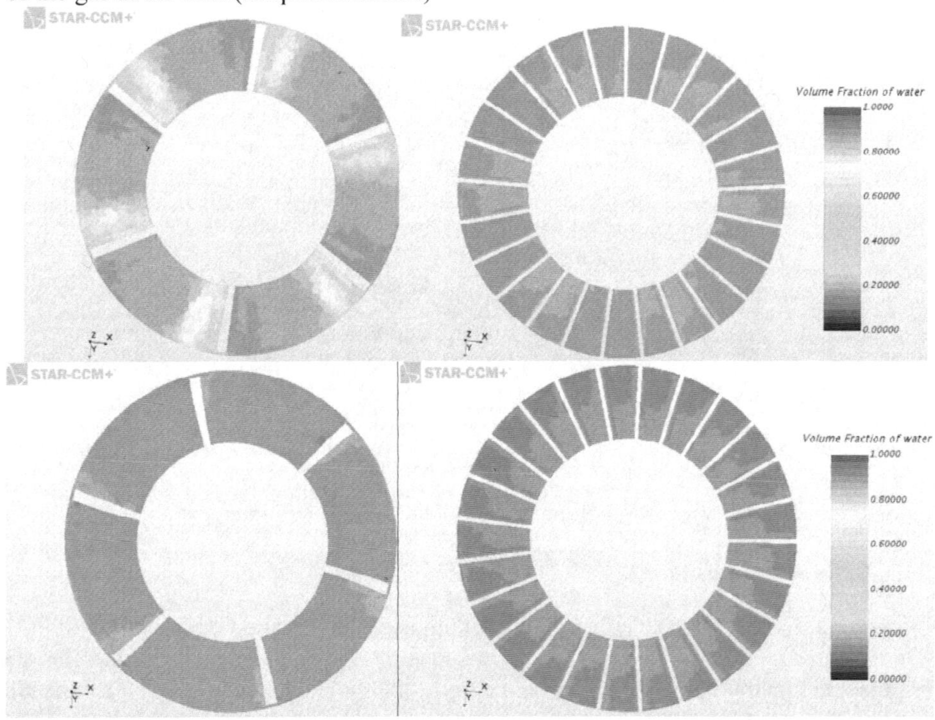

Fig. 7. The volume fraction of water in the cross sections of the pump at 20% of the gas at the inlet (segregated flow model)

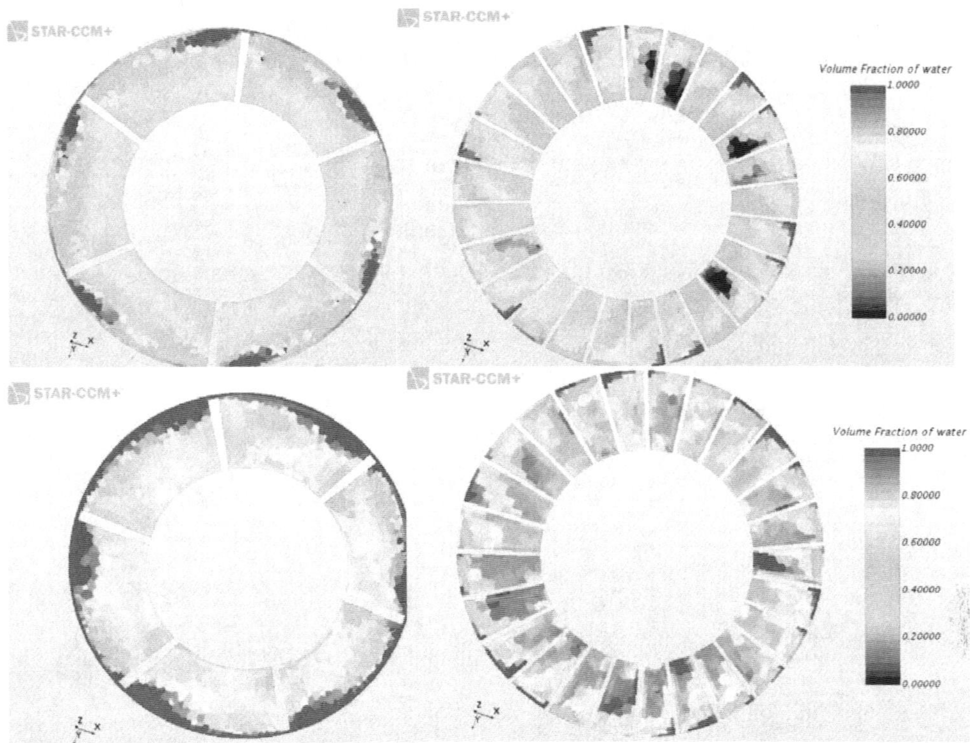

Fig. 8. The volume fraction of water in the cross sections of the pump with 50% of the gas at the inlet (simplified model)

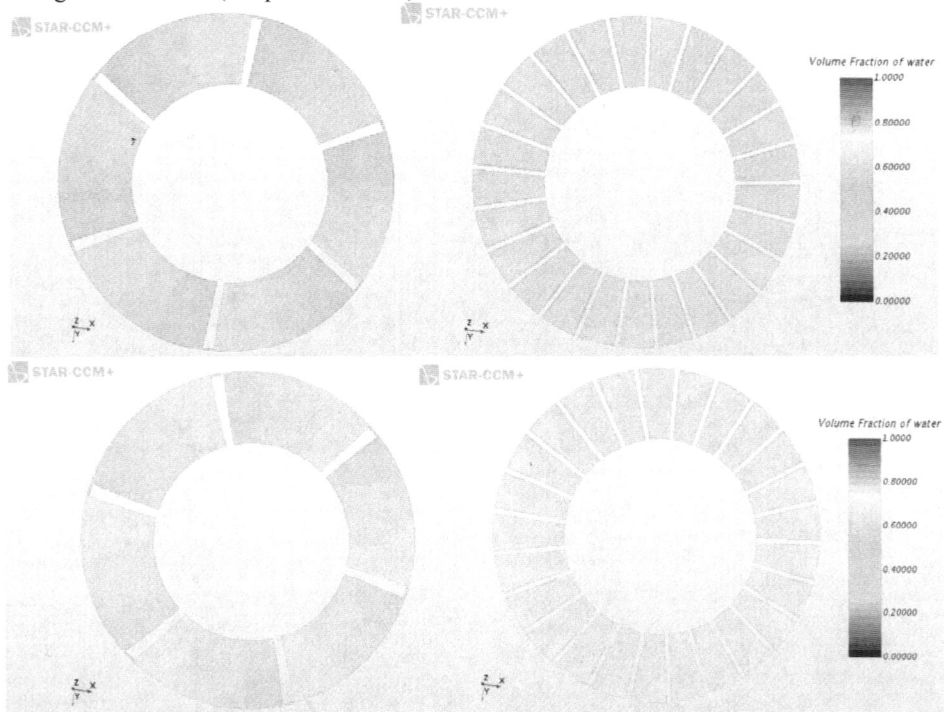

Fig. 9. The volume fraction of water in the cross sections of the pump with 50% of the gas at the inlet (segregated flow model)

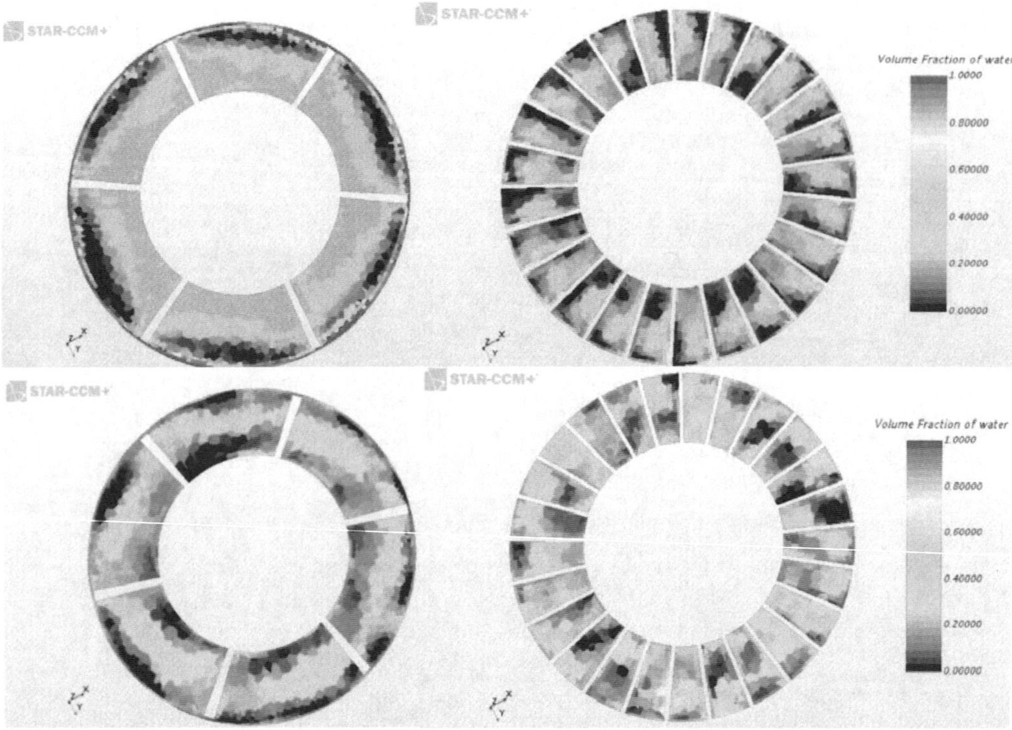

Fig. 10. The volume fraction of water in the cross sections of the pump with 80% of the gas at the inlet (simplified model)

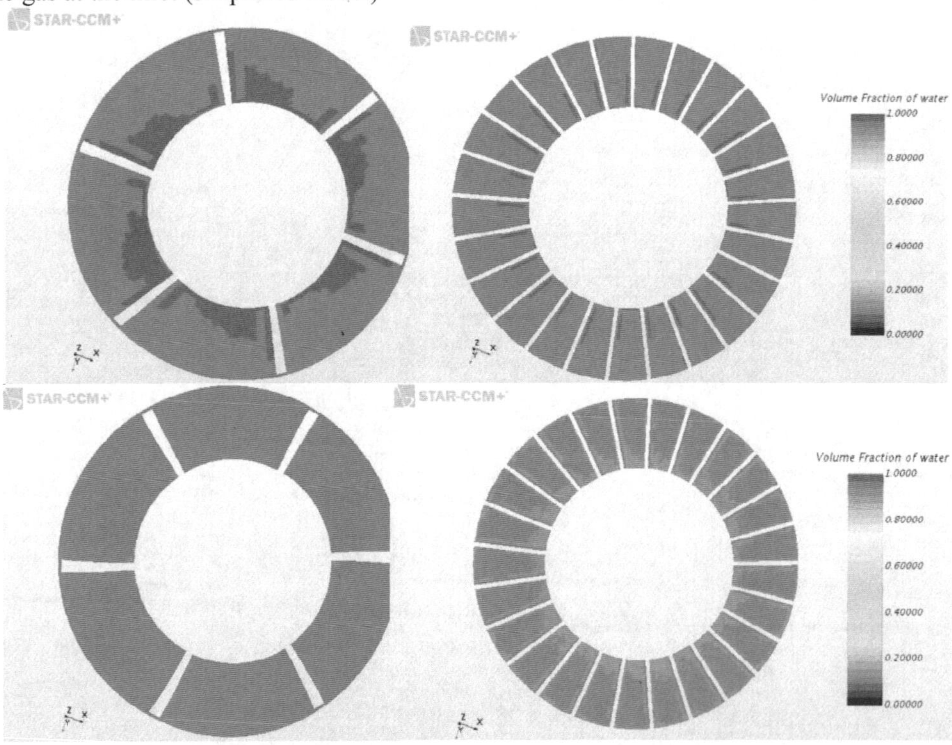

Fig. 11. The volume fraction of water in the cross sections of the pump with 90% of the gas at the inlet (segregated flow model)

Comparing Fig. 6, 8, 10 with fig. 7, 9, 11, it can be seen that the flow distributionsignificantlydiffersin the simplified model and in the segregated flow model. This can be explained by the fact that in full model appear diffusion of gas and liquid, which corresponds to reality. However, in the simplified model, the physics of the process is clearly visible: the liquid in impeller isthrown away to the periphery by centrifugal force and gas spreads throughout the flow, but accumulates closer to the center. Also in the guide vanes gas accumulates on the back side of the blades. But in reality, most likely, such an obvious phase separation in the pump does not occur. In both models, the amount of gas in the second stage is less than in the first.

The simulation results are summarized in table 1.

Table 1. SimulationResults

Gas content α, %	Homogeneous multiphase modelVOF			Segregatedmultiphaseflowmodel		
	H, м	Q, м^3/ч	η, %	H, м	Q, м^3/ч	η, %
0	6,9	80,2	58,6	6,53	81,05	56,2
10	6,93	72	79,6	6,5	72,3	53,1
20	7	64,6	84,2	6,43	64,24	48
30	6,9	55,9	73,5	6,2	56,7	43,1
40	6,7	48,52	38,7	5,7	48,2	36,2
50	6	40	32,4	4,9	39,9	30,9
60	4,78	32,3	25,8	3,7	33	25,7
70	2,8	24,3	18,8	2,6	24,3	18,5
80	1,6	15,8	11,5	1,58	16,4	12,5
90	-	-	-	0,6	7,2	6,2

The dependences of pump head, efficiency and productivity on gas content are presented in Fig. 12-14 (dependencies obtained using a homogeneous multiphase VOF model are shown in blue, red shows a multiphase segregated flow model).

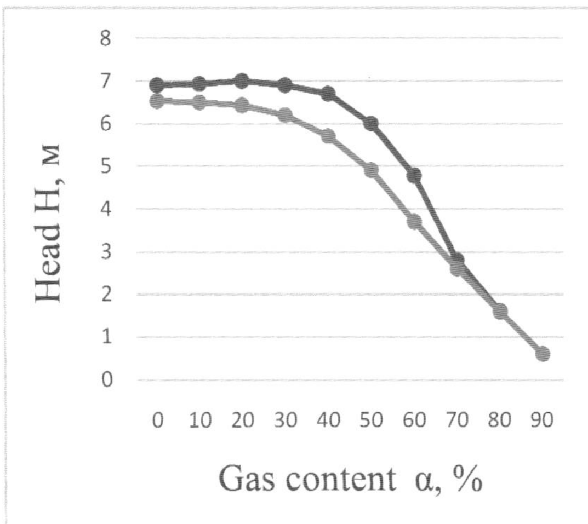

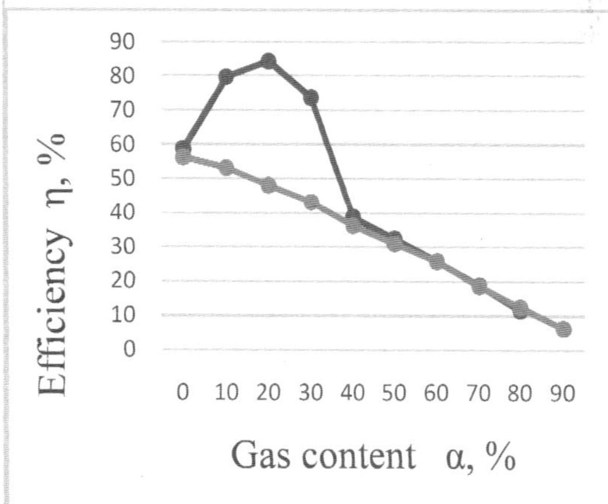

Fig.12. Dependencies of head and pump efficiency on gas content

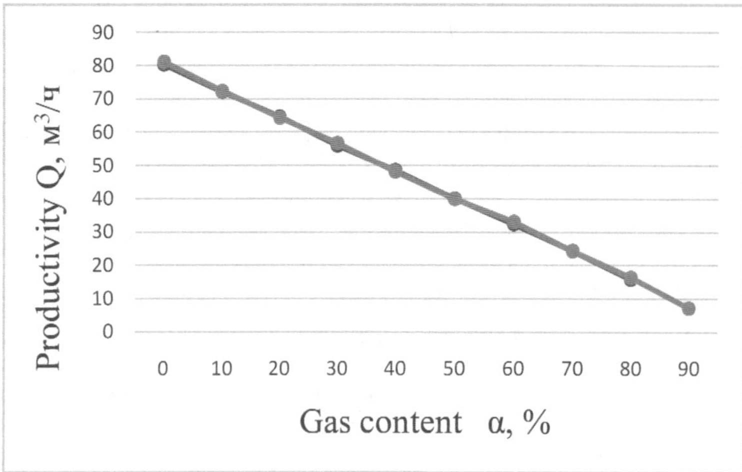

Fig.13. Dependence of pump flow on gas content

From Fig.12-13, it can be concluded that pump head calculated by the VOF method is overestimated relative to the other model, and the efficiency with gas content from 10% to 40% is unreasonably high.

Conclusion

The paper describes a mathematical model of a multi-phase flow, which was implemented by two methods of calculation in the STAR CCM + software package.

The performed simulation showed physicaly results in both cases. As a first approximation the homogeneous multi-phase VOF model can be used. Its advantage is that the calculation requires significantly less computing power. But still, this is a simplified model and errors in it are no exception, for example, too high efficiency in the calculated flow section of the pump. Using a multi-phase segregated flow model, it was possible to calculate fluid flow with gas content up to 95%; in turn, the simplified model showed the result only up to 80% of the gas phase.

In fact, a more complex two-phase flow structure in a multiphase pump is possible. The multiphase flow modelscontains many empirical coefficients, so an experiment is required to validation of this model.

Bibliography

[1] Gläser, D., Flemisch, B., Helmig, R., Class, H. A hybrid-dimensional discrete fracture model for non-isothermal two-phase flow in fractured porous media (2019) GEM - International Journal on Geomathematics, 10 (1), № 5.

[2] Mikelić, A., Wheeler, M.F., Wick, T. Phase-field modeling through iterative splitting of hydraulic fractures in a poroelastic medium (2019) GEM - International Journal on Geomathematics, 10 (1), № 2.

[3] Pinilla, J.A., Guerrero, E., Pineda, H., Posada, R., Pereyra, E., Ratkovich, N. CFD modeling and validation for two-phase medium viscosity oil-air flow in horizontal pipes (2019) Chemical Engineering Communications, 206 (5), pp. 654-671.

[4] Feng, T., Wang, M., Song, P., Liu, L., Tian, W., Su, G.H., Qiu, S. Numerical research on thermal mixing characteristics in a 45-degree T-junction for two-phase stratified flow during the emergency core cooling safety injection (2019) Progress in Nuclear Energy, 114, pp. 91-104.

[5] Lomakin, V.O. Investigation of two-phase flow in axial-centrifugal impeller by hydrodynamic modeling methods (2015) Proceedings of 2015 International Conference on Fluid Power and Mechatronics, FPM 2015, статья № 7337302, pp. 1204-1206.

[6] Putra, R.A., Schäfer, T., Neumann, M., Lucas, D. CFD studies on the gas-liquid flow in the swirl generating device (2018) Nuclear Engineering and Design, 332, pp. 213-225.

[7] Cao, S., Peng, G., Yu, Z. Hydrodynamic design of rotodynamic pump impeller for multiphase pumping by combined approach of inverse design and CFD analysis (2005) Journal of Fluids Engineering, Transactions of the ASME, 127 (2), pp. 330-338.

[8] Shi, G., Wang, Z., Wang, Z., Li, Z., Liu, X. Research on the pressurization performance of an impeller in a multi-phase pump under different working conditions (2019) Advances in Mechanical Engineering, 11 (3), .

[9] El Hajem, M., Morel, R., Champagne, J.Y., Vilagines, R., Pagnier, P. The study of the flow in anhelico-axial pump using laser Doppler velocimetry [Article@Exploration par anémométrie laser Doppler de l'écoulementdansunepompehélico-axiale] (2001) Houille Blanche, (3-4), pp. 34-39.

[10] Shi, G.-T., Wang, Z.-W., Luo, K. Analysis of the Turbulent Flow Intensity and Dissipation Characteristics of an Oil-gas Multiphase Pump in Its Compression Stages [Article@油气混输泵压缩级内湍流强度及湍流耗散特性分析] (2018) RenengDongliGongcheng/Journal of Engineering for Thermal Energy and Power, 33 (6), pp. 115-121.

[11] Falcimaigne, J., Brac, J., Charron, Y., Pagnier, P., Vilagines, R. Multiphase pumping: Achievements and perspectives [Article@Pompagepolyphasique: Réalisationsetperpectives] (2002) Oil and Gas Science and Technology, 57 (1), pp. 99-107.

[12] ANIN Framo Engineering Multiphase pumping - a technology and field testing status report (1994) World Pumps, 339, pp. 29-30.

[13] Jiang, J., Rui, Z., Hazlett, R., Lu, J. An integrated technical-economic model for evaluating CO 2 enhanced oil recovery development (2019) Applied Energy, 247, pp. 190-211.

[14] Shi, Y., Zhu, H., Zhang, J., Yin, B., Xu, R., Zhao, J. Investigation of condition parameters in each stage of a three-stage helico-axial multiphase pump via numerical simulation (2017) Proceedings of the International Offshore and Polar Engineering Conference, pp. 156-162.

[15] Zhang, J.Y., Li, Y.J., Cai, S.J., Zhu, H.W., Zhang, Y.X. Investigation on the gas pockets in a rotodynamic multiphase pump (2016) IOP Conference Series: Materials Science and Engineering, 129 (1), № 012007.

[16] Zhang, Y., Di, Y., Liu, P., Li, W. Simulation of tight fluid flow with the consideration of capillarity and stress-change effect (2019) Scientific Reports, 9 (1), № 5324.

[17] Lomakin, V.O., Kuleshovav, M.S., Bozh'eva, S.M. Numerical Modeling of Liquid Flow in a Pump Station (2016) Power Technology and Engineering, 49 (5), pp. 324-327.

[18] Lomakin, V.O., Kuleshova, M.S., Kraeva, E.A. Fluid Flow in the Throttle Channel in the Presence of Cavitation (2015) Procedia Engineering, 106, pp. 27-35.

[19] Brennen, C.E. Hydrodynamics of pumps (2011) Hydrodynamics of Pumps, 9781107002371, pp. 1-270.

[20] Brennen, C.E. Fundamentals of Multiphase Flows (2005) Cambridge:Cambridge University Press, pp. 1-410.

[21] Gouskov, A.M., Lomakin, V.O., Banin, E.P., Kuleshova, M.S. Minimization of Hemolysis and Improvement of the Hydrodynamic Efficiency of a Circulatory Support Pump by Optimizing the Pump Flowpath (2017) Biomedical Engineering, 51 (4), pp. 229-233.

[22] V Lomakin et al 2019 *IOP Conf. Ser. Eng.* **492** 012012

[23] Lomakin, V.O., Chaburko, P.S., Kuleshova, M.S. Multi-criteria Optimization of the Flow of a Centrifugal Pump on Energy and Vibroacoustic Characteristics (2017) Procedia Engineering, 176, pp. 476-482.

IOP Conf. Series: Materials Science and Engineering **589** (2019) 012015 doi:10.1088/1757-899X/589/1/012015

Measurements of acoustic flow parameters in the orifice on non-linear regimes

A Bykov[1,2], **A Komkin**[1] **and V Moskalenko**[1]

[1]Bauman Moscow State Technical University, 5 Second Baumanskaya Street, Moscow, 105005, Russian Federation

[2]E-mail: alecksbykov@mail.ru

Abstract. This research presents the results of measurements of the parameters of the acoustic jet from the orifice in the partitions on nonlinear regimes. The resulting dependences of jet velocity on the distance from the measurement point to the partition with the orifice are presented.

1. Introduction

In acoustics, nonlinear waves propagation processes at high sound pressure levels take place starting from certain amplitude of oscillating velocity in the orifice, and it can cause considerable changes in acoustic impedanceю. These changes can significantly affect the acoustic characteristics of the Helmholtz resonators [1, 2].

The study of the nonlinear properties of the orifice is devoted to a large number of works that are mainly experimental [3 – 12]. In these studies main focus was on the real part of the orifice impedance, defined as the acoustic impedance R.

Additionally to direct measuring of the orifice impedance it is necessary to describe a shape of the gas-dynamic processes when a high-intensity sound passes through orifice in the partition to understand the mechanisms that forms the nonlinear impedance of the orifice. A proper form of jet-vortex motions in the air near the orifice was obtained in [5] by applying smoke particles to the measuring tube. Similar results for jet-vortex motions in a liquid were obtained in [8]. After then the studies of partition velocity in the orifice radius by mean of hot-wire anemometer took place. In [10] the partition velocity in the orifice was measured and it was found that velocity is related to the sound pressure in the chamber by a quadratic dependence. A number of studies on the characteristics of acoustic jets were performed in [13 – 15], it was shown that the process of jet formation with acoustic waves by leading them through an orifice has a complex character. Similar studies were carried out in [16 – 20] in the study of synthetic jets generators. Recently, methods of laser visualization for oscillatory processes were used in experimental studies of jet-vortex motions near the orifice [20 – 23]. At the same time, computer simulation methods for studying acoustic processes in the orifice area [24 – 26] find its application. However, most of these calculations were done with a constant flow. Moreover, with the help of computational methods it has not been possible to identify general properties of the orifice nonlinear impedance formation so far.

2. Problem definition

This study is devoted to obtaining reliable experimental data on the effect of the orifice diameter on the amplitude of the oscillation velocity in the orifice at high levels of sound pressure.

3. Experimental setup

In the measurements, we used an impedance tube with inner diameter d = 99 mm. At one end of the impedance tube partition with an orifice of certain diameter, d_0, was installed, at the other end – the source of sound, 450-watt loudspeaker. A narrow chirp signal that was formed by a signal generator with a bandwidth of 10 Hz around central frequency of 150 Hz and was applied to the speaker by mean of a power amplifier. The amplitude of the chirp signal varied, so that the sound pressure levels in the impedance tube could vary in the range from 85 dB to 150 dB. The sound pressure in the impedance tube was measured by two 1/4-inch condenser microphones. Signals from microphones after passing a signal amplifier were subjected to spectral analysis. Further on, the obtained spectra were processed in a computer by the method of transfer functions [27, 28] in order to obtain the characteristics of sound pressure directly at the partition with an orifice. Simultaneously with the acoustic measurements and processing, there wasthe measurement was performed using the Pitot tube of the oscillatory velocity in the opening of the partition and its vicinity along the longitudinal axis of symmetry.

4. Results

Figure 1a shows the dependences of the axial longitudinal velocity, V_0, in the area of 10 mm diameter orifice from distance to the partition expressed in calibers, κ, at various sound pressure levels. A characteristic feature of the curves presented in this figure is that in the initial section, at distances of up to ½ caliber, there is a gradual increase in the measured speed. At a distance of the caliber, the longitudinal axial velocity reaches its maximum and further on gradually reduces. Additionally, as curves show with the sound pressure level increase in the impedance tube, the values of the longitudinal velocity increases through the entire length of the measuring section. Moreover, the higher sound pressure level, the greater range of changes of the longitudinal velocity on the measuring section.

Figure 1b shows the dependences of the longitudinal axial velocity, V_0, on the caliber distance, κ, to the measurement point from the partition for three orifice diameters of 5, 10 and 15 mm. The curves presented here corresponds with the curves in fig. 1a. With the decrease of orifice diameter, the corresponding curve shifts upward along the ordinate axis. However, the value of the axial velocity in the cross section of the orifice remains approximately constant.

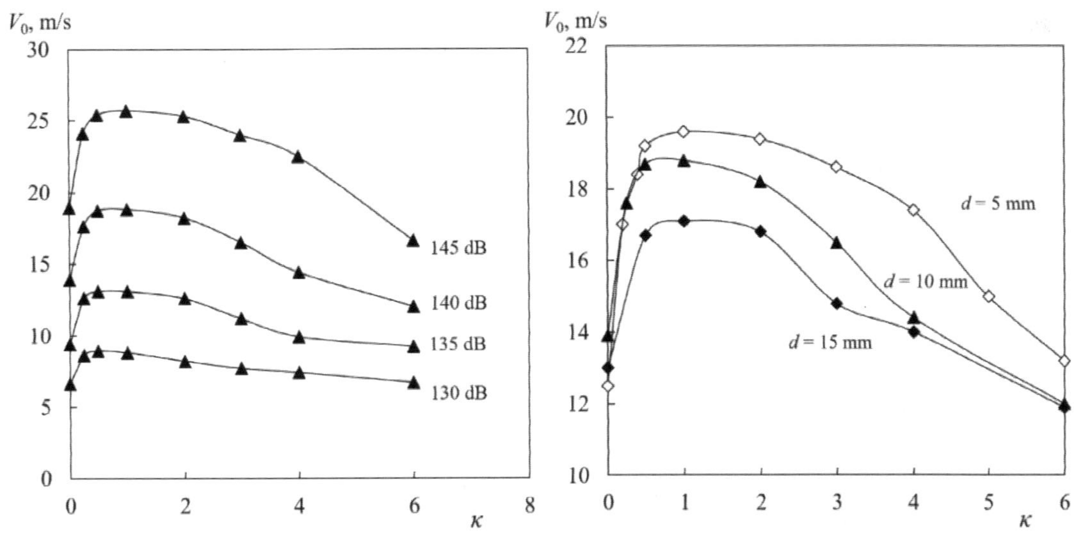

a) b)

Figure 1. Dependences of the longitudinal axial velocity, V_0, on the distance to the partition,

κ - calibers: a – for different sound pressure levels at the partition with an orifice of 10 mm diameter; b – for partitions with different diameters at a sound pressure level of 140 dB.

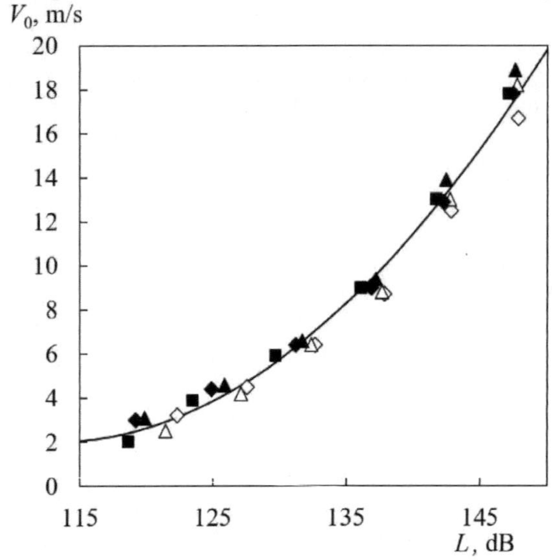

Figure 2. Experimental results of the effect of sound pressure level, Ln, at the orifice on axial longitudinal velocity V_0 for different diameters: ◇-d=5mm, △-d=7mm, ▲-d=10mm, ■-d=12mm, ◆-d=15mm.

Figure 2 shows the dependence of the axial longitudinal velocity, V_0, in the orifices on the sound pressure level, L, at the partition. As expected from data in figure 2, the axial velocity in the cross section of the orifice does not depend on the diameter and is determined only by the sound pressure level. Presented in figure 3 dependence can be approximated by the following function:

$$V = 0.013(L_n - 113)^2 + 2.$$

5. Conclusion

The conducted studies allowed us to establish the characteristics of the oscillatory motion of the medium in the vicinity of a partition with a hole on nonlinear operating regimes. It is shown that the velocity in the hole does not depend on its diameter, but is determined only by the level of sound pressure at the partition.

References

[1] Wu J., Rudnick I. Measurements of the nonlinear tuning curves of Helmholtz resonators // J. Acoust. Soc. Am. 1986. V. 80, № 5. P. 1419–1422.

[2] Komkin A.I., Mironov M.A., Bykov A.I. Sound Absorption by a Helmholtz Resonator // Acoustical Physics. 2017. v. 63. № 4. P. 385–392.

[3] Sivian L.J. Acoustic impedance of small orifices // J. Acoust. Soc. Am. 1935. V. 7, № 3. P 94–101.

[4] Ingard, U., Labate S. Acoustic circulation effects and the non-linear impedance of orifices // J. Acoust. Soc. Am. 1950. V. 22, № 2. P 211–218.

[5] Ingard U. On the theory and design of acoustic resonators // J. Acoust. Soc. Am. 1953. V. 25, № 6. P. 1037–1061.

[6] Thurston G.B., Martin C.E. Periodic Fluid Flow through Circular Orifices // J. Acoust. Soc. Am. 1953. V. 25, № 1. P. 26–31.

[7] Ingard U., Ising H. Acoustic nonlinearity of an orifice // J. Acoust. Soc. Am. 1967. V. 42, № 1. P. 6–17.

[8] Panton R.L., Goldman A.L. Correlation of nonlinear orifice impedance // J. Acoust. Soc. Am. 1976. V. 60, № 6. P. 1390–1396.

[9] Dickey N.S., Selamet A., Novak J.M. The effect of high-amplitude sound on the attenuation of perforated tube silencer // J. Acoust. Soc. Am. 2000. V.108, № 5. P. 1068 –1081.

[10] Komkin A.I., Bykov A.I., Mironov M.A. Inertial attached length of an orifice at high sound pressure levels // Acoustical Physics. 2018. v. 64. № 3. P. 287–292.

[11] Komkin A.I., Bykov A.I., Mironov M.A. Acoustic Resistance ofan orifice at high sound pressure levels // Acoustical Physics. 2018. v. 64. № 5. P. 563–566.

[12] Bykov A.I., Komkin A.I., Mironov M.A. Nonlinear acoustic impedance of orifices // Proceedings of the 25-th International congress on sound and vibration. Hiroshima, Japan, 8-12 July 2018. P.1–7.

[13] Lebedeva I.V. Experimental investigation of acoustic flow in the vicinity of the orifice // Acoustical Physics. 1980. T. 26. №.4. C. 599–602.

[14] Lebedeva I.V., Grushin A.E. Development of acoustic jet in space at the open end of the waveguide // Vestnik MGY. Ser.3. 2002. №.2. C. 49–52.

[15] Lebedeva I.V., Grushin A.E. Amplitude and frequency characteristics of acoustic jets // Acoustical Physics. 2003. v. 49. №.3. C. 300–304.

[16] Smith, B.L. Glezer, A. The formation and evolution of synthetic jets // Phys. Fluids. 1998, V. 10, № 9. P.2281-2297.

[17] Mallinson, S. G., Reizes, J. A., and Hong, G. An Experimental and Numerical Study of Synthetic Jet Flow // The Aeronautical Journal, V. 105, №. 1043. 2001. P. 41–49.

[18] Holman R., Utturkar Y., Mittal R., Smith B.L, Cattafesta L.: Formation criterion for synthetic jets, AIAA J. 2005. V.43. P. 2110-2117.

[19] Chaudhari M., Verma G., Puranik B., Agrawal A. Frequency response of a synthetic jet cavity // Exp. Therm. Fluid Sci.2009, V. 33, P. 439-448.

[20] Gil P. Morphology of synthetic jet // ZN PRzMechanika. 2017. V. 89 (2/17). P. 183-191.

[21] Sharp J. P., Greated C. A., Gray C., Campbell D. M. // The Measurement of Acoustic Streaming using Particle Image Velocimetry // Acustica. 1989. V. 68, N 2. P. 168–172.

[22] Valiere J.-C. Acoustic Particle Velocity Measurementsusing lasers. Principles, signal processing and applications.Wiley, 2014. – 157 p.

[23] Buick J.M, Atig M., Skulina D. J., Campbell D. M., Dalmont J. P., Gilbert J. Investigation of non-linear acoustic losses at the open end of a tube // J. Acoust. Soc. Am. 2011. V.129, № 3. P. 1261 –1272.

[24] ScarpatoA. Linear and nonlinear analysis of the acoustic response of perforated plates traversed by a bias flow // Ph.D. dissertation, EcoleCentrale Paris, 2014. 178 p.

[25] Akamatsu K. Tsutahara M. Acoustic properties of perforated plates with bias flow // Proceedings of the INTER-NOISE-2016, Hamburg, Germany, 2016. P. 6610-6621.

[26] Miró A., Soria M., Moulinec C., Cajas J. C., Fournier Y. Numerical investigations on rectangular and circular synthetic jet impingement // Tenth International Conference on Computational Fluid Dynamics (ICCFD10), Barcelona, Spain, July 9-13, 2018. 18 p.

[27] ISO 10534-2:1998. Acoustics - Determination of sound absorption coefficient and impedance in impedance tubes - Part 2: Transfer-function method.

[28] Komkin A.I. Methods of measuring the acoustic characteristics of sound-absorbing materials // Measurement Techniques. 2003. V. 46. № 3. P. 290–295.

IOP Conf. Series: Materials Science and Engineering **589** (2019) 012016 doi:10.1088/1757-899X/589/1/012016

Mathematical modeling and analysis of the electrohydraulic actuator with the throttle control

N Sosnovsky[1,2] and D Ganieva[1]

[1]Bauman Moscow State Technical University, 5 Second Baumanskaya Street, Moscow, 105005, Russian Federation

[2]Email: sosn60@mail.ru

Annotation

High demands to modern electrohydraulic actuators are placed. This applies to the static and dynamic characteristics of the drives, which is achieved by using different correcting devices (controllers). In the article analysis of the most effective regulator for an electro-hydraulic actuator is conducted during the comparison of its dynamic characteristics at introduction of various types of correcting devices: the introduction of state controller, the introduction of additional pressure feedback and a proportional controller. Block diagrams of the electrohydraulic actuator are constructed according to the corresponding mathematical models. The optimization of the control parameters was carried out to improve the static characteristics of the actuator.

Introduction.

High demands to modern electrohydraulic actuators are placed. This applies to the static and dynamic characteristics of the drives, which is achieved by using different correcting devices (controllers). In this case, the design parameters of the hydraulic drive are determined on the basis of energy, power and static characteristics [1]-[3]. To reduce the influence of external and parametric disturbances on the hydraulic drive, as well as to improve its dynamic performance, it is necessary to increase the quality factor of the electro-hydraulic actuator by increasing the gain of the proportional controller (P-controller) in a forward-path. The hydraulic drive is characterized by relatively low natural damping, and an increase in the quality factor in the presence on the hydraulic cylinder rod large inertial loads reduces the stability and dynamic quality of the drive [4]-[7]. One of the controllers that can solve this problem is the state controller, the principle of which is the introduction of additional state variable feedback.

Feedbacks on the drive state coordinates are entered into the control loop. The transmission coefficients of these feedbacks are adjusted so as to be able to achieve the required properties, dynamic and static characteristics of the electro-hydraulic actuator [8]-[12]. It should be noted that it is advisable to use state controllers only in hydraulic actuators with electrohydraulic amplifier, the natural frequencies of which are at least one and a half times higher than the natural frequency of a loaded hydraulic motor [13]-[16].

Negative speed feedback increases the natural frequency and reduces the relative damping of the hydraulic cylinder. The introduction of feedback on the speed and acceleration of the output link of the electro-hydraulic drive allows to set the parameters allows to set the parameters of the natural frequency and damping of the hydraulic motor independently of each other; therefore, it provides any speed-of-response behavior of the system with any gain in a forward-path [17]-[19].

In the articleanalysis of the most effective regulator for an electro-hydraulic actuator is conducted during the comparison of its dynamic characteristics at introduction of various types of correcting devices: the introduction of state controller, the introduction of additional pressure feedback and a proportional controller.

Mathematical model of electrohydraulic actuator.

Linear mathematical models can only be used for preliminary evaluation of the dynamic characteristics of real systems. In reality, there are nonlinearities that ambiguously affect the operation of the actuator. In particular, it is necessary to take into account the nonlinearity of the pressure loss/flow characteristics of the hydraulic amplifier:

$$Q_З = k_з' \cdot x_з \cdot \sqrt{\left| p_{пит} - p_н \cdot sign(x_з) \right|}\ , \tag{1}$$

where $k_з' = \mu_з \cdot \pi \cdot d_з \cdot k_п \cdot \sqrt{\frac{2}{\rho}}$,

$p_н = p_1 - p_2$ – the pressure difference in the hydraulic cylinder cavities;

$Q_З$ – flow through spool; $x_з$ – displacement of servovalve spool; $d_з$ – spool diameter.

Equation of motion of a loaded hydraulic cylinder:

$$m \cdot \frac{d^2}{dt^2} y + k_{тр} \cdot \frac{d}{dt} y = F_ц \cdot (p_1 - p_2) - P_н, \tag{2}$$

where y – displacement of hydraulic cylinder rod; $k_{тр}$ – friction coefficient; $F_ц$ – hydraulic cylinder piston area.

The pressure difference in the hydraulic cylinder cavities is found from the equation (2), which takes into account dependences on the force of viscous friction and external loading:

$$p_н = \frac{P_н}{F_ц} + \frac{m}{F_ц} \cdot \frac{d^2}{dt^2} y + \frac{k_{тр}}{F_ц} \cdot \frac{d}{dt} y. \tag{3}$$

To improve the static characteristics of the actuator (increase the static stiffness of the actuator), as well as to obtain a transient with the least oscillation, the optimization of the control parametersbased on the ITAE functionality (integral time of absolute error)was carried out. The basis of this functional is the integral of the product of the transient time t by the absolute run time error:

$$I = \int_0^\infty t \cdot \left| y_{зад} - y \right| dt,$$

where $y_{зад}$ – set position of a rod of a hydraulic cylinder.

The results of mathematical modeling.

According to the standard mathematical model of a linear system, a block diagram of an electro-hydraulic actuator was made in the program "ПКМВТУ" [20] with additional feedback on state variables (Fig. 1).

The optimization of the control parameters was carried out.The transient graphs on the position of the hydraulic cylinder rod was obtained (Fig. 2).

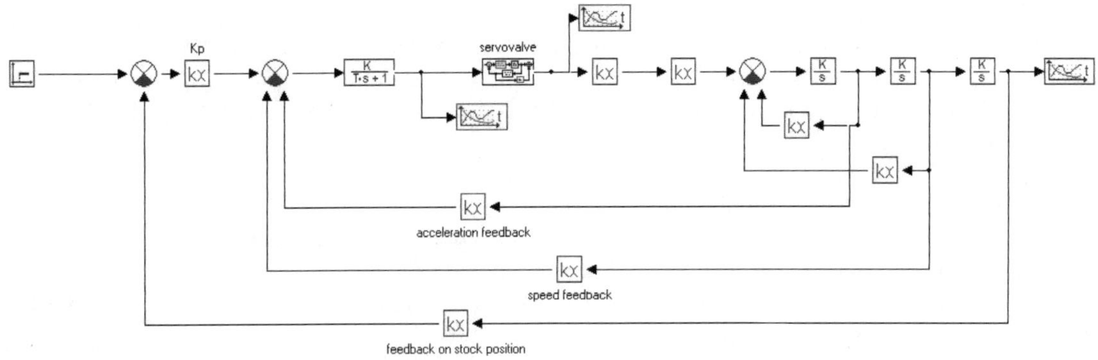

Fig. 1. Block diagram of an electro-hydraulic actuator with a state controller.

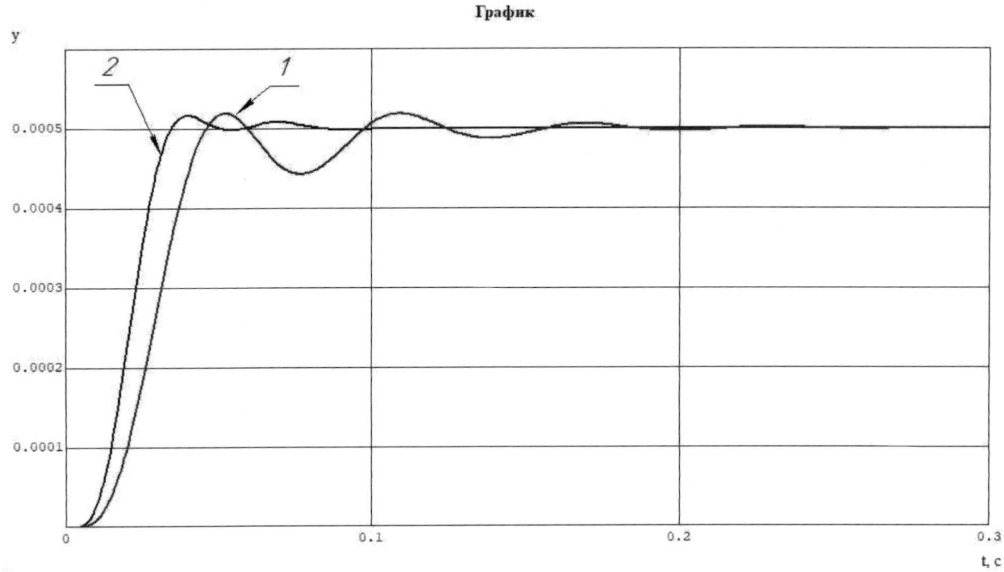

Fig. 2. Transient graphs of the electrohydraulic actuator.
1 – with a proportional controller; 2 – with state controller.

Indicators of the quality of transient processes are presented in table 1:
σ – run time error, %
$t_{п/п}$ – transition time, c

Table 1.

Correcting device	σ, %	$t_{п/п}$, c
proportional controller	4.12	0.115
state controller	3.35	0.042

According to the developed mathematical model, taking into account the non-linearity of the pressure loss/flow characteristic, a block diagram of the electrohydraulic drive was constructed, according to which, by optimization, the coefficients of the regulators were refined and the transients were calculated. Transient graphs with the introduction of different types of controllers (Fig. 3):

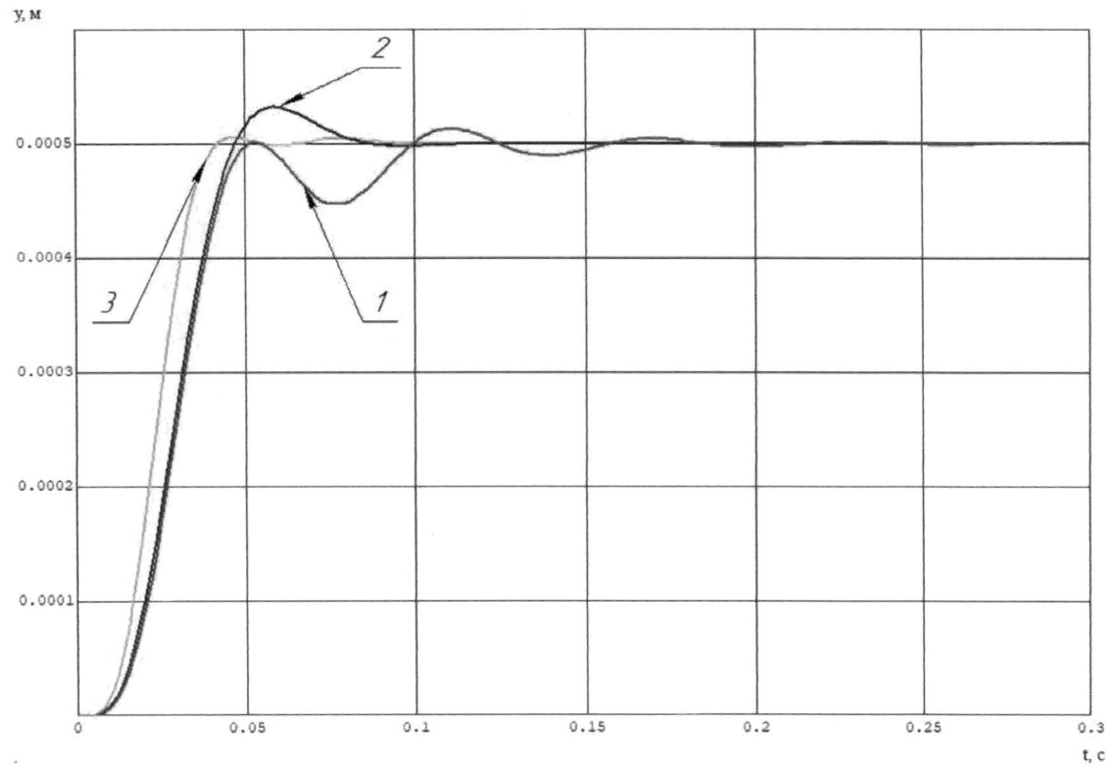

Fig. 3. Transient graphs of the electrohydraulic actuator with the introduction of different types of controllers.1 – with a proportional controller;2 – with additional pressure difference feedback; 3 – with state controller.

The results are presented in table 2:
k_p – control factor in a forward-path (proportional controller).

Table 2.

Correcting device	σ, %	$t_{п/п}$, c	k_p
proportional controller	2.46	0.094	1.3375
additional pressure difference feedback	6.39	0.072	1.6
state controller	1.05	0.038	3.0364

With the help of the ITAE functional, the most optimal values of feedback coefficients were found: for differential pressure - 2 * 10-8; for speed - 6.671 and for acceleration - 0.073.

Conclusion.

The dynamic characteristics of an electro-hydraulic actuator with three types of correcting devices are obtained: with a conventional proportional controller in a forward-path; with additional pressure difference feedback in the cavities of the hydraulic drive; with additional feedback on speed and acceleration (state controller).

The introduction of additional pressure difference feedback led to a significant reduction in transition time, to an increase in the stability of the system and to an increase in the quality factor, however the run time error also increased.

The best controller for our actuator turned out to be a state controller. With the introduction of additional feedbacks on state variables, the best dynamic characteristics of the actuator were obtained: the transition time was significantly reduced, and the run time error became significantly less, the quality and stability of the system were increased. Using ITAE functionality for optimizing control parameters of an electrohydraulic actuator leads to an improvement in the dynamic and static characteristics, including by increasing the control factor in a forward-path.

Reference.

[1] Popov, D.N., Sosnovskiy, N.G., Siukhin, M.V. Control of synergetic processes to ensure the asymptotic stability of hydraulic systems (2017) Herald of the Bauman Moscow State Technical University, Series Natural Sciences, (3), pp. 37-51.

[2] Zhao, J., Wang, M., Wang, Z., Grekhov, L., Qiu, T., Ma, X. Different boost voltage effects on the dynamic response and energy losses of high-speed solenoid valves (2017) Applied Thermal Engineering, 123, pp. 1494-1503.

[3] Popov, D.N., Sosnovskiy, N.G., Siukhin, M.V. Wave Processes Regulators Optimization in Hydraulic Systems (2018) IOP Conference Series: Materials Science and Engineering, 468 (1), статья № 012014.

[4] Krut'ko, P.D. Algorithms of terminal control of linear dynamic systems (1998) ShuiyunGongcheng/Port & Waterway Engineering, (9), p. 859.

[5] Oleg, T. The experience of creating compact energy efficiency hydrostatic servo drive (2015) Proceedings of 2015 International Conference on Fluid Power and Mechatronics, FPM 2015, статья № 7337187, pp. 603-609.

[6] Semenov, S., Kulakov, D. Mathematical modeling of the mechanisms of volumetric hydraulic machines (2019) IOP Conference Series: Materials Science and Engineering, 492 (1), статья № 012042.

[7] Kulakov, D., Kulakov, B. Study of the dynamics of an electro-hydraulic servo drive physical model on FESTO learning system (2019) IOP Conference Series: Materials Science and Engineering, 492 (1), статья № 012029.

[8] Krut'ko, P.D. Robustly stable structures of control systems of high dynamic precision: Algorithms and dynamics of control of model objects (2005) Journal of Computer and Systems Sciences International, 44 (2), pp. 273-292.

[9] Bushuev, A.Yu., Ivanov, M.Yu., Korotaev, D.V. Minimization of a mismatch time of movement of actuators of a throttle synchronization system * (2018) Journal of Physics: Conference Series, 1141 (1), статья № 012090.

[10] Fursov, A.S., Emel'yanov, S.V., Kapalin, I.V., Sagadinova, E.S. Stabilization of Multiple-Input Switched Linear Systems with Operation Modes of Different Dynamical Orders (2018) Differential Equations, 54 (11), pp. 1517-1523.

[11] Semenov, S.E. Mathematical modeling of the electro-hydraulic actuation systems of the machines with tree-like kinematic structure (2015) Proceedings of 2015 International Conference on Fluid Power and Mechatronics, FPM 2015, статья № 7337184, pp. 583-592.

[12] Krut'ko, P.D., Chernous'ko, F.L. Decomposing algorithms for the motion control of nonlinear dynamic systems (2001) IzvestiyaAkademiiNauk. TeoriyaiSistemyUpravleniya, (4), pp. 8-24.

[13] Meretskaya, E.R., Chernyshev, A.V. Calculating and theoretical research and determination of the functional parameters of a pneumo-mechanical device actuator (2018) AIP Conference Proceedings, 2007, статья № 030045.

[14] Popov, E.P. Control and software of complex robotengineering systems (1995) IzvestiyaAkademiiNauk. TeoriyaiSistemyUpravleniya, (1), pp. 193-204.

[15] Krut'ko, P.D., Chkheidze, G.A. Synthesis of algorithms for controlling tracking systems of high dynamic accuracy (1993) Journal of Computer and Systems Sciences International, 31 (5), pp. 1-16.

[16] Ostroukh, A., Surkova, N., Varlamov, O., Chernenky, V., Baldin, A. Automated process control system of mobile crushing and screening plant (2018) Journal of Applied Engineering Science, 16 (3), pp. 343-348.

[17] Belousov, B., Demik, V., Kozlova, A., Ksenevich, T., Naumov, S., Medvedev, E., Lyushnin, S., Kuzminkov, K. The schematic diagrams of actuators of an mechatronic wheel steering system (2018) FISITA World Automotive Congress 2018.

[18] Bazinenkov, A.M., Mosina, M.S., Mikhailov, V.P. The Investigation of the Work-fluid Characteristics of Precision Hydraulic Active Vibration Control Module (2017) Procedia Engineering, 176, pp. 296-300.

[19] Timofeev, G.A., Kuzenkov, V.V. Features of the dynamics of a servo drive with wave gear (2015) Journal of Machinery Manufacture and Reliability, 44 (6), pp. 513-519.

[20] Kozlov OS, Kondakov D.E., Skvortsov L.M. et al. Software package for the study of the dynamics and design of technical systems. Information Technologies. 2005. № 9.

IOP Conf. Series: Materials Science and Engineering **589** (2019) 012017 doi:10.1088/1757-899X/589/1/012017

The evaluation of the effect of gas content on the characteristics of a Centrifugal Pump

M Saprykina[1,2] and V Lomakin[1]

[1]Bauman Moscow State Technical University, 5 Second Baumanskaya Street, Moscow, 105005, Russian Federation

[2]E-mail: marias108@mail.ru

Abstract
The effect of gas content on the changing ofhead of a centrifugal pump isresearched. The method of numerical hydrodynamic modeling is used for solving this problem. The using mathematical model is described. The results of the simulation of fluid flow,which are depended on the air content, are presented. The data on the changingof head and the termination of its work are obtained.

Introduction

In these latter days, many people have to deal with a rather large number of complicated factors with oil production. These include the greater curvature of the wellbore, small diameters of the casing, high gas content, removal of mechanical impurities, salt deposition, high temperature of the liquid and others [1].

The effect of the gas factor complicating the operation of a centrifugal pump is regarded in this article. The gas factor is one of the most important one while you are choosing the method of operation and designing the optimal operation mode of the reservoir-well system. The presence of gas in the oil / water mixture also changes the behavior of the pump. The value of the optimal gas content will additionally depend on the properties of the oil and the water content in the mixture. The submersible centrifugal pump is quite sensitive to the presence of free gas in the pumped liquid. The characteristics of the pump are deformed because of its dependence on quantity of gas content.

The presence of free gas limits the work of the ECP, because:

- free gas reduces the volume of fluid in the ECP stage;
- separation of liquid and gas in the centrifugal force field reduces the pressure of the stage;
- "gas impurity" is forming in the first steps of the pump and accumulating at the center. It blocks the flow of liquid, which leads to disruption of the pump supply. [2] - [4]

Numerous and long-term studies of the work of the ESP system made it possible to identify three qualitatively different areas of the work of centrifugal pump. It is pumping out the gas-liquid mixture. In the first area, which is characterized by a small content of free gas in the pumped liquid, the actual characteristics of the pump do not differ from the bench characteristics for a pure liquid (there is no free gas), and the efficiency of the pump is maximum. The pressure at the pump intake, corresponding to small gas content in the pumped liquid, is the optimum pressure at the inlet.The second area of work of the ECP is characterized by increasing of free gas quantity in the pumped liquid. As a result the actual characteristics of the pump differ from the bench ones when the pump works without free gas.

Yet, the pump maintains stable operation with an acceptable efficiency. The inlet pressure in this area of the work of free gas pump is called the admissible inlet pressure. The third area of work of the ECP is characterized by a significant amount of free gas in the pumped liquid, as a result the stable operation of the pump is disturbed until the supply is interrupted. In this case, the efficiency of the pump is significantly reduced (down to zero when the supply is disrupted), and long-term operation of the ECP in this area becomes impossible. [5]

When the existing literature analysis on this topic was conducted, it was revealed that there is no complete research on this topic. Thus, in [6], the effect of the gas factor was studied in conjunction with others one during the overhaul period of operation of the reservoir-well-pump system. In [7] - [8] the main attention is paid to the review and development of dispersing devices for pumping gas liquid mixture from wells.To conclude, the scientific literature has no studies and articles covering all aspects of the effect of gas content on the pump operation.

Mathematical model of multiphase fluid flow

In this paper, we use the model of a multiphase incompressible fluid flow (ρ = const). Numerical simulation is based on solving discrete analogs of the basic hydrodynamic equations.

The calculation is carried out on the basis of a mathematical model of a divided multiphase flow. That is, the equations of mass and momentum are solved separately for each phase. Yet, the pressure field is the same for all phases. The developed pump is used in the system of formation water purification. It supplies the working fluid to the flotation machine. It pumps a two-phase medium: water-air. Numerical hydrodynamic modeling was performed in the StarCCM + software package. [9] - [11]

The mathematical model consists of differential and algebraic equations:

1. The volume of the i-th phase in computional cell is calculated as:

$$V_i = \int_V \alpha_i dV,$$

where α_i – the concentration of the i-th phase in the cell.

The sum of the concentrations of all phases in the cell is one.

$$\sum_{i6}^{n} \alpha_i = 1.$$

2. The equation for conservation of mass (continuity equation):

$$\frac{\partial}{\partial t}\int_V \alpha_i \rho_i dV + \int_A \alpha_i \rho_i \vec{V_i} d\vec{a} = 0,$$

where ρ_i – the i-th phase density; $\vec{V_i}$ – the i-th phase velocity (in the case of turbulent flow modeling by a RANS-type).[12]-[15]

3. The equation of the changing in the amount of motion:

$$\frac{\partial}{\partial t}\int_V \alpha_i \rho_i \vec{V_i} dV + \int_A \alpha_i \rho_i \left(\vec{V_i}\vec{V_i}\right) d\vec{a} = -\int_V \alpha_i \nabla p dV +$$

$$+\int_V \alpha_i \rho_i \vec{g} dV + \int_A \left[\alpha_i \left(T_i + T_i^t\right)\right] d\vec{a} + \int_V \vec{M_i} dV,$$

where $\left(\vec{V_i}\vec{V_i}\right)$ – the tensor product of the velocity vectors of the i-th phase; p– the pressure; $\vec{g}$ –the mass intensity vector (in this case, the gravity force is 9.81 m / s2 and the inertial force due to the rotation of the computational area); T_i - the molecular viscosity stress tensor; T_i^t – the turbulent stress tensor; $\vec{M}$ –the vector of the total intensity of interfacial interaction forces per unit volume, for the vector $\overrightarrow{M_i}$ the equality is:

$$\sum_i \overrightarrow{M_i} = 0.$$

The vector $\overrightarrow{M_i}$ characterizes all the forces that separated phases interact with each other.[16]-[20]

$$\overrightarrow{M_i} = \sum_{i \neq j}\left(\overrightarrow{F_{ij}^{D}} + \overrightarrow{F_{ij}^{VM}} + \overrightarrow{F_{ij}^{L}} + \overrightarrow{F_{ij}^{TD}} + \overrightarrow{F_{ij}^{WL}}\right),$$

where $\overrightarrow{F_{ij}^{D}}$ – the resisting force; $\overrightarrow{F_{ij}^{VM}}$ – the power of the virtual mass; $\overrightarrow{F_{ij}^{L}}$ – the ascensional power; $\overrightarrow{F_{ij}^{TD}}$ – the turbulent dispersive force; $\overrightarrow{F_{ij}^{WL}}$ – the force, which is caused by wall effects.

In fig. 1 the computational grid for hydrodynamic modeling is shown.

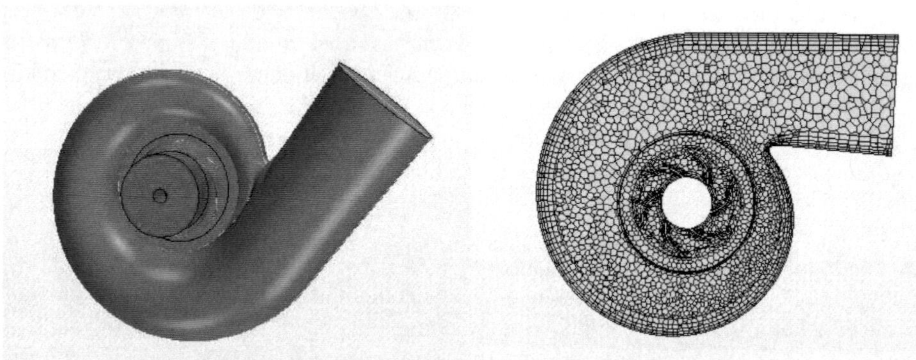

аб

Fig. 1. The computational grid: a-3d model; b-section.

The pump was calculated with a changing of the volume fractions of water and air in order to determine the time when the head will drop and its operation will stop. The head was calculated by the following formulas:

$$H = \frac{p_2 - p_1}{\rho_w \cdot g},$$

где p_1— the pressure in 1 dimensional section, Pa; p_2 — the pressure in 2 dimensional section, Pa; ρ_w — the water density, 1000 kg/м^3; g —the acceleration of gravity, м/s^2.

When a two-phase medium are being pumped, the head should be considered to take into account changes in the density of the pumped medium, which is calculated by the formula:

$$\rho = \frac{m}{V} = \frac{\alpha_1 \cdot V \cdot \rho_1 + \alpha_2 \cdot V \cdot \rho_2}{V} = \alpha_1 \cdot \rho_1 + \alpha_2 \cdot \rho_2,$$

whereα_1 — the water volume fraction; ρ_1 — the water density, кg/м^3; α_2 — the water volume fraction; ρ_2 — the air density, кg/м^3.

Therefore, the head, which was received by the StarCCM + software, was adjusted for the calculated density:

$$H_1 = H \cdot \frac{\rho_w}{\rho},$$

where$H$ — the head, which was received by the StarCCM+, м; ρ_w — the water density, кg/м^3; ρ — calculated density of the two-phase medium, кg/м^3.

For the convenience of evaluation the effect of gas content on the pump head, the relative head was also calculated:

$$\bar{H} = \frac{H_1}{H_u},$$

where $\bar{H}$ — the relative head, м; H_u— the ideal head, 10,5 м; H_1 — the correct head,м.

Calculation results

In fig. 2 the visualized data on the volume of air in the calculated pump are shown.

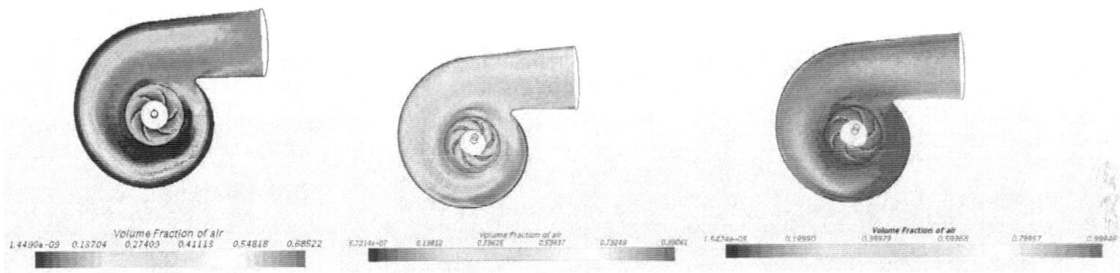

абв

Fig. 2. The air volume visualization
а —φ_{air} = 10%; б— φ_{air} = 60%; в — φ_{air} =90%.

From the STAR CCM + software package the necessary graphs were obtained. They are shown in fig. 3 and 4.

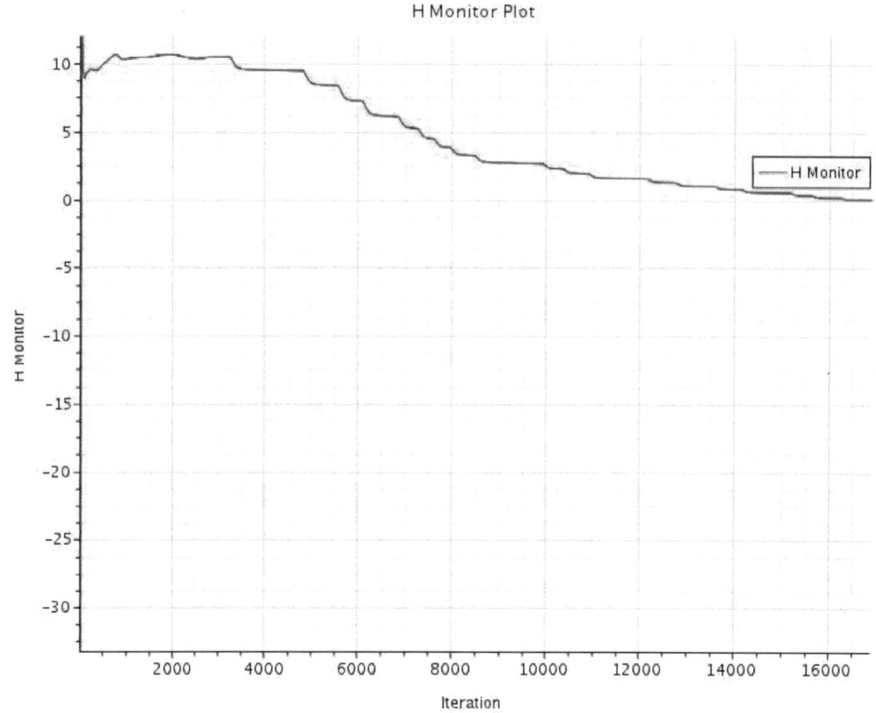

Fig. 3. The head graph

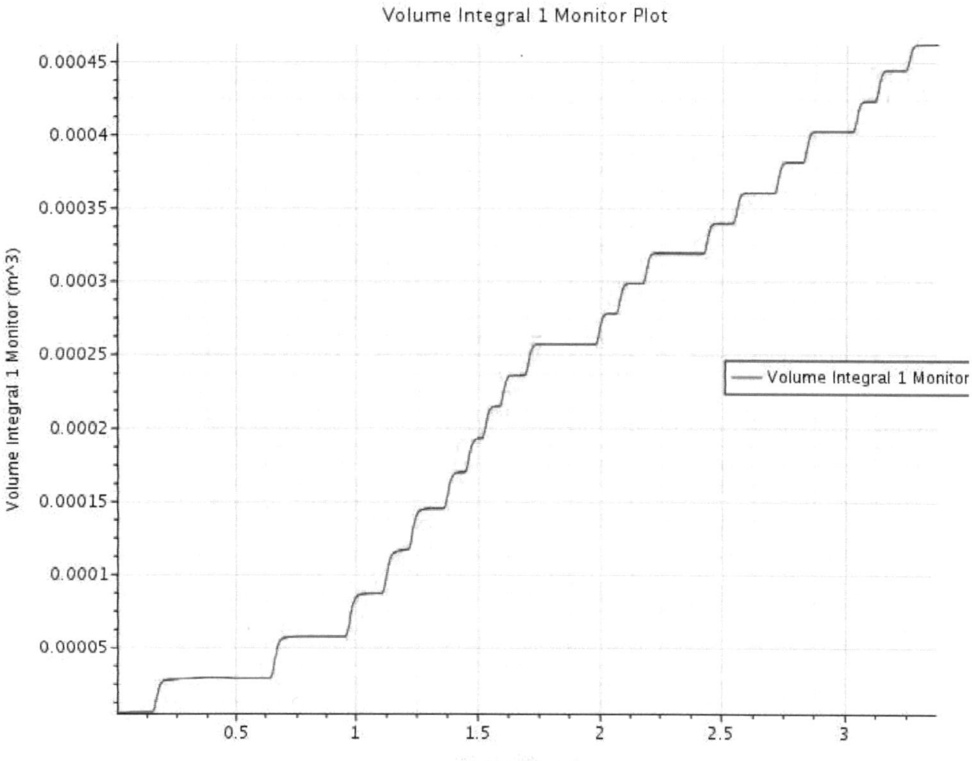

Fig. 4.The volume of air graph

The data on the head and volume of air in the developed pump was taken from the graphs. According to the formula the relative head was calculated. The results are presented in table 1.

Table 1. The results

φ_{air}	φ_{water}	H	H_1
0,01	0,99	10,551	10,656
0,05	0,95	10,368	10,598
0,1	0,9	9,473	10,523
0,15	0,85	8,391	9,871
0,2	0,8	7,251	9,062
0,25	0,75	6,124	8,162
0,3	0,7	5,271	7,525
0,35	0,65	4,548	6,992
0,4	0,6	3,879	6,461
0,45	0,55	3,187	5,789
0,5	0,5	2,676	5,346
0,55	0,45	2,345	5,104
0,6	0,4	1,924	4,801
0,65	0,35	1,565	4,461
0,7	0,3	1,292	4,295
0,75	0,25	1,018	4,057

Окончание табл. 1

φ_{air}	φ_{water}	H	H_1
0,8	0,2	0,785	3,906
0,85	0,15	0,553	3,662
0,9	0,1	0,366	3,621
0,95	0,05	0,183	3,578
0,99	0,01	0,046	3,501

According to the data of the table, the dependence of the head on the amount of air in the pump under study was plotted and performed on fig.5.

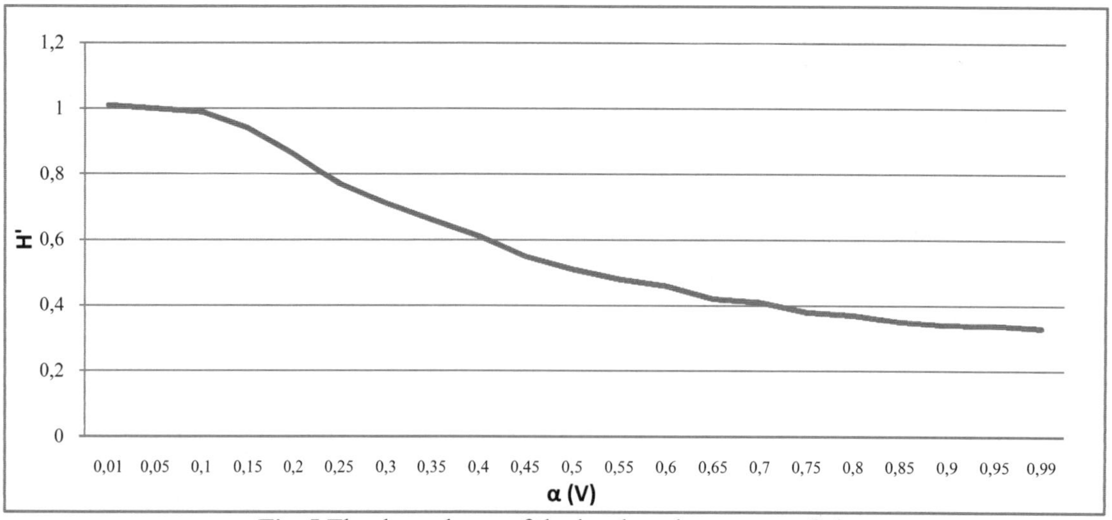

Fig. 5 The dependence of the head on the amount of air

Conclusion

The analysis showed the effectiveness of the application of this mathematical model. The calculation showed that the head of the pump is a function of the density of the pumped medium, in the case of pumping a two-phase medium. The head is dropping with increasing gas content. That is associated with an uneven distribution of phases in the blade channel.

References

[1] Ivanovsky V.N., Sabirov A.A., Donskoy U.A. The effect of dissolved carbon dioxide on carbonate precipitation during oil production using electric centrifugal pumps. Oil industry, 2010

[2] Bellarya S.A.I., Husain A., Samadc A., KanaiaR.A.. Performance Optimization of Centrifugal Pump for Crude Oil Delivery. TheJournalofEngineeringResearch (TJER), Vol. 15, No. 1 (2018) 88-101

[3] Antonio C. Bannwart. Experimental Study of Centrifugal Pump Handling Viscous Fluid and Two-Phase Flow. SPE production&operations 30(02), 2013

[4] Larry Bachus, Angel Custodio. Know and Understand Centrifugal Pumps. ElsevierInc., 2003, 270 p.

[5] Gülich J.F. Centrifugal Pumps. Springer-Verlag Berlin Heidelberg, 2010, 964 p.

[6] Kosteniuk S.A., Podyapolsky A.I. The technology of well operation with electric centrifugal pumps with increased gas content. Oil industry, 2010 (in Russian)

[7] Mirsaetov O.M., VyoshevK.I.. Time to failure of the reservoir-well-pump system. Earth Sciences, 2005, No. 11.(in Russian)

[8] Ageev S., Druzhinin E., ESP with increased gas content at the inlet. Drilling and oil, 2003.

[9] Lomakin, V.O., Kuleshovav, M.S., Bozh'eva, S.M. Numerical Modeling of Liquid Flow in a Pump Station (2016) Power Technology and Engineering, 49 (5), pp. 324-327.
DOI: 10.1007/s10749-016-0623-9

[10] Lomakin V.O. Investigation of two-phase flow in axial-centrifugal impeller by hydrodynamic modeling methods Proceedings of 2015 International Conference on Fluid Power and Mechatronics, FPM 2015 no 7337302 pp 1204–6
DOI: 10.1109/FPM.2015.7337302

[11] Lomakin V O, Petrov A I and Kuleshova M S 2014 Investigation of a two-phase flow in an axial centrifugal wheel by methods of hydrodynamic modeling Science and education: electronic scientific edition of Bauman MSTU (9) pp 45–64 (in Russian)

[12] Brennen C.E. Fundamentals of Multiphase Flows. Cambridge: Cambridge University Press, 2005. 410 p.

[13] Wilcox D.C. Turbulence Modeling for CFD. 3rd ed. DCW Industries, 2006. 515 c.

[14] Ding H., Visser F.C., Jiang Y., Furmanczyk M. Demonstration and Validation of a 3D CFD Simulation Tool Predicting Pump Performance and Cavitation for Industrial Applications // Journal of Fluids Engineering. 2011. Vol. 133, no. 1. P. 011101-011101-1
DOI: 10.1115/1.4003196

[15] Lomakin V.O. Numerical simulation of flow parts of pump models and verification of simulation results by comparison of obtained values with experimental data. Naukaiobrazovanie MGTU im. N.E. Baumana = Science and Education of the Bauman MSTU, 2012, no. 5, pp. 52-62. DOI: 10.7463/0512.0356070 (in Russian).

[16] Lomakin V.O., Petrov A.I. Verification of the calculation results using hydrodynamic modeling package STAR-CCM + for flow channel of the centrifugal pump AX 50-32-200. Izvestiyavysshikhuchebnykhzavedeniy. Mashinostroenie [Proceedings of Higher Educational Institutions. MachineBuilding], 2012, no. S, pp. 6–9 (inRuss.).

[17] Zhu J, Zhu H, Zhang J and Zhang H-Q 2019 A numerical study on flow patterns inside an electrical submersible pump (ESP) and comparison with visualization experiments Journal of Petroleum Science and Engineering 173 pp 339–350
DOI: 10.1016/j.petrol.2018.10.038

[18] Timushev S F,Knyazev V A,Panaiotti S S,Soldatov V A and Rohatgi U S 2009 Numerical procedure for assessment of centrifugal pump cavitation erosion ASME International Mechanical Engineering Congress and Exposition, Proceedings 14 pp 73-78
DOI: 10.1115/IMECE2008-67177

[19] Arefyev, K.Y., Voronetsky, A.V., Suchkov, S.A., Ilchenko, M.A. Computational and experimental study of the two-phase mixing in gas-dynamic ignition system (2017) Thermophysics and Aeromechanics, 24 (2), pp. 225-237.
DOI: 10.1134/S086986431702007X

[20] Lozovetskii, V.V., Pelevin, F.V. Resistance of spherical fills in the case of flow of single- and two-phase media (2009) Journal of Engineering Physics and Thermophysics, 82 (2), pp. 277-282.
DOI: 10.1007/s10891-009-0194-9

IOP Conf. Series: Materials Science and Engineering **589** (2019) 012018 doi:10.1088/1757-899X/589/1/012018

Mathematical representation of pressure regulator with variable characteristic

I Kolodin[1,2] and M Ryabinin[1]

[1]Bauman Moscow State Technical University, 5 Second Baumanskaya Street, Moscow, 105005, Russian Federation

[2]E-mail: kolodin@student.bmstu.ru

Abstract
One of major pressure regulator features is its static characteristic. It is necessary to choose parameters fulfilling static characteristic requirements while designing a regulator. The article observes possible requirements imposed on static characteristic and shows mathematical description of balance of pressure regulator with variable characteristic. Static characteristics are plotted for some of parameters being varied and others fixed. Influence of parameters is analyzed on the base of obtained data and recommendations for its choosing are given.

Introduction

Variable-displacement pumps with pressure regulator became widespread in application for hydraulic servos of portable machines and other industrial fields. Mathematical modeling of such devices is an important problem [1]-[9] as well as of hydraulic servo and other devices included into it [10]-[17].

Choo regulator is a complex and problematic task [18]-[20], therefore influence of each the parameters should be considered and trends should be revealed.

Existing hydrosystems working at constant pressure require pressure level to be changed according to operation performed by technological hydraulic hardware. We will use NP-96 high pressure pump arrangement for analytical research. The pump is a variable-displacement plunger pump, which has a functional ability of regulator pressure setup correction using service liquid under pressure generated by control unit under pressure formed by control unit.

This pump design shaft considers that the bar rotates between two roller bearings and it is connected with driving prings by splines while adjustable angle cam plate is fixed in body frame on half axles. Adjustable angle cam plate is controlled by control plunger which is reacted against by a spring. Service liquid under reduced pressure is supplied to control plunger. Reduced pressure level depends on control valve position. The liquid is driven to draining channel through a plunger hole. Pump delivery pressure is applied to one face of control valve, while spring force and regulator setting correcting pressure is applied to the other one. Control valve position in steady mode provides equality of liquid flow through filling and draining openings between valve and shell casings edges. Pressure under control plunger is constant, control plunger effort is equal to spring effort, and adjustable angle cam plate position provide pump supply and pressure required by technological equipment.

While regulator setting correcting pressure increases, control valve moves and closes service liquid supply under control plunger. Pressure under plunger decreases, angle of cam plate grows, while pump supply and pressure also grow. Growth of pump supply pressure stops control valve movement.

Pressure under plunger becomes stable but less than the initial one. Control plunger effort is equal to spring effort. Adjustable angle cam plate is set in a position providing required pressure behind the pump.

Mathematical model of the regulator.

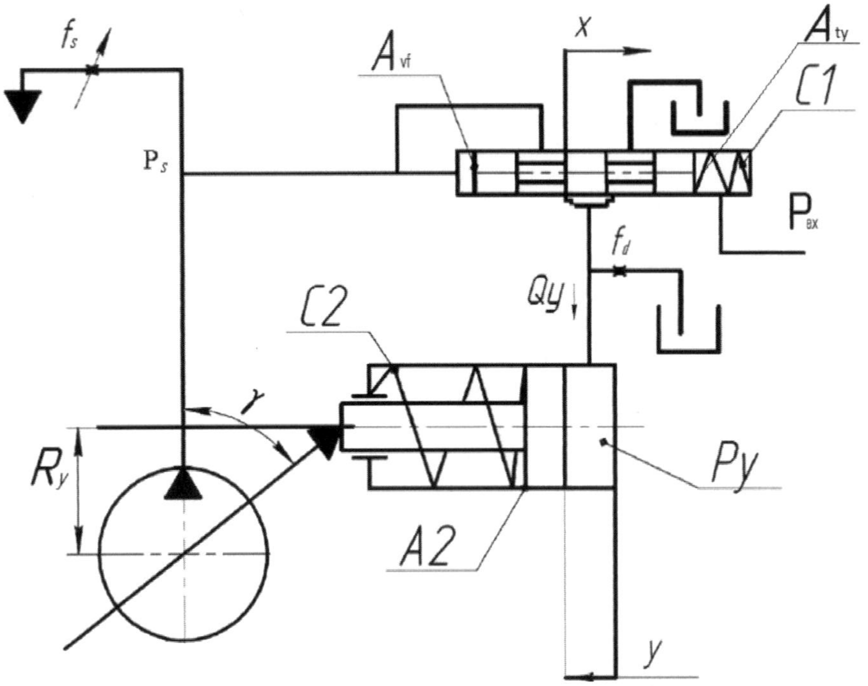

Fig.1. Principal scheme of the regulator.

Piping balance equation:

$$c_2 \cdot (y_0 + y) - \frac{T_{ky}}{R_y} = p_y \cdot A_2 \tag{1}$$

Where c_2 — control piping spring stiffness, y_0 — initial spring prepressure, y — piping position coordinate, T_{ky} — adjustable angle cam plate moment, R_y — distance between plate and actuating cylinder steering axes, p_y — control pressure, A_2 — control piping end face area.

Initial spring prepressure must compensate force applied to the piping from plate side:

$$c_2 \cdot y_0 = \frac{T_{ky}}{R_y} \tag{2}$$

We get

$$c_2 \cdot y = p_y \cdot A_2 \tag{3}$$

Valve balance equation:

$$c_1 (x_0 + x) + p_{ex} \cdot A_{ty} = p_s \cdot A_{vf} \tag{4}$$

IOP Conf. Series: Materials Science and Engineering **589** (2019) 012018 doi:10.1088/1757-899X/589/1/012018

Where, c_1- stiffness of spring affecting on valve,
x_0– initial spring prepressure,
x– displacement between valve and neutral point,
p_{ex}– external pressure,
A_{ty}– area to which external pressure is applied,
p_s– supply pressure,
A_{vf}– valve end face area.
Connection between $p_y$andp_s :

$$p_s = p_y \cdot \frac{\left(A_{sl} + f_d\right)^2 + A_s^2}{A_s^2} \quad (5)$$

Where $A_{sl} = b \cdot X_{sl}$,
$A_s = b \cdot X_s$,
b– summarized width of distributor windows
f_d— area of throttle draining control chamber.
Following model of valve geometry is used:

$$S + 2R + x + S + 2R + x \vee$$
$$\left(\delta + 2R\right)^2 + 0.25 \cdot \quad (6)$$
$$X_s = -2R + \sqrt{}$$

$$S + 2R - x + S + 2R - x \vee$$
$$\left(\delta + 2R\right)^2 + 0.25 \cdot \quad (7)$$
$$X_{sl} = -2R + \sqrt{}$$

Where
R- valve edge radius,
δ- radial gap of valve,
S- valve edges overlap.
Pump supply:

$$Q_s = Q_{max} - k_y * y \quad (8)$$

Where
Q_{max} - maximal pump supply.

$$k_y = \frac{Q_{max}}{y_{max}} \quad (9)$$

Wherey_{max}– maximal shift of control piping.
Therefore mathematical description of pressure regulator balanceis as follows:

$$c_2 \cdot y = p_y \cdot A_2$$

$$c_1\left(x_0 + x\right) + p_{ex} \cdot A_{ty} = p_s \cdot A_{vf}$$

$$p_s = p_y \cdot \frac{\left(A_{sl} + f_d\right)^2 + A_s^2}{A_s^2}$$

$$S + 2R + x + S + 2R + x \vee$$

$$(2)$$

$$S + 2R - x + S + 2R - x \vee \qquad (10)$$

$$(2)$$

$$-2R + \sqrt{\left(\delta + 2R\right)^2 + 0.25 \cdot Q_s = Q_{max} - k_y * y}$$

$$\left(\delta + 2R\right)^2 + 0.25 \cdot A_{sl} = b \cdot$$

$$-2R + \sqrt{}$$

$$A_s = b \cdot$$

Mathematical modeling results

Equation system (10) was solved with the help of Python programming language. Matplotliblibrary was used for plotting. Required solution is correspondenceQ(p). Static characteristics for regulator parameters chosen for design calculated for two different values p_{ex} = 1MPaand p_{ex} =5.5 MPa.

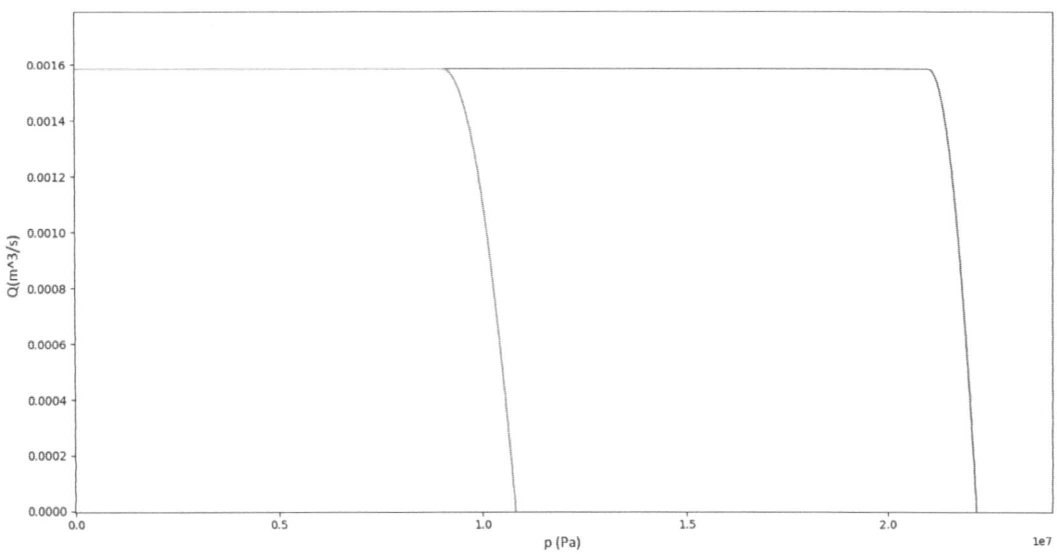

Fig. 2. Static characteristic

Following requirements were put on static characteristic of the regulator:

1. Certain position of sloping part of the static characteristic must correspond to certain value of external pressure.
2. Sharp slope angle of sloping part of the characteristic provides almost constant supply pressure level.
3. Slope angle of sloping part must remain sharp at different levels of external pressure.

Let`s consider influence of regulator parameters on static characteristic. To achieve that we will vary one of the parameter's value while keeping other parameters constant. In order to check correspondence to the 3d requirement put on the regulator static characteristic, we will carry external pressure level, choosingvalues of p_{ex} =1 MPa and p_{ex} =5.5 MPa. The results are shown on one plot.

Q_{max} – pumping unit parameter defining regulator operating range. This parameter influences the height of horizontal part of the static characteristic.

Let`s consider parameters influencing sloping part of the static characteristic.

R_y Is reversely proportional to k_y. Variations of k_y give us the followingplots.

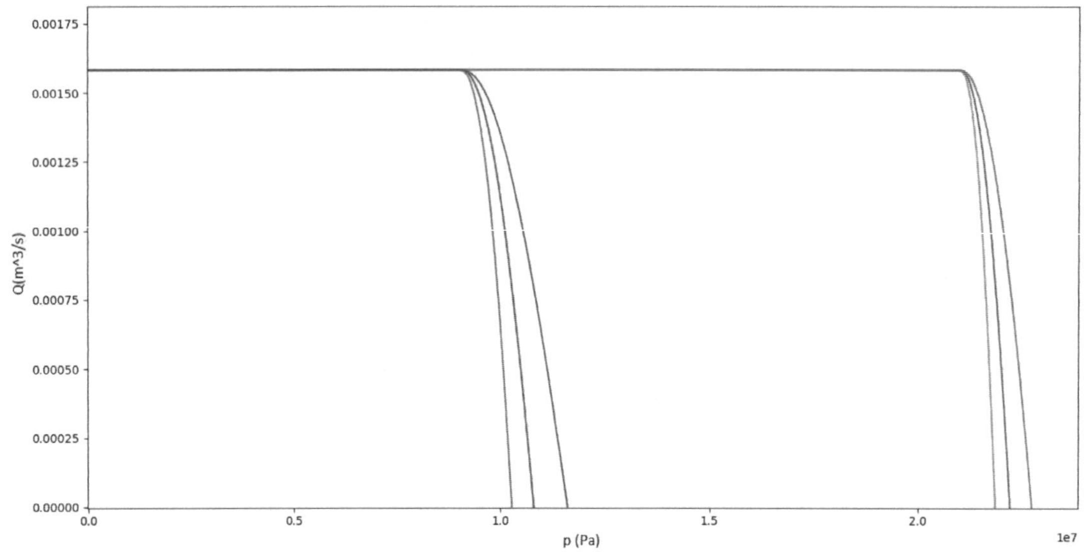

Fig. 3. Influence of R_y on static characteristic

The plots allow to conclude that decreasing parameter R_y leads to increase of slope angle of static characteristic.

Influence of parametersc_2, $\frac{1}{A_p}$ on static characteristic is the same. In order to increase slope of theplot those parameters values should be as low as possible. However, it`s important to note that moment on the regulating unit has to be compensated by increasing values of parameters c_2, R_y, y_0.

Parameters$b$and $f_d$also influence the slope. Value of parameter$b$is limited by technological requirements. The plots show that parameter$b$ should as big as possible. It allows to imporove static characteristic and increase speed of control of pressure control piston. However, since the speed of control of control of pressure control piston often is an initial parameter for regulator design,change ofwindows width may cause a need to increase control of control of pressure control piston spring stiffness. It`s not desirable according to previous conclusions regarding static characteristic.

Design process allows to use quite wide range of values of parameter f_d. By decreasing the parameter it`s possible to improve static characteristic and compensate conflict of choice of $b$and $c_2$parameter values.

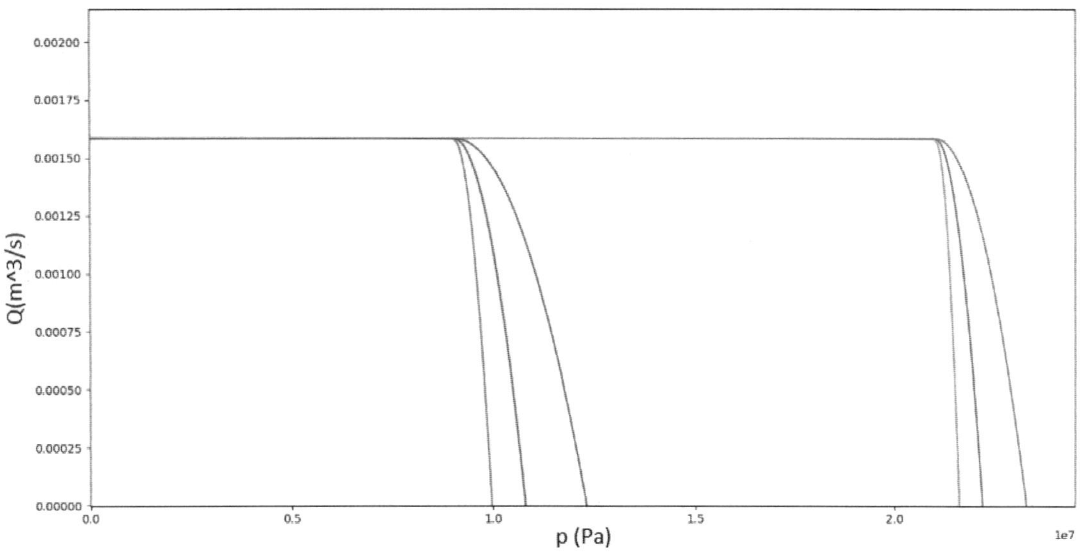

Fig. 4. Influence of *b*on static characteristic

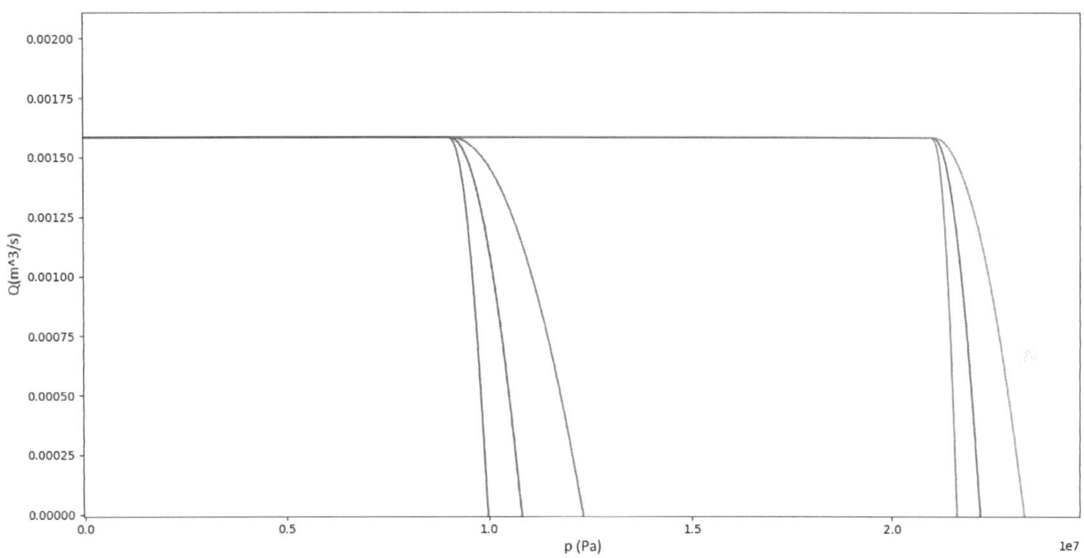

Fig. 5. Influence off_d on static characteristic

Let`s consider parameters influencing initial regulating pressure level. Such parameters are c_1, x_0, A_{ty}, A_{vf}. During design process these parameters are defined basing on initial data.

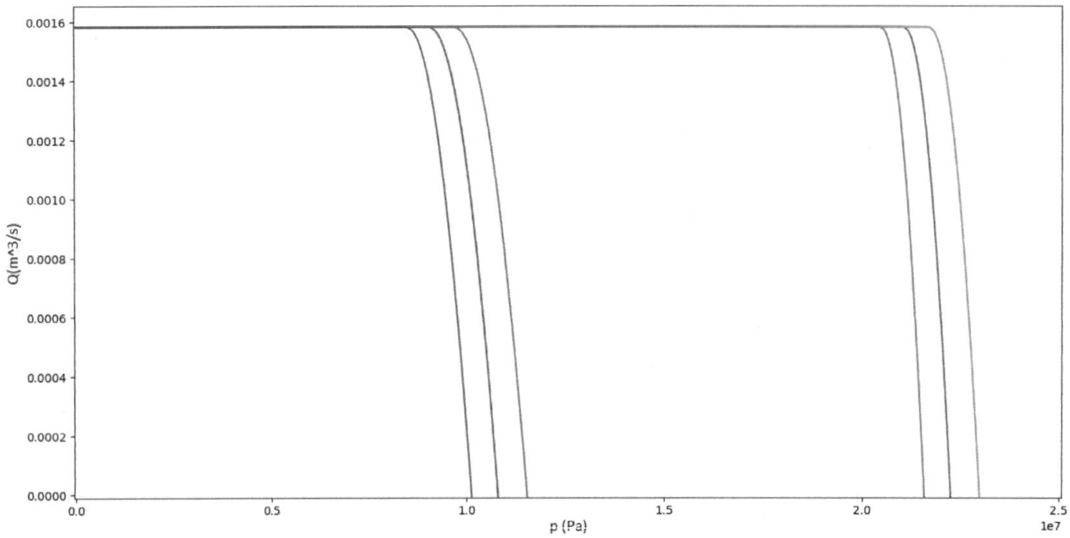

Fig. 6. Influence of c_1 on static characteristic

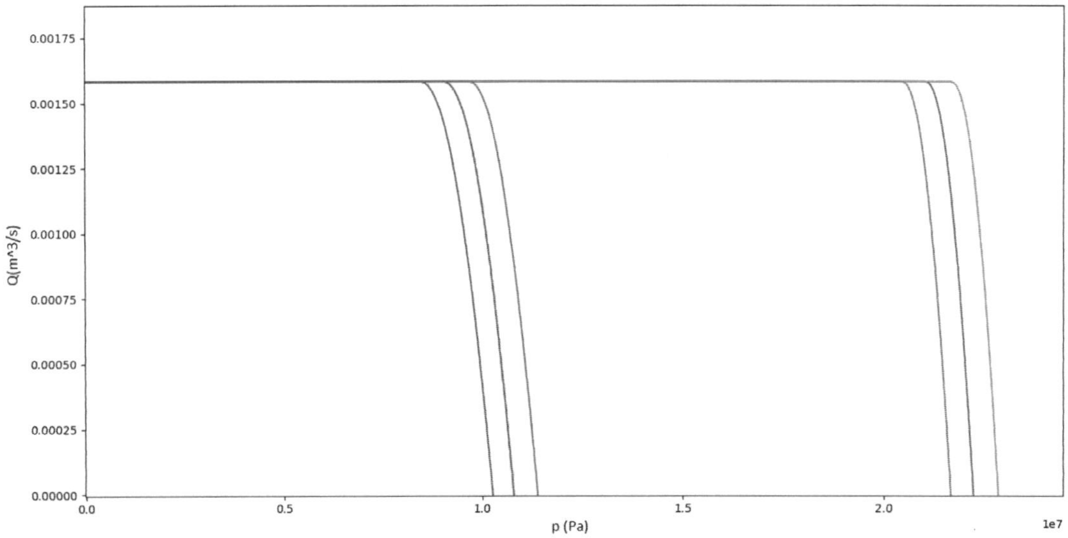

Fig. 7. Influence of x_0 on static characteristic

IOP Conf. Series: Materials Science and Engineering **589** (2019) 012018 doi:10.1088/1757-899X/589/1/012018

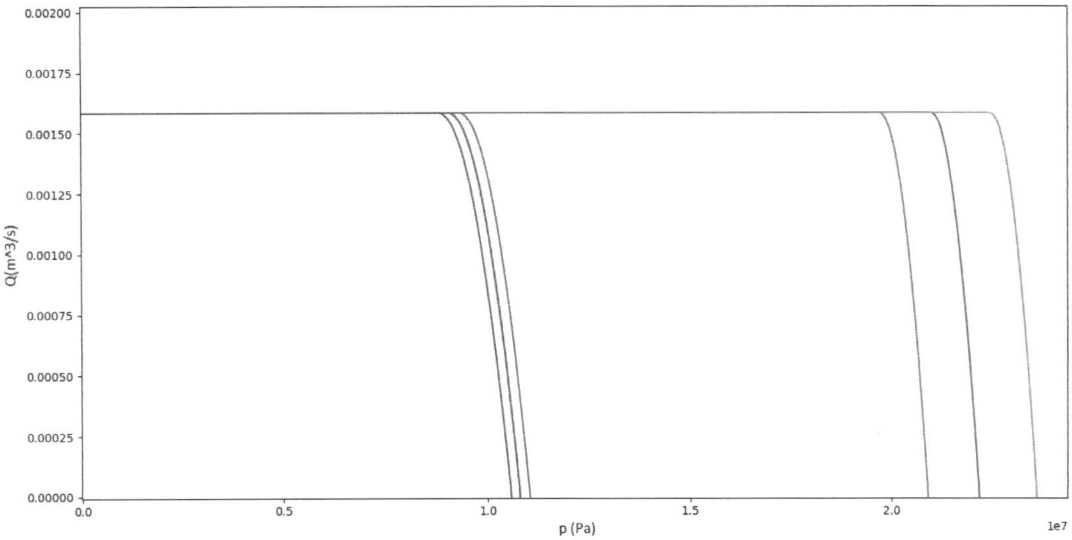

Fig. 8. Influence ofA_{ty}on static characteristic

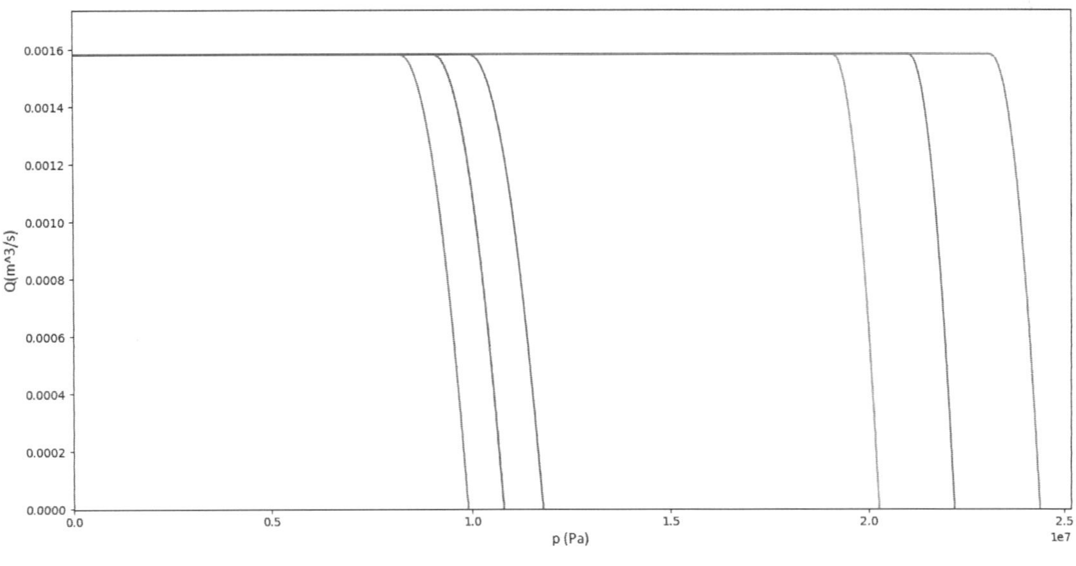

Fig. 9. Influence of A_{vf}on static characteristic

Let's consider affection of valve microgeometry on statistic characteristics of controller.

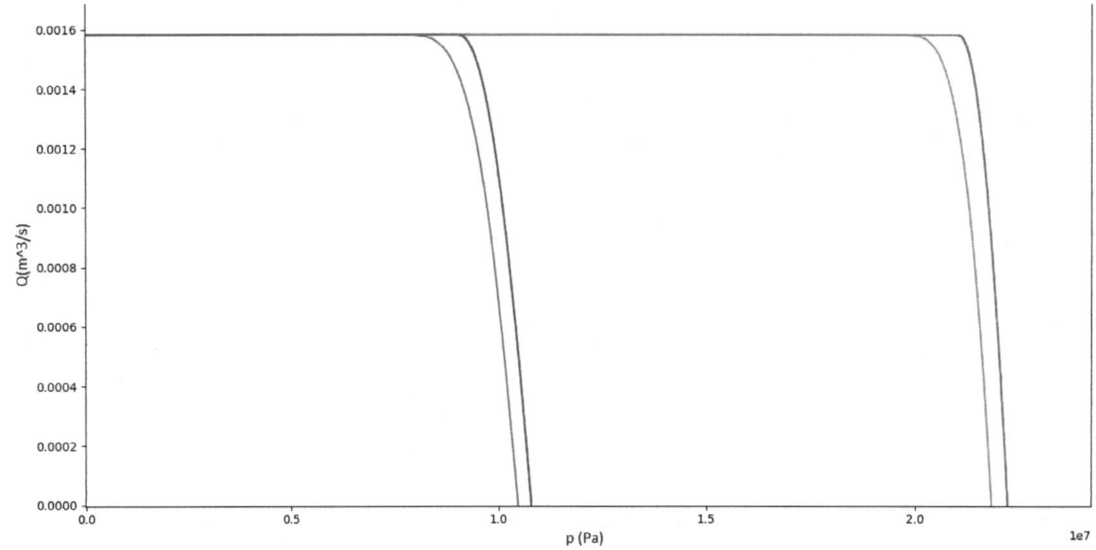

Fig. 10. Influence of *R*on static characteristic

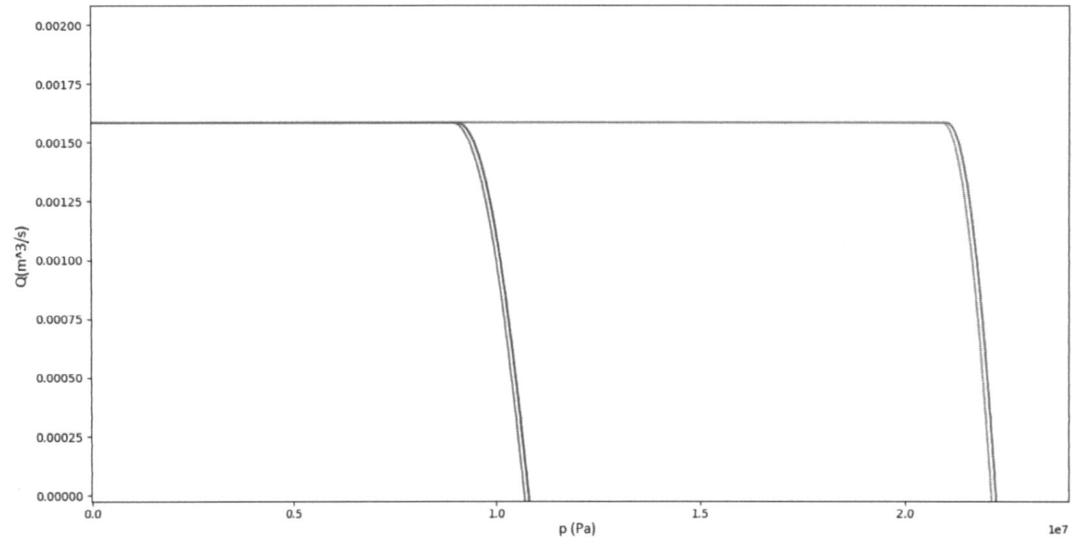

Fig. 11. Influence of *S*on static characteristic

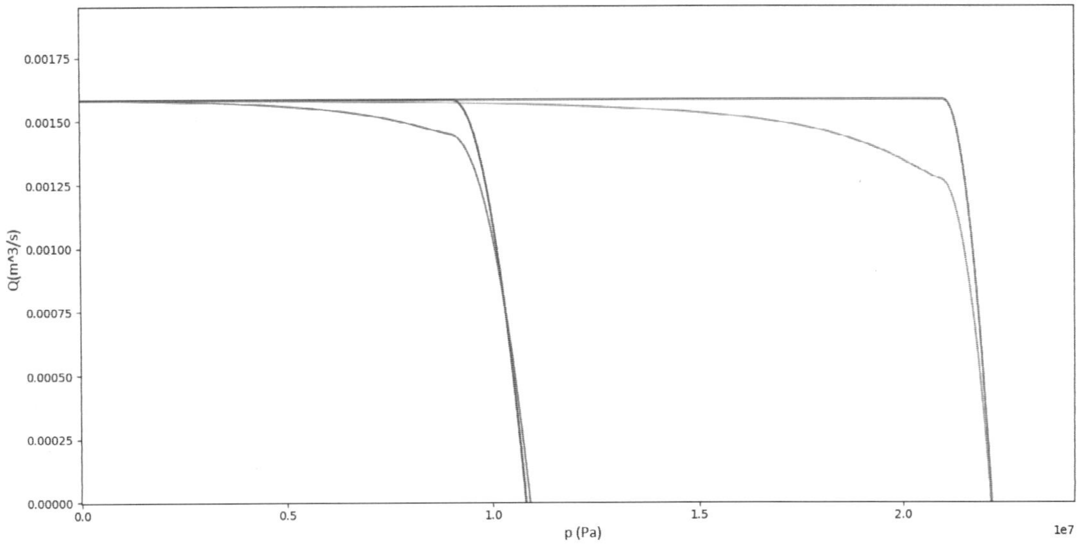

Fig. 12. Influence of δon static characteristic

These parameters are limited by technological requirements. The plots allow to conclude that the parameters values should be as low as possible, but if they are low enough, further decrease doesn`t change static characteristic remarkably.

Conclusions.

Pressure regulator static characteristic changeable by external pressure level was obtained. Represented mathematical representation takes into account affection of microgeometry of valve on static characteristic. Analysis of regulator parameter affection on its static characteristic is held. Also in this article were shown demands on static characteristics and recommendations about its support. Some allowances have been made in this model. For example, moment on piston cam plate changes in a minor way.

Reference

[1] Fulbright, N.J., Van De Ven, J.D. Dynamic response of pressure compensated variable displacement linkage pump (2018) BATH/ASME 2018 Symposium on Fluid Power and Motion Control, FPMC 2018, .

[2] Fornarelli, F., Lippolis, A., Oresta, P., Posa, A. Investigation of a pressure compensated vane pump (2018) Energy Procedia, 148, pp. 194-201.

[3] Wisch, J.K., Manring, N.D., Fales, R.C. Dynamic Characteristics of a Pressure-Compensated Inlet-Metered Pump (2017) Journal of Dynamic Systems, Measurement and Control, Transactions of the ASME, 139 (6), article № 064502, 4 quotations.

[4] Back to the basics part IV: Variable displacement and pressure-compensated pumps (2000) Diesel Progress North American Edition, 66 (9), p. 90.

[5] Bouzard, G.R. Pressure compensated pump control for forestry log skidders (1994) SAE Technical Papers, 1 quotation.

[6] Zarotti, Luca G., Paoluzzi, Roberto Contrast of pressure and flow compensated pumps (1994) American Society of Mechanical Engineers, Petroleum Division (Publication) PD, 64 (8-2), pp. 421-429.

[7] Kim, S.D., Cho, H.S., Lee, C.O. ON THE ORDER REDUCTION OF THE DYNAMIC MODEL OF A PRESSURE-COMPENSATED VARIABLE DISPLACEMENT

HYDRAULIC PUMP. (1987) American Society of Mechanical Engineers (Paper), 9 p. 1 quotation.

[8] Henke, R.W. UNDERSTANDING PRESSURE-COMPENSATED PUMPS. (1985) Hydraulics and Pneumatics, 28 (11), pp. 80-83.

[9] Pressure-compensated high-pressure pump. (1984) CHART. MECH. ENGR., 31 (10, Oct. 1984), p. 21

[10] Semenov, S., Kulakov, D. Mathematical modeling of the mechanisms of volumetric hydraulic machines (2019) IOP Conference Series: Materials Science and Engineering, 492 (1), article № 012042, .

[11] Semenov, S.E. Mathematical modeling of the electro-hydraulic actuation systems of the machines with tree-like kinematic structure (2015) Proceedings of 2015 International Conference on Fluid Power and Mechatronics, FPM 2015, article № 7337184, pp. 583-592. 1 quotation.

[12] Kulakov, D.B., Semenov, S.E., Kulakov, B.B., Shcherbachev, P.V., Tarasov, O.I. Hydraulic Bipedal Robots Locomotion Mathematical Modeling (2015) Procedia Engineering, 106, pp. 62-70. 2 quotations.

[13] D Kulakov and B Kulakov 2019 *IOP Conf. Ser. Mater. Sci. Eng.* **492** 012029

[14] Oleg, T. The experience of creating compact energy efficiency hydrostatic servo drive (2015) Proceedings of 2015 International Conference on Fluid Power and Mechatronics, FPM 2015, article № 7337187, pp. 603-609.

[15] Ulrich, H., Kleist, A., Grebe, K.-C. Comparison of control concepts for an electrohydraulic pressure- compensated variable-displacement pump [Article@Vergleich von Regelungskonzepten fur eineelektro-hydraulischdruckgeregelteVerstellpumpe] (1992) OLHYDRAULIK & PNEUMATIK, 36 (9), pp. 602-608+611-613. 1 quotation.

[16] Iyengar, S.K.R. Simulation of multi-actuator systems using pressure-compensated pumps. (1980) BASIC FLUID POWER RES. J., 13 (1 , Jan. 1980), pp. 87-90.

[17] You, Z., Burton, R.T., Ukrainetz, P.R. Interactions between two pressure compensated pumps in a single load circuit (1989)

[18] Baz, A. OPTIMIZATION OF THE DYNAMICS OF PRESSURE-COMPENSATED AXIAL PISTON PUMPS. (1983) Journal of Fluid Control, 15 (2), pp. 64-81.9 quotations.

[19] Jenkins, R.P., Ivantysynova, M. Investigation of instability of a pressure compensated vane pump (2016) 9th FPNI Ph.D. Symposium on Fluid Power, FPNI 2016, 1 quotation.

[20] Pery, Arie EXPLANATION OF LOW FREQUENCY PEAK IN PRESSURE COMPENSATED PUMPS DISCHARGE PRESSURE FREQUENCY SPECTRUM. (1984) Proceedings of the National Conference on Fluid Power, Annual Meeting, 38, pp. 119-125.

IOP Conf. Series: Materials Science and Engineering **589** (2019) 012019 doi:10.1088/1757-899X/589/1/012019

Optimal geometric parameters selection of walking rig individual elements

E Kolotvin

Bauman Moscow State Technical University, 5 Second Baumanskaya Street, Moscow, 105005, Russian Federation

E-mail: egor.kolotvin.95@gmail.com

Abstract
The main types of rigs with which it is possible to conduct geological exploration under water are listed. Their disadvantages are indicated. For geological exploration of the seabed surface, it is proposed to use the double-support drilling rig with stepper propulsion engine developed in 2016. A technique for selecting the optimal geometric parameters of the cantilever sections of the walking rigbearing frame is presented.

1. Introduction

Caterpillar, wheel and auger thrusters can be used as vehicles on the bottom. The study and industrial development of the seabed resources is impossible without technical means - primarily without underwater mining equipment. An important role among underwater vehicles is assigned to bottom units that carry mining and exploration working bodies in the form of suction tips, rippers, buckets, pick-ups, dumps, soil pumps, airlift heads, probes, drilling rigs. [1] - [5].

It is known that the passability of the wheel is limited by a ledge with a height of one third of its radius and a trajectory width of two thirds. The wheel, the auger and the caterpillar deform the soil, creating a continuous gauge, which in turn leads to environmental damage and additional energy consumption. The walking propulsor is most suitable for operation in the seabed conditions, which allows overcoming obstacles while leaving a discrete gauge. [6] - [10].

The underwater platform is known, which includes a sampler connected to a floating cable, mounted on a vertical frame movably, which, in turn, is rigidly mounted on the frame (see certificate of authorship USSR 920438, E 02 D 1/04, 1984). The main disadvantage of the known technical solution is the possibility of its overturning during a storm and the drift of the craft, which leads to the loss of the device.

A device for rotary underwater drilling is known, comprising a support frame, a rotator including a shaft-drum with a rope wound on the drill pipe, an actuator and a tension mechanism (certificate of authorship SU 1139818 A, Bull. No. 6, 15.02.85). The disadvantage of the considered device is that it cannot independently move from the drilled well to a new drilling site.

Known drilling machine, including a support frame, on which is installed the drive of the core pipe, made in the form of a coil with coils wound on it of a flexible element. One end of the flexible element is connected to the float of the other element with the lifting mechanism of the maintenance vessel. The coil has a profile axial bore to accommodate the driving rod, which is rigidly connected to the core tube (certificate of authorship N 1701884, M. Kl^6, E 21 B 7/12, 1990). The disadvantages of this

IOP Conf. Series: Materials Science and Engineering **589** (2019) 012019 doi:10.1088/1757-899X/589/1/012019

underwater drilling machine is the impossibility of its use at great depths, large energy costs when embedding the float when lifting a flexible element on the vessel.

It is also known machine for underwater drilling exploration wells (Russian Federation Patent No. 2260664; E21B 7/124 publ. September 20, 2005, Bull. No. 26). The disadvantage of this device is inaccurate positioning of the machine when setting on the point of drilling.

To carry out geological exploration on the seabed, it is proposed to use a two-support walking unit (utility model patent No. 166446 "Walking drilling rig", utility model priority: 07/04/2016). (Figure 1). The installation includes a farm, supports with platforms, hydraulic cylinders of supports, working body, control system. The support frame is made of two parallel tubes with longitudinal guides and is equipped with earrings pivotally connected to the platforms of the supports, while transverse beams rigidly interconnect the ends of the tubes with blocks placed on them. The working body is made in the form of a trolley with rollers that interact with the longitudinal guides of the truss, with a drilling machine rigidly attached to it and two winches equipped with flexible traction elements enclosing blocks of transverse beams, one end of which is fixed on the winch drum and the other on the worker's trolley body. [11]; [12].

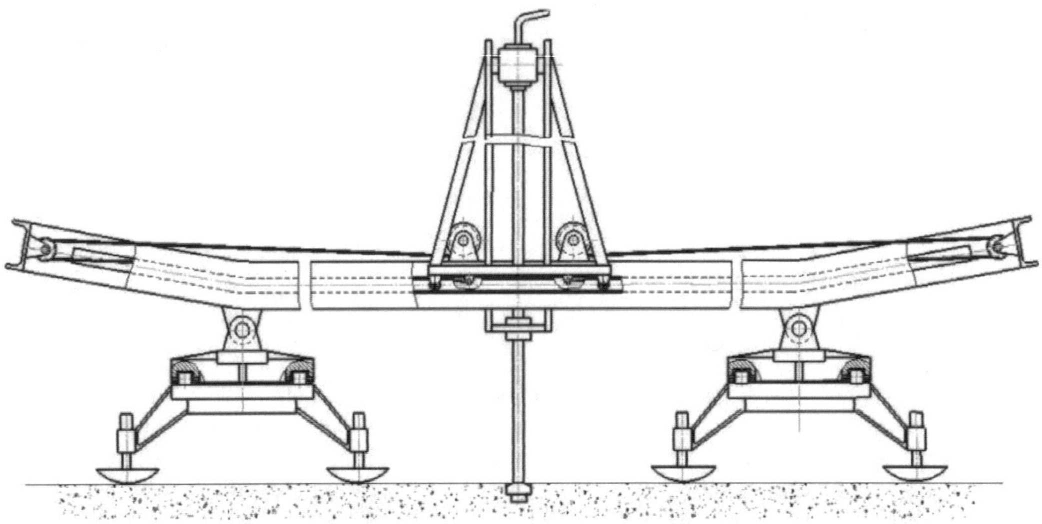

Figure 1. Constructive scheme of the walking rig[12].

2. Methods

To determine the minimum weight of the machine, it is necessary to build the dependence of the counterweight on the length of the cantilever section of the supporting frame. The equation of moments of forces acting on the installation relative to point A [13] - [16] was compiled (Figure 2).

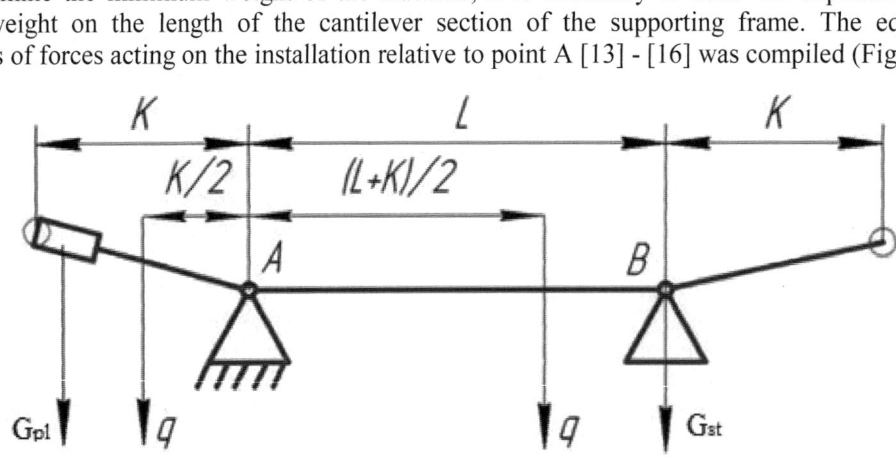

Figure 2. Design scheme.

$$M_A = G_{PL}K + 2q\frac{K^2}{2} = 2q\frac{(L+K)^2}{2} + G_{ST}L;$$

$$G_{PL}K = q(L+K)^2 + G_{ST}L - qK^2;$$

$$G_{PL}K = q((L+K)^2 - K^2) + G_{ST}L;$$

$$G_{PL} = q\frac{(L+K)^2 - K^2}{K} + \frac{G_{ST}L}{K};$$

$$G_{PL} = q\frac{(L^2 + 2LK + K^2 - K^2)}{K} + \frac{G_{ST}L}{K};$$

$$G_{PL} = q\frac{L(L+2K)}{K} + \frac{G_{ST}L}{K};$$

Где: q is the pipe linear mass, (kg / m);
K is the length of the cantilever frame,(m);
L is the section length between pts. A and B,(m);
$G_{ПЛ}$ is the platform weight (counterweight),(kg);
G_{CT} is the weight of the support table,(kg).

$$G_{PL} = \alpha_{PL}q + \beta_{PL}G_{ST}\ *. [17]$$

Where: α_{PL} is the angular equation coefficient;*
$\beta_{PL}G_{ST}$ is the initial ordinate.

$$\gamma_{PL} = \frac{\alpha_{PL}}{L};\ \alpha_K = \frac{K}{L}.$$

$$G_{PL} = \gamma_{PL}qL + \beta_{PL}G_{ST}\ |: L;$$

$$g_{PL} = \gamma_{PL}q + \beta_{PL}g_{ST};$$

$$\gamma_{PL} = \frac{1+2\alpha_K}{2\alpha_K};\ \beta_{PL} = \frac{1}{\alpha_K}.$$

g_{PL} and g_{PL} are specific gravity values of the counterweight and the support table per unit length of the working section of the bearing frame;

α_K is the console coefficient, which characterizes consoles length relative to the length of the working section of the frame (when $\alpha_K = 0$ there are no consoles, when $\alpha_K = 1$ the length of the console is equal to the length of the working section). [18] - [20].

$$g_{PL} = q\frac{1+2\alpha_K}{\alpha_K} + 5q\frac{1}{\alpha_K}; \qquad q = 276,4(kg/m);\ L = 10(m).$$

$$g_{PL} = q\frac{1+2\alpha_K}{\alpha_K} + 5q\frac{1}{\alpha_K};$$

$$g_{PL} = q\frac{1+2\alpha_K+5}{\alpha_K};$$

$$g_{PL} = q(2 + \frac{6}{\alpha_K});$$

$$G_{PL} = qL(2 + \frac{6}{\alpha_K}).$$

IOP Conf. Series: Materials Science and Engineering **589** (2019) 012019 doi:10.1088/1757-899X/589/1/012019

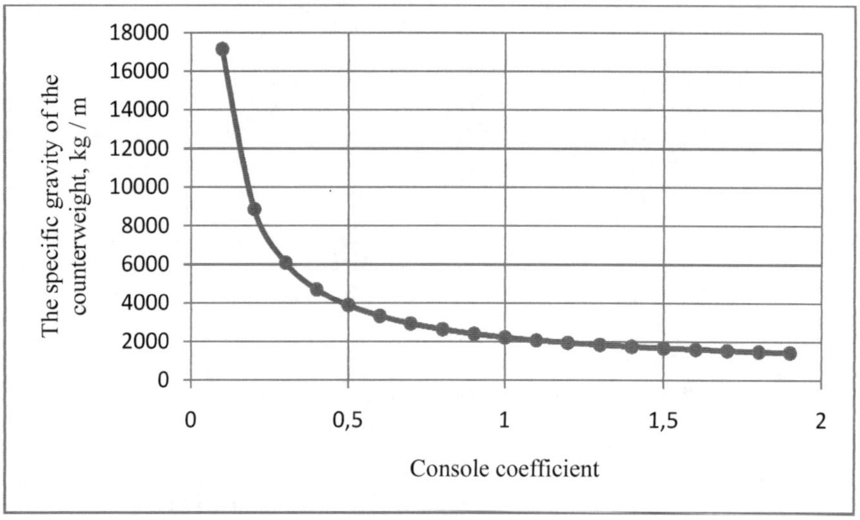

Figure 3. Specific gravity of the counterweight depending on the console coefficient.

With an increase in the length of the cantilevers (α_K coefficient), the counterweight decreases sharply. The specific mass of the counterweight should be selected depending on the ratio of the length of the console to the length of the working section of the supporting frame. The criterion for choosing the coefficient α_K is the minimum of the drilling rigmass.

$$G_{Total} = 2G_{ST} + G_{Frame} + G_{PL};$$
$$K = \alpha_K L;$$
$$G_{Total} = 10qL + 2qL + 4\alpha_K qL + qL(2 + \frac{6}{\alpha_K});$$
$$G_{Total} = qL(14 + 4\alpha_K + \frac{6}{\alpha_K}).$$

3. Results

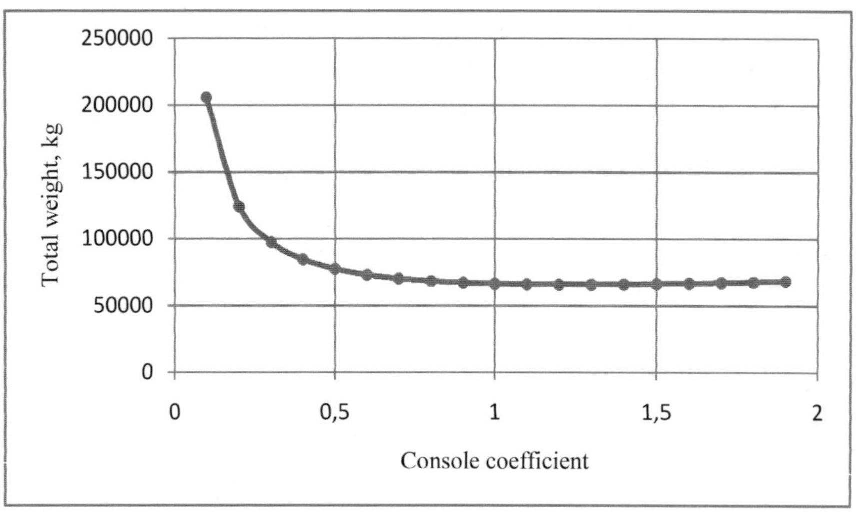

Figure 4. Total weight of the rig depending on the console coefficient.

Dependencies $G_{Total}(\alpha_K)$ show that the minimum mass of the installation is achieved with the following (α_K) values:

$$G_{Total(MIN)} = 65783,2(kg)\big|when\ \alpha_K = 1,2\big|$$

$$G_{Total(1,2)} = qL(14 + 4\alpha_{K(1,2)} + \frac{6}{\alpha_{K(1,2)}}) = 276,4 \cdot 10(14 + 4 \cdot 1,2 + \frac{6}{1,2}) = 2764 \cdot 23,8 = 65783,2(kg).$$

Table 1.

α_K	G_{Frame}, kg	G_{PL}, kg	G_{Total}, kg
0,1	6633,6	171368	205641,6
0,2	7739,2	88448	123827,2
0,3	8844,8	60808	97292,8
0,4	9950,4	46988	84578,4
0,5	11056	38696	77392
0,6	12161,6	33168	72969,6
0,7	13267,2	29219,43	70126,63
0,8	14372,8	26258	68270,8
0,9	15478,4	23954,67	67073,07
1,0	16584	22112	66336
1,1	17689,6	20604,36	65933,96
1,2	18795,2	19348	65783,2
1,3	19900,8	18284,92	65825,72
1,4	21006,4	17373,71	66020,11
1,5	22112	16584	66336
1,6	23217,6	15893	66750,6
1,7	24323,2	15283,29	67246,49
1,8	25428,8	14741,33	67810,13
1,9	26534,4	14256,42	68430,82

4. Conclusions

To determine the minimum weight of the rig, the mass of the counterweight was determined as a function of the cantilever section length of the bearing frame. Counterweights and carrier frames weights were determined for different console coefficients. The rig total mass depending on the console coefficient shows that the minimum mass is achieved when the value α_K= 1.2. It is economically feasible to choose console sections with a coefficient equal to 1.2.

The walking rig is important to use for geological exploration at the sea bed and ocean bed because of economic feasibility, high efficiency, high permeability and environmental friendliness.

References

[1] Grigoriev S N, Romanov R I, Fominski V Y, Volosova M A and Zhukova E A 2014 Structure and tribological behavior of nanocomposite C-Ti-WSex coatings Journal of Friction and Wear 35 (4) pp 263–9;

[2] Kalashnikov I E, Bolotova L K, Bykov P A, Kobeleva L I, Katin I V, Mikheev R S and Kobernik N V 2016 Tribological properties of the babbit B83–based composite materials fabricated by powder metallurgy Russian Metallurgy (Metally) (7) pp 669–74;
[3] Kovalchuk A K, Kulakov D B and Semenov S E 2008 Block-matrix equations of robots executive mechanisms with tree-like kinematic structureIzvestiyavuzov.ysshikhuchebnykhzavedeniyMashinostroenie — Proceedings of Higher Educational Institutions. Machine Building 5-21 in Russian;
[4] Kulakov D B, Semenov S E, Kulakov B B, Shcherbachev P V and Tarasov O I 2015 Hydraulic Bipedal Robots Locomotion Mathematical Modeling Procedia Engineering **106** 62-70;
[5] Semenov S E 2015 Mathematical modeling of the electro-hydraulic actuation systems of the machines with tree-like kinematic structure Proceedings of 2015 International Conference on Fluid Power and Mechatronics, FPM 2015 583-92 article no 7337184;
[6] Semenov S E, Kovalchuk A K and Kulakov D B 2010 Mathematical modeling of mechanisms with a tree-arborescence kinematic structure Drive Technology;
[7] S Semenov and D Kulakov 2019 Mathematical modeling of the mechanisms of volumetric hydraulic machines;
[8] Borisov B P 2018 Volumetric hydraulic machines: study guide (Moscow: BMSTU Publ.) 237 p;
[9] Nikolenko I V and Zheglova V M 2018 Design and calculation of axial piston hydraulic machines: monograph (Simferopol: ARIAL Publ.) 300 p;
[10] Fomin M V 2015 Seating selection for slip bearings with hydrodynamic lubrication Russian Engineering Research35 (1) pp 25–9;
[11] RussianFederationPatent № 2648365,13.06.2017. Timofeev I.P., Sokolova G.V., Bolshunov A.V., Vasilyev N.I., Ignatiev S.A., Kolotvin E.V., Avdeev A.M. Shagayushchayaustanovkadlyatransportirovaniyaiukladkineftegazovyhtrubnamorskomdne [Walking installation for transportation and laying of oil and gas pipes on the seabed];
[12] RussianFederationPatent№ 166446, 04.07.2016. I.P. Timofeev, G.V. Sokolova, G.A. Colton, I.A. Korolev, E.V. Kolotvin. Shagayushchayaburoayaustanovka[Walking drilling rig];
[13] Kobernik N V, Mikheev R S, Kalashnikov I E, Kobeleva L I and Bolotova LK 2017 Tribological properties of Babbitt alloy coatings modified with carbon nanotubes Inorganic Materials: Applied Research 8 (3) pp 428–33;
[14] Berta G L, Casoli P and Vacca A 2002 Proceedings of Fluid Power Systems and Technology Division (FPSTD) ASME International Mechanical Engineering Congress and Exposition [Conference] Simulation mode od axial piston pumps inclusive of cavitation (New Orleans, Louisiana, USA : [s.n.], 2002, November 17–22);
[15] Hao M and Qi X Y2012 Modeling Analysis and Simulation of Hydraulic Axial Piston Pump Advanced Materials Research 430–432 pp 1532–35;
[16] Zalogin O V, Noskova S and Cherevatova G 2016 Experimental study of the dynamics of regulation of an axial piston pump with proportional electric control Hydraulics (2) Available at: http://hydrojournal.ru/item/43-eksperimentalnoe-issledovanie-dinamikiregulirovaniya-aksialno-porshnevogo-nasosa-s-proportsionalnym-elektroupravleniem;
[17] Deeken M 2003 Simulation of the tribological contacts in an axial piston machine Ölhydraulik und Pneumatik 47 Nr 11–12;
[18] Zloto T 2017 Simulation of the Hydrostatic Load of the Valve Plate-cylinder Block System in an Axial Piston PumpProcedia Engineering177 pp 247–54 ;
[19] Wayne D, Milestone S A and Olsen G. 1983 The kinematic analysis of axial piston pumps Mechanism and Machine Theory18 (6) pp 475–479;
[20] Perfilev A V and Mukhin A A 2018 Improving the energy efficiency of aircraft pumping installations Hydraulics(6) Available at: http://hydrojournal.ru/item/84-povyshenieeffektivnosti-aviatsionnykh-nasosnykh-ustanovok.

IOP Conf. Series: Materials Science and Engineering **589** (2019) 012020 doi:10.1088/1757-899X/589/1/012020

Definition of positive displacement pump output regulation mechanism parameters with providing of constant flow capacity in output

S Barmin[1] and O Nikitin[1,2]

[1]Bauman Moscow State Technical University, 5 Second Baumanskaya Street, Moscow, 105005, Russian Federation

[2]E-mail: nof1936@yandex.ru

Abstract

This paper surveys mechanism of supply regulation for axial-piston pump with adjustable-angle swash plate operating with constant power. The mechanism includes a piston, whose working space is connected to delivery port, regulation unit and two springs. Least square method and successive approximations method are used to calculate parameters of the mechanism: cut-in pressure level for the second spring and springs stiffness. Regulation effectiveness is estimated by calculation of relative error of power delivered.

Introduction

Pump units allowing hydraulic servo to operate at constant power are widely spread in manufacturing industry. [1, 2] While designing this type of pump units, it is necessary to make pump output regulation unit provide constant power of pressure fluid flow (multiplication of pump output and pressure in pressure pipe values) in all of modes, so that $N = Qp = const.$ In $Q — p$ coordinates the plot of the relation is hyperbolic. Hyperbolic law of supply variation can hardly be sustained mechanically or hydromechanically, which causes necessity to approximate the law by two linear functions. Constructionally it`s relatively easy to provide linear relation between supply and pressure. Connection of pressure pipe and stock cavity of shiftable ram combined with availability of spring group enables us to provide flow control depending on pumping pressure. It can be provided by a mechanism including a piston, whose working space is connected to delivery port, regulation unit and two springs. Practically applied method of spring selection doesn`t provide necessary effectiveness of energy transfer. [3,4]

Mathematical description of approximation process

While conducting approximating of hyperbolic law of supply conversion in conditions of supply pressure conversion, the problem of implementation of two sequential linear laws which provide the fullest power supply. In other words, $N_{\text{hyp}} = N_{app}$ so that the energy is fully supplied and the moment of the second spring p_{st} switching is defined. Approximating straight lines are drawn through edge pointsp_0 and $p_{\text{max}}$of pressure change range. [5,6]

Supply regulation process is shown on the example of axial-piston pump with adjustable-angle swash plate. On the base of it preparative prepression of first spring z_0, stiffness of springs $c_1$and c_2, diameter of rod d_{r}, shoulders a and b are defined. On fig.1 construction scheme of mechanism of supply regulation for axial-piston pump with adjustable-angle swash plate is shown. The swash plate changes pump operating volume from maximum to zero while changing angle in range of 20…0° ', and it corresponds to linear law in this range.

Constant power $N=Qp=const$ in coordinates Q - p theoretically defined by hyperbolic law $Q_T(p)=N/p$. If the pump is equipped with a leverage mechanism providing operating volume change, supply change law will be of the form of $Q(p_i)$, where i corresponds to a certain part of approximation (here i = 1and 2). In the considered case supply regulation mechanism includes two springs of different stiffness and a rod. Control signal passed to rod side is pressure level of hydraulic fluid from delivery line defined by external load of hydraulic fluid discharged flow consumer. First spring operates in range of $p_0 \le p_1 \le p_{st}$ of discharge pressure level, and the second one operates in the range of $p_{st} \le p_2 \le p_{max}$ (fig.2).

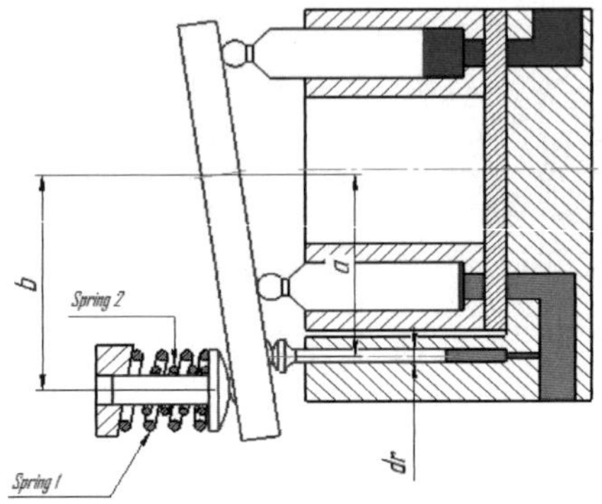

Fig. 1. Construction scheme of mechanism of pump supply regulation

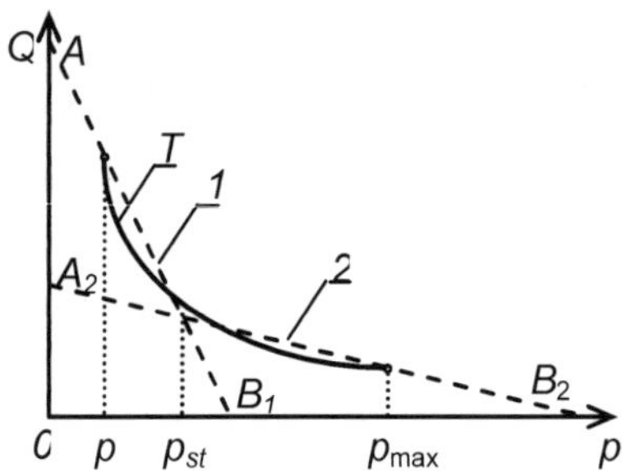

Fig.2. Operating characteristic of pump with supply regulation mechanism, T - theoretical, *1* and *2* — approximating lines .

Initial pre-pressure of the spring holds the swash plate at position of maximal angle to initial pressure of supply regulation p_0. The other spring starts operating when hydraulic fluid discharge pressure $p_{st}$is achieved (it corresponds to a certain angle of the swash plate), whose level is defined by law $Q(p_i)$ shown on fig. 2. Further both springs operate together in the range of $p_{st} \le p_2 \le p_{max}$ of discharge pressure. The piston and springs 1, 2 of supply regulation mechanism may have different shoulders relative to the swash plate rotation axis. This type of construction allows to provide constant power of the pump with lower error. Operating parameters corresponding to each pressure change range get indexes 1 and 2. [7,8]

In the theoretical variant of regulation pump supply Q and pressure p are connected by relation $Q_T\, p = N = const$ in operating range of $p_0 \le p \le p_{max}$. In case of approximation we get the following system:

$$Q_1(p_1) = A_1 - B_1 p_1 \quad at\ p_0 \le p \le p_{st} \quad (1)$$

$$Q_2(p_2) = A_2 - B_2 p_2 \quad at\ p_{st} \le p \le p_{max} \quad (2)$$

In order to increase regulation parameters calculation effectiveness we will derive a formula for p_{st}, which would allow to deliver pressure with the lowest possible error while taking into account pressure change range between values of p_0 and p_{max}. For that we will use non-dimensional values.[9, 10]

Maximum pump supply $Q_{max}= A_1 - B_1 p_0$ is provided in the beginning of regulation process p_0. Dividing values for pump supply $Q_1(p_1)$ and $Q_2(p)$ *by* Q_{max}, we obtain the following system used to calculate pump supply in a non-dimensional form:

$$\begin{cases} \overline{Q}_1(\overline{p}) = \overline{A}_1 - \overline{B}_1 \overline{p} \text{ at } 1 \le \overline{p} \le \overline{p}_{st} \\ \overline{Q}_2(\overline{p}) = \overline{A}_2 - \overline{B}_2 \overline{p} \text{ at} \le \overline{p} \le \overline{p}\max(2) \text{ and } \overline{Q}_T(\overline{p}) = 1/\overline{p}. \end{cases} \quad (3)$$

At first approximation $\overline{p}_{st}$ is defined as follows. Tangentials to $\overline{Q}_T$ are drawn through edge points of regulation range with coordinates of $\overline{p}_0=1$, $\overline{Q}_0=1$ and $\overline{p}_{\max}$, , $\overline{Q}(p_{\max})$ Point of crossing of the tangentials has coordinates of $\overline{p}_{st}$ and $\overline{Q}(\overline{p}_{st})$, which are defined as follows:

$\overline{B}_1=\overline{Q}_T(1)$ – derivative of function $\overline{Q}_T(\overline{p})$ at the moment pf start of the pump regulation start;

$\overline{B}_2=\overline{Q}_T(\overline{p}_{\max})$ – derivative of function $\overline{Q}_T(\overline{p})$ at a point with coordinates of $\overline{p}=\overline{p}$ max(final moment of pump supply regulation process);

$\overline{A}_1=\overline{Q}_T(1)-\overline{B}_1\cdot 1$; $\overline{A}_2=\overline{Q}_T(\overline{p}_{\max})-\overline{B}_2\overline{p}_{\max}$; $\overline{A}_1-\overline{B}_1\overline{p}_{st}=\overline{A}_2-\overline{B}_2\overline{p}_{st}$

and we obtain: $\overline{p}_{st}=(\overline{A}_1-\overline{A}_2)/(\overline{B}_1-\overline{B}_2)$. (4)

In order to calculate $p_0 \le p_1 \le p_{st}$ more precisely approximation by least square method is used. [9, 10] In order to do that range of 1... $p_{st}$is broken into n random parts. Number of parts should be as big as possible. Calculation of differential derivatives of the function for variables $\overline{A}_1$ and $\overline{B}_1$ and equating them to zero allows to calculate values of coefficients A_1 and B_1 for the first part.

Coefficients A_2 and B_2 for the second part are obtained the same way by breaking the range of $p_{st} \le p_2 \le p_{\max}$ into m random parts.

$$\begin{cases} \overline{A}_1=\dfrac{\sum_{i=1}^{n}\overline{Q}_T(\overline{p}_1)\cdot\sum_{i=1}^{n}\overline{p}_1^2-\sum_{i=1}^{n}\overline{Q}_T(\overline{p}_1)\cdot\overline{p}_1\cdot\sum_{i=1}^{n}\overline{p}_1}{n\cdot\sum_{i-1}^{n}\overline{p}_1^2-(\sum_{i=1}^{n}\overline{p}_1)^2} \\ \overline{B}_1=\dfrac{n\cdot\sum_{i=1}^{n}\overline{Q}_T(\overline{p}_1)\cdot\overline{p}_1-\sum_{i=1}^{n}\overline{p}_1\cdot\sum_{i=1}^{n}\overline{Q}_T(\overline{p}_1)}{(\sum_{i=1}^{n}\overline{p}_1)^2-n\cdot\sum_{i=1}^{n}\overline{p}_1^2} \end{cases}. \quad (5)$$

Coefficients A_2 and B_2 for the second part are obtained the same way. For that the range of $\overline{p}_{st}$ ... $\overline{p}_{\max}$ is broken into m random parts. And then the following system is obtained:

$$\overline{A}_2=\frac{\sum_{i=1}^{m}\overline{Q}_T(\overline{p}_{\max})\cdot\sum_{i=1}^{m}\overline{p}_{\max}^2-\sum_{i=1}^{m}\overline{Q}_T(\overline{p}_{st})\cdot\overline{p}_{\max}\cdot\sum_{i=1}^{m}\overline{p}_{\max}}{m\cdot\sum_{i-1}^{m}\overline{p}_{\max}^2-(\sum_{i=1}^{m}\overline{p}_{\max})^2}$$

$$\overline{B}_2=\frac{m\cdot\sum_{i=1}^{m}\overline{Q}_T(\overline{p}_{\max})\cdot\overline{p}_{\max}-\sum_{i=1}^{m}\overline{p}_{\max}\cdot\sum_{i=1}^{m}\overline{Q}_T(\overline{p}_{\max})}{(\sum_{i=1}^{m}\overline{p}_{\max})^2-m\cdot\sum_{i=1}^{m}\overline{p}_{\max}^2} \quad (6)$$

In case of approximation by least square method and successive approximation method correspondences of start pressure level $\overline{p}_{st}$in range of $\overline{p}_0=1$ and $\overline{p}_{\max}=p_{\max}/p_0$ define coefficients A_1, B_1, A_2 and B_2 $\overline{A}_1$, $\overline{B}_1$, $\overline{A}_2$ and $\overline{B}_2$ by formulas (5) and (6). The results are shown in table №1.

The analysis shows that definition of second spring starting moment by least square method and successive approximation methods takes quite long time because a sum of large number of components is required, and also because change in intimal data (regulation start moment p_0, final regulation moment $p_{\max}$) leads to repeated calculation of second spring start moment p_{st}..

Table 1. Values of approximation parameters for the corresponding values.

Values		$\overline{p}_{st}$	Δ, %	$\overline{A}_1$	$\overline{B}_1$	$\overline{A}_2$	$\overline{B}_2$
$\overline{p}_{max}$	1,5	1,116	0,008	2,075	1,067	1,549	0,595
	1,75	1,275	0,008	1,803	0,809	1,34	0,445
	2	1,363	0,007	1,725	0,738	1,217	0,366
	2,25	1,44	0,011	1,677	0,696	1,116	0,306
	2,5	1,516	0,012	1,632	0,657	1,03	0,259
	2,75	1,538	0,037	1,636	0,66	0,98	0,233
	3	1,615	0,027	1,591	0,621	0,913	0,201

Equality of areas (power) under plots $\overline{Q}_1(\overline{p})$, $\overline{Q}_2(\overline{p})$ and $\overline{Q}_T(\overline{p})$ (fig.2) is used as approximation accuracy criteria , which means $S_1 + S_2 = S$,

where $S_1 = \int_1^{\overline{p}_{st}} \overline{Q}_1(\overline{p})d\overline{p}$ – area under plot $\overline{Q}_1(\overline{p})$; $S_2 = \int_{\overline{p}_{st}}^{\overline{p}_{max}} \overline{Q}_2(\overline{p})d\overline{p}$ –area under plot $\overline{Q}_2(\overline{p})$;

$S_T = \int_1^{\overline{p}_{max}} \overline{Q}_T(\overline{p})d\overline{p}$ – area under plot $\overline{Q}_T(\overline{p})$.

Relative error in this case is: $\Delta = \frac{S_T - (S_1 + S_2)}{S_T} \cdot 100.$ (7)

Correspondence $\overline{p}_{st} = f(\overline{p}_{max})$ in form of $\overline{p}_{st} = 0,71 + 0,311\,\overline{p}_{max} = 1,021 + 0,311(\overline{p}_{max} - 1)$, (8)

is obtained taking into account $\overline{p}_0 = 1$ and parameters used for calculating errors Δ by formula (7).

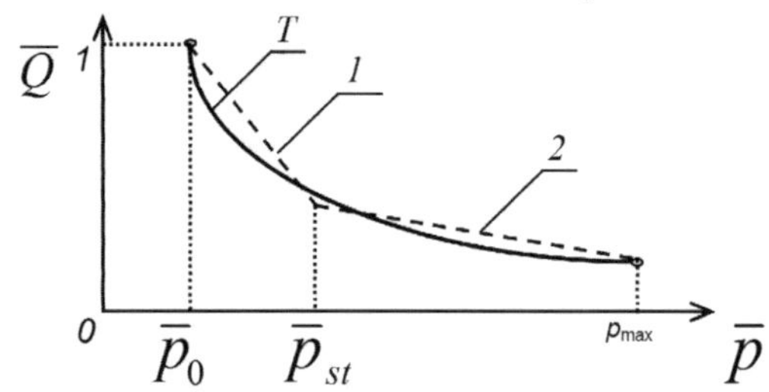

Fig.3. Dimensionless operating characteristics of pump with supply regulation mechanism: *T* – theorertical and *1* and *2* - approximated

On fig. 3 non-dimensional operating characteristics are shown: *T* – theoretical and *1* and *2* - approximated for pump with supply regulation mechanism and providing constant power at any pressure level in accepted change range. Approximating non-dimensional lines are not obligated to be drawn through points at which non-dimensional values of pump supply are equal to 1 and $\overline{Q}_T(\overline{p}_{max})$. Approximating lines don't have to pass through points in which non-dimensional value of pump output is equal to 1 and $\overline{Q}_T(\overline{p}_{max})$.

It is connected with the specificity of least square method.[11,12]

Fig.3. Dimensionless operating characteristics of pump with supply regulation mechanism: *T* – theorertical and *1* and *2* - approximated

Checking calculation of chosen mechanism parameters

Taking into account calculations conducted to obtain $p_{st} = f(p_{max})$, calculations were performed in order to check parameters values of mechanism, which regulates supply providing constant capacity of hydraulic fluid flow for a pump with a capacity of 15kW and operating in the pressure range of 18MPa..40MPa. Structural scheme of pump supply regulation mechanism and operating characteristic of a pump with regulation mechanism are similar to ones shown on fig.1, while functional diagrams are shown on fig.2. Checking calculations were performed taking into account geometrical parameters (A, b, V_0, γ) and using system (1). [13,14]

On the first part $p_0 \le p_1 \le p_{st}$force applied to the swash plate by rod, acts against the force applied by first spring to the swash plate and defined by initial pre-pressure of spring z_0, its stiffness c_1, diameter of d_r,, shoulders a and b (fig.1).

Equations of forces balance relative to the swash plate rotation axis are as follows:

$$F_{01} \cdot b = p_0 a \pi d_r^2 / 4, \quad (9)$$

where $F_{01} = c_1 z_0$ – force applied to the swash plate by first spring and provided by its initial pre-pressure z_0,c_1 – first spring stiffness, p_0 – discharge line pressure before movement start, or start of regulation process.

Further pressure growth $p_1 > p_0$ causes spring deformation and increase of force applied to the swash plate by it.

$$F_{sp1} = C_1[z_0 + z_1(p)] \quad (10)$$

where$z_1(p)$- first spring deformation, caused by change of pressure in the range of $p_0 < p_1 < p_{st}$, and it can be derived through swash plate angle as follows:

$$z_1(p) = b[\text{tg}\,\gamma_{max} - tg\gamma(p)] \quad (11)$$

where γ_{max} and $\gamma(p)$ – maximal value and current value of swash plate angle defined by pressure p.

Pump supply Q is in direct proportion to swash plate angle tangent:

$$Q = \frac{V_o}{2\pi}\omega\frac{\text{tg}\gamma(p)}{\text{tg}\gamma_{max}} \quad (12)$$

whereω – rotational speed of pump shaft V_0 – pump operating volume.

Using equations (9), (10), (11) in equation (12) and taking into account contsruction parameters of operating volume regulator we obtain relation between pump supply Q_1 and supply pressure in the range of $p_0 \le p_1 \le p_{st}$ with only first spring operating as follows:

$$Q_1(p_1) = A_1 - B_1 p_1 \quad (13)$$

where $A_1 = \frac{V_o}{2\pi}\omega(\frac{z_o}{b \cdot \text{tg}\gamma_{max}} + 1)$. $B_1 = \frac{V_0}{2\pi}\omega\frac{\pi d_r^2}{4c_1 \text{tg}\gamma_{max}} \cdot \frac{a}{b^2}$.

When supply pressure level equal to p_{st} and corresponding angle of swash plate is achieved, the second spring with a stiffness c_2 of starts deforming . Therefore, in the range of $p_{st} \le p_2 \le p_{max}$ two springs operate simultaneously. Equations of forces balance relative to the swash plate rotation axis can be descripted in the following form:

$$F_{s.1} \cdot b + F_{s.2} \cdot b = p\frac{\pi d_r^2}{4}a, \quad (14)$$

where $F_{s.2} = c_2 b[\text{tg}\gamma_{st} - \text{tg}\gamma(p)]$ – force applied by second spring to the swash plate, c_2– second spring stiffness, γ_{st} – swash plate angle at the pressure of p_{st}, which is pressure level for second spring to start operate. It depends on second spring size and defined during pump design.

After transforming equation (12) taking into account equations (9), (10), (11), and (14), we will obtain relation between pump supply Q_2 and supply pressure when two springs operate simultaneously in the following form:

$$Q_2(p_2) = A_2 - B_2 p_2 \quad (15)$$

where $A_2 = \frac{V_0}{2\pi}\omega\frac{c_1(z_0 + b \cdot tg\gamma_{max}) + c_2 b \cdot \text{tg}\gamma_{st}}{b(c_1 + c_2)\text{tg}\gamma_{max}}$ and $B_2 = \frac{V_0}{2\pi}\omega\frac{\pi d_r^2}{4} \cdot \frac{a}{b^2(c_1 + c_2)\text{tg}\gamma_{max}}$.

After calculating coefficients A_1, B_1, A_2 and B_2 by formulas (5) and (6) in dimensional form (p_{st} for first approximation defined same as for non-dimensional $\overline{p}_{st}$) all parameters of regulating unit are

IOP Conf. Series: Materials Science and Engineering **589** (2019) 012020 doi:10.1088/1757-899X/589/1/012020

defined (c_1 and c_2 —spring stiffness, z_0 – initial pre-pressure of first spring, γ — swash plate angle causing the second spring to work, d_r— rod diameter, shoulders a and b diameter).

Number of parts n and m *is* chosen to make pressure value change with a step of 0,001 MPa.

After defining coefficients A_1, B_1, A_2 and $B_2 p_{st}$ is calculated by formula (4) for the second approximation and coefficients A_1, B_1, A_2 and B_2 are defined repeatedly by formulae (13) and (15). After each approximation relative error Δ is calculated by formula (7).

Results of calculation of p_{st} and Δ for successive approximations for the pressure change range of 18..40Mpa are shown in table 2:

Table 2. Results of calculation of p_{st} and Δ values performed during checking calculations for pump.

Approximation number	1	2	3	4	5	6
p_{st}, MPa	24.828	25.082	25.268	25.403	25.5	25.569
Δ, %	4.9952	0.0255	0.0238	0.0217	0.0245	0.0205

Values of relative error and cut-in pressure p_{st}show high convergence with the results of theoretical calculations. After 4-5 approximations limΔ→0 for concerned variant of output regulation design we have received $p_{st} = 25{,}27...25{,}4\ MPa$ ($\Delta_{min} = 0.0217$ %) in terms of transmission of energy $\Delta = N_{app} - N_{hyp.}$

On the basis of table 2 data we conclude that second spring start moment p_{st}corresponds to one third of range of regulated pressure from the moment of regulation start p_0. In this case we obtain the following:

$$p_{st} = p_0 + \frac{1}{3}(p_{max} - p_0) \quad (16)$$

In a non-dimensional form this relation can be described as follows:

$$\overline{p}_{st} = 1 + \frac{1}{3}(\overline{p}_{\max} - 1).(17)$$

Discussion and analysis of the results

Expressions (16) and (17) describe characteristics and values of parameters quite accurate in correspondence to formula (8). It allows to conclude that expressions (16) and (17) can be used for defining second spring start moment with high precision and applied for relative pressure change range of any length in a relative range of 1...3,0.

Relative error of power transfer in separate specific points of supply change characteristic (for example, two edge points and one middle point for each of 3 ranges obtained after approximation of hyperbolic function by two lines)a aren`t higher than ± 1 %.

Therefore, possibility of usage of least square method for defining supply regulation mechanism for axial-piston pump providing operating flow supply at constant capacity at the output with low loss (almost full usage of available input power)

Practical checking of chosen parameters of supply regulation mechanism and providing constant hydraulic fluid flow intensity for pump with capacity of 15 kW operating in the pressure range of 18..40MPa, verified the results of mathematical calculations.

Conclusion

1. Proposed method of calculation of second spring start pressure in consideration of pressure change range allows to create and define parameters and size of highly effective pump supply regulation mechanism.

2. Results of research and approximation with the help of least square method and successive square method showed the following:

- second spring start pressure level in the range of obtained errors depends on value of supply pressure change range;

- the following expression can be practically used to define second spring start pressure $p_{st} = p_0 + \frac{1}{3}(p_{max} - p_0)$.

3. Other parameters of the mechanism (spring stiffness, initial deformation of first spring, swash plate angle providing second spring to start, rod diameter, shoulders a and b diameter) should be defined on the base of obtained expressions by formulae (13) and (15), taking into account construction size of regulation mechanism.

Reference

[1] Borisov, B.P. Positive-displacement machines. M.: Moscow State Technical University named after N. E. Bauman, 2018. – p.240.
[2] Nikitin, O.F. Hydraulics and air-oil circuit, second issue., M.: Moscow State Technical University named after N. E. Bauman, 2012. – p.418.
[3] Kalitkin, N.N. Digital methods. M.: Nauka, 1978. – p. 512.
[4] Vlasova E.A., Zarubin V.S., Kuvyrkin G.N. Approximate methods of mathematical physics. / Eds. V.S. Zarubin, A.P. Kryshcnenko. - M.: BMSTU publishing house, 2004. - 704 p. (Mathematics in technical university Series; Issue XIII).
[5] ShvayogiSalutage, Mathew Crewicr, Oing Yuan, Hui Dind. Axial Piston Pump Performance Prediction using 3D CFD Simulation. Eaton Corp. SimeriesInc/ 2015
[6] Zakharov, M.N., Laryushkin, P.A., Erastova, K.G. Stable Geometry of a Plane Five-Link Mechanism (2018) Russian Engineering Research, 38 (2), pp. 72-76. DOI: 10.3103/S1068798X1802020X
[7] Ivanov, A.S., Fomin, M.V. Classification and Calculation of Drive Loads (2017) Russian Engineering Research, 37 (12), pp. 1017-1022. 1 quotation. DOI: 10.3103/S1068798X17120061
[8] Sayapin, S.N. Control system for the Stirling engine (2017) Russian Engineering Research, 37 (10), pp. 841-844. DOI: 10.3103/S1068798X17100185
[9] Lomakin, V.O., Kuleshova, M.S., Bozh'eva, S.M. Numerical Modeling of Liquid Flow in a Pump Station (2016) Power Technology and Engineering, 49 (5), pp. 324-327. DOI: 10.1007/s10749-016-0623-9
[10] Fedoseev, V.N.Triboengineering of friction couples of braking devices operating with repeated brief actuations (1998) TrenieiIznos, 19 (2), pp. 200-206.
[11] Tsai, Shir-Tau, "Simulation of the control characteristics of an axial piston pump and a flapper-nozzle valve combination" (1988).Retrospective Theses and Dissertations. 8895.
[12] Zarotti, G.L. and Nervegna, N. – "Pump efficiencies approximation and modelling" – Int Fluid Power Symp, 6th; Cambridge, Engl; 8-11 April 1981, Pages 145-164.
[13] Noh D. Manring, "Fluid Power Pump and Motors: Analysis, Design and Control" (2013), 1st Edition, Kindle Edition, Page 320.
[14] Jong-Hycok Kim, Chag-SooJeon, Yeh-Sun Hong. "Constant pressure control of a swash plate type axial piston pump by varying both volumetric displacement and shaft speed", International Journal of Precision Engineering and Manufacturing, October 2015, Volume 16, Issue 11, pp 2395-2401.

IOP Conf. Series: Materials Science and Engineering **589** (2019) 012021 doi:10.1088/1757-899X/589/1/012021

3D-printed polyether ether ketone samples mechanical properties estimation

A Sviridov[1], Iu Lopatina[1] and Iu Kurganova[2,3]

[1]Federal State Budgetary Scientific Institution «Federal scientific agroengineering center VIM» (FSBSI FSAC VIM)
[2]Bauman Moscow State Technical University, 5 Second Baumanskaya Street, Moscow, 105005, Russian Federation

[3]E-mail: kurganova_ya@mail.ru

Abstract
This paper is dedicated to study of mechanical characteristics (tensile limit, elasticity modulus and relative elongation) of samples made of polyether ether ketone by additive method FDM (FFF). The paper includes results of mechanical tests on tension of sample party and comparison to results of other authors. Conclusions on reasons of the results obtained are made.

Introduction

Modern trends of technical development show growing usage of new materials having high specific strength properties and unique technological properties [1-5]. One of such materials is high temperature structural polymer polyether ether ketone (PEEK), which has gained widespread use in different manufacturing industries due to its unique set of properties [6-10]. Chemical structure of this polymer consists of phenylene rings connected in-series with ester and carbonyl groups by pair bonds and provides high strength and flexibility of a material, and retaining these properties during heating up to remarkably high temperature (about 300 °C). Among other issues the material is flame resistant and gives few smoke and toxic substances while burning. In spite of high cost, this set of properties makes PEEK a high demand material for aircraft, machine, instrument and other industries. PEEK is used for manufacturing of sealing washers, bearings, sensors bodies, coils and other details contacting fuel, lubricant and cooling fluids.

Polyether ether ketone is a thermoplastic polymer, and it makes it possible to use PEEK in a FDM (FFF) 3D-printing, which is based on depositing of polymer material layer by layer in accordance to advance prepared program. Usage of additive technologies for details manufacturing allows to reduce their cost while piece or low volume production; in comparison to CNC milling resulting in 90% of initial material becoming metal chips, 3D-printing allows to cut expenses on raw materials. Obvious advantages of 3D-printing as a method of details manufacturing make processing of polyether ether ketone by FDM (FFF) methods a subject of researchers focus in recent years [11-15].

In spite of the benefits, details manufacturing by FDM (FFF) methods has a number of shortcomings, one of which is degradation of mechanical characteristics of 3D-printed components in comparison to components manufactured by traditional methods (casting and extrusion). The purpose

IOP Conf. Series: Materials Science and Engineering **589** (2019) 012021 doi:10.1088/1757-899X/589/1/012021

of this paper is estimation of mechanical properties of samples made of polyether ether ketone by 3D-printing.

Materials and methods

Tension testing is the most universal among other kinds of testing, because allow to define mechanical properties of a material on each stage of its deformation (from elastic deformation to destruction). In this paper mechanical characteristics of the material at tension are defined according to the standard ASTM D638-14 "Standard Test Method for Tensile Properties of Plastics". Sample of type 4 (fig. 1) was chosen for testing as recommended by the standard as suitable for estimation of properties of materials of different nature during comparative tests. Samples testing were carried out on electromechanical universal tensile testing machine TIME WDW-50E equipped with longitudinal deformation sensor.

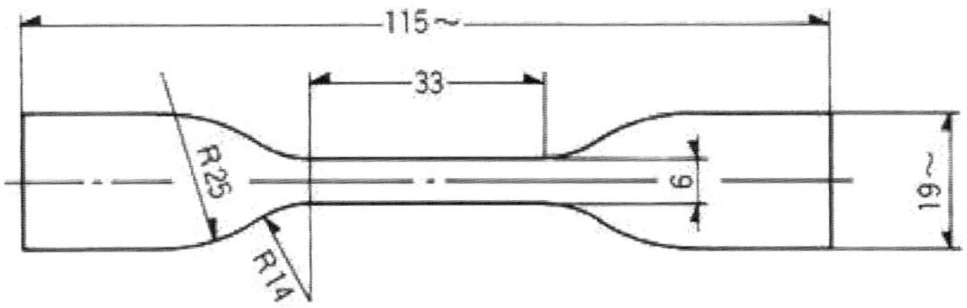

Figure 1. Sample of type 4 by standard ASTM D638-14, thickness is 4 mm

Subject of the research is 3D-printing filament made of granules Victrex® PEEK 381G manufactured by Russian-based company U3Print. Samples of this material were manufactured by FDM (FFF) method on high temperature 3D-printer TotalZAnyForm 950 PRO HOT+ at modes shown in table 1.Printing was implemented on polycarbonate list of 0.8mm thickness with the use of vacuum table and 20 minutes long preliminary heating of the camera. Geometry of laying of the layers is as follows: periphery walls of one-layer thickness along the whole perimeter of the sample, 100% inner filling, laying of layers at angles of 45°, -45°.

Table 1. Printing modes parameters for samples tension testing

Parameter	Value
Polymer type	PEEK
Extruder bore diameter	0.7 mm
Layer height	0.25 mm
Layer thickness	0.75 mm
Supply coefficient	100
Extruder temperature	450 °C
Camera temperature	290 °C
Printing speed	40 mm/c

Results and discussion

Tensile testing results for a party of 3 samples of the material under research are shown in table 2. As it can be concluded from testing results scattering of obtained values of tensile limit and elasticity modulus is not high and only 10% of average value. Relative elongation is varied more widely – from 3.0 to 5.5 %. Average value of tensile limit is about 67 MPa, while elasticity modulus – about 3.2 GPa.

Table 2. Mechanical properties of samples according to testing results

№ образца	Tensile limit (σ_B), MPa	Elasticity modulus (E), GPa	Relative elongation (δ), %
1	68.0	3.2143	3.0
2	64.7	3.4354	3.4
3	68.0	3.1428	5.5
Average value	66.9	3.2641	3.96

Figure 2 shows machine diagram of tension of sample №1 having a form typical for all three tested samples. As it can be seen on the figure, yield plateau is absent on the diagram, what allows to make a conclusion about brittle destruction of samples. This conclusion corresponds to the data on relative elongation which is not high and is about 4% on average.

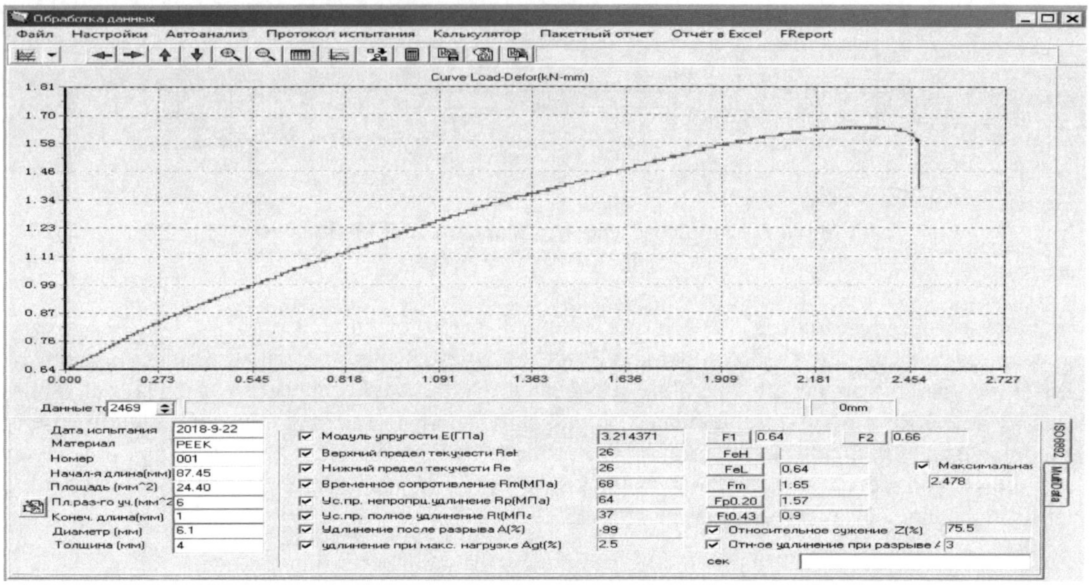

Figure 2. Machine diagram of tension of sample №1 in coordinates of kN-mm

Obtained experimental results were compared to data from different sources (table 3). It can be seen that tensile limit of samples obtained in this paper is 1.3 times lower than PEEK samples manufactured by pressure casting. Values of relative elongation is one order lower (4% in opposite to 34%).Degradation of strength properties of 3D-printed sample is caused by usage of FDM (FFF) technology, which produces samples consisting of layers having low adhesion relative to each other. Interlayer adhesion can be regulated in certain limits by speed of extruder and thickness of layers: those parameters influence spreading speed of heat from bore to material and, consequently, adhesion of layers. It appears that authors of papers [17] and [18] managed to find printing modes providing better interlayer adhesion of the tested material, at which point experimental data obtained in those papers are closer to characteristics of cast sample than those obtained in this paper. On the contrary, authors of papers [19] and [20] don't show corresponding parameters of printing modes, which causes mechanical characteristics of samples to be remarkably different from those of cast samples.

Table 3. Mechanical properties of samples made of PEEK according to different sources

	Manufacturing method and sample properties						
	Current experiment	Pressure casting[16]	FDM (FFF), filling 100%, laying of layers $+45^0/-45^0$ [17]	FDM (FFF), filling 100%, laying of layers 0^0 [18]	FDM (FFF), filling 100%, laying of layers 90^0 [18]	FDM (FFF), filling 100%, laying of layers $+45^0/-45^0$ [19]	FDM (FFF), filling 60 % [20]
Tensile limit (σ_B), MPa	66.9	100.0	98.9	82.6	72.9	~50	39.1
Elasticity modulus (E), GPa	3.2	3.7	3.9	3.8	3.5	~3	0.5
Relative elongation (δ), %	4.0	34.0	~50	110.0	2.9	~2-3	10.4

Conclusions

Therefore, polyether ether ketone as kind of aromatic polyarylene ether ketones are important polymeric material in applications, and different spheres of manufacturing industry show growing interest in it every year. Conducted research showed that strength of samples manufactured by FDM (FFF) 3D-printing technology is about 1.3 times lower than cast samples strength, what most likely caused by specificity of the technology involving creating details layer by layer. However, literature sources analysis showed existence of possibility of sample printing modes correction tending to improve their mechanical properties. Further it is planned to conduct research involving different sorts of polymers of polyetether ether ketone for 3D-printing.

Reference

[1] Gorodetskii, M.A., Nelyub, V.A., Malysheva, G.V., Shaulov, A.Y., Berlin, A.A. Technology of Forming and the Properties of Reinforced Composites Based on an Inorganic Binder // Russian Metallurgy (Metally). 2018. Volume 2018, Issue 13. pp. 1195–1198. DOI: 10.1134/S0036029518130074

[2] Korotchenko, A.Yu.,Khilkov, D.E., Tverskoy, M.V. The Development of New Materials and Modes of Casting Metal Powder Mixtures (MIM Technology) // Materials Science Forum. 2018 . Volume 945 . pp. 538 – 542. DOI: 10.4028/www.scientific.net/MSF.945.538

[3] Dumansky, A.M., Alimov, M.A., Terekhin, A.V. Experiment- and computation-based identification of mechanical properties of fiber reinforced polymer composites // Journal of Physics: Conference Series. 2019. Volume 1158, Issue 2. DOI: 10.1088/1742-6596/1158/2/022037

[4] Smerdova, O., Cayer-Barrioz, J., Le Bot, A., Sarbaev, B. Analytical model and experimental validation of friction laws for composites under low loads // Tribology Letters. 2012. Volume 46, Issue 3. pp. 263–272. DOI: 10.1007/s11249-012-9947-2

[5] Azarov, A.V., Antonov, F.K., Golubev, M.V., Khaziev, A.R., Ushanov, S.A. Composite 3D printing for the small size unmanned aerial vehicle structure // Composites Part B: Engineering. 2019. Volume 169. pp. 157-163. DOI: 10.1016/j.compositesb.2019.03.073

[6] Smith, J.A., Mele, E., Rimington, R.P., Capel, A.J., Lewis, M.P., Silberschmidt, V.V., Li, S. Polydimethylsiloxane and poly(ether) ether ketone functionally graded composites for

biomedical applications // Journal of the Mechanical Behavior of Biomedical Materials. 2019. Volume 93. pp. 130-142. DOI: 10.1016/j.jmbbm.2019.02.012

[7] He, M., Chen, X., Guo, Z., Qiu, X., Yang, Y., Su, C., Jiang, N., Li, Y., Sun, D., Zhang, L. Super tough graphene oxide reinforced polyetheretherketone for potential hard tissue repair applications // Composites Science and Technology. 2019. Volume 174. pp. 194-201. DOI: 10.1016/j.compscitech.2019.02.028

[8] Barile, C., Casavola, C., De Cillis, F. Mechanical comparison of new composite materials for aerospace applications // Composites Part B: Engineering. 2019. Volume 162. pp. 122-128. DOI: 10.1016/j.compositesb.2018.10.101

[9] Karthikeyan, L., Mathew, D., Robert, T.M. Poly(ether ether ketone)-bischromenes: Synthesis, characterization, and influence on thermal, mechanical, and thermo mechanical properties of epoxy resin // Polymers for Advanced Technologies. 2019. Volume 30, Issue 4. pp. 1061-1071.

[10] Lhymn, C., Lhymn, Y.O. Wear behaviour of silicone-impregnated poly(ether etherketone) (PEEK) composites // Advances in Polymer Technology. 1988. Volume 8, Issue 4. pp. 417-430. DOI: 10.1002/adv.1988.060080406

[11] Singh, S., Prakash, C., Ramakrishna, S. 3D printing of polyether-ether-ketone for biomedical applications // European Polymer Journal. 2019. Volume 114. pp. 234-248. DOI: 10.1016/j.eurpolymj.2019.02.035

[12] Luo, M., Tian, X., Shang, J., Zhu, W., Li, D., Qin, Y. Impregnation and interlayer bonding behaviours of 3D-printed continuous carbon-fiber-reinforced poly-ether-ether-ketone composites // Composites Part A: Applied Science and Manufacturing. 2019. Volume 121. pp. 130-Geng, P.,

[13] Geng, P., Zhao, J., Wu, W., Ye, W., Wang, Y., Wang, S., Zhang, S. Effects of extrusion speed and printing speed on the 3D printing stability of extruded PEEK filament // Journal of Manufacturing Processes. 2019. Volume 37. pp. 266-273. DOI: 10.1016/j.jmapro.2018.11.023

[14] Stepashkin A., Chukov, D.I., Senatov, F.S., Salimon, A.I., Korsunsky, A.M., Kaloshkin, S.D. 3D-printed PEEK-carbon fiber (CF) composites: Structure and thermal properties // Composites Science and Technology. 2018. Volume 164. pp. 319-326. DOI: 10.1016/j.compscitech.2018.05.032

[15] Basgul, C., Yu, T., Macdonald, D.W., Siskey, R., Marcolongo, M., Kurtz, S.M. Structure-property relationships for 3D-printed PEEK intervertebral lumbar cages produced using fused filament fabrication // Journal of Materials Research. 2018. Volume 33, Issue 14. pp. 2040-2051. DOI: 10.1557/jmr.2018.178

[16] Hoskins, T.J., Dearn, K.D., Kukureka, S.N. Mechanical performance of PEEK produced by additive manufacturing // Polymer Testing. 2018. Volume 70. pp. 511-519. DOI: 10.1016/j.polymertesting.2018.08.008

[17] Rinaldi, M., Ghidini, T., Cecchini, F., Brandao, A., Nanni, F. Additive layer manufacturing of poly (ether ether ketone) via FDM // Composites Part B: Engineering. 2018. Volume 145. pp. 162-172. DOI: 10.1016/j.compositesb.2018.03.029

[18] Arif, M.F., Kumar, S., Varadarajan, K.M., Cantwell, W.J. Performance of biocompatible PEEK processed by fused deposition additive manufacturing // Materials and Design. 2018. Volume 146. pp. 249-259. DOI: 10.1016/j.matdes.2018.03.015

[19] Zhao, F., Li, D., Jin, Z. Preliminary investigation of poly-ether-ether-ketone based on fused deposition modeling for medical applications // Materials. 2018. Volume 11, Issue 2. Article number 288. DOI: 10.3390/ma11020288

[20] Deng, X., Zeng, Z., Peng, B., Yan, S., Ke, W. Mechanical properties optimization of poly-ether-ether-ketone via fused deposition modeling // Materials. 2018. Volume 11, Issue 2. Article number 216. DOI: 10.3390/ma11020216

Manufacturing technology stamp-welded workpieces based on porous-net material with desired properties

A Tretiakov[1,2] and M Serezhkin[1]

[1]Bauman Moscow State Technical University, 5 Second Baumanskaya Street, Moscow, 105005, Russian Federation

[2]E-mail: tretyakov@bmstu.ru

Abstract
The manufacture of stamp-welded and porous elements of porous-net material (PNM) from stainless steel 12H18N10T (0,12% C, 18% Cr, 10% Ni, 1% Ti) is difficult due to the low ductility under tension and the lack of weldability. It limits the use of PNM, although the fact that they may have the necessary technological and operational properties. The effect of porosity and heat conductivity anisotropy on the strength of welded splices of parts made of PNM have been established. Technologies of electron-beam welding (EBW) and sheet stamping of filter elements, which are implemented during the manufacture of prototypes of filter elements with a cleaning precision 10-12 microns, are presented.

Introduction

For the implementation of different technological processes, it is necessary to create new equipment, which includes hydraulic and pneumatic systems. To ensure the required service life of such installations it is necessary to use reliable filter elements. Thus, the creation of reliable filter elements is an important task of modern engineering.

In a number of sources [1, 2], it has been shown that for efficient refrigeration of the walls of power plants caused by temperature increase and velocities of energy carriers increase, a method of gas supply to the boundary layer through a porous shell is used. Also, for increase of modern pneumatic and hydraulic systems reliability, the requirements for purity of the liquids and gases used in them are increased due to their filtration. In the process of creating a design of filters, an important step is an informed choice of permeable elements [3, 4], which should provide the necessary cleaning precision, energy carrier consumption, withstand the required pressure and temperature, and also have the necessary corrosion resistance [5].

Reliable operation of such structures largely depends on the right choice of parameters for the properties of materials (PM). To PM include a fabric of man-made and natural fibers [6], the grid of metal wires, blanks, obtained from metal powders [3, 4, 7] and fibers, perforated sheets [8] and some others.

As a result of the analysis of the electric power plant structures and their working conditions, it was revealed that the most forward-looking are porous materials on a metal base. Woven nettings made from corrosion-resistant steels have been used as blanks for filter elements. The cleaning precision of liquids and gases by one layer of the mesh reaches 20...40 microns [3], some improvement of filter characteristics can be achieved using multilayer briquettes of grids.

IOP Conf. Series: Materials Science and Engineering **589** (2019) 012022 doi:10.1088/1757-899X/589/1/012022

The creation of permeable materials based on metal fibers (metal matrix composites - MMC) is interestingly, studies [9, 10, 11] show that these materials have excellent hydraulic and mechanical properties. However, the lack of data about the stampability and weldability of fiber materials does not currently allow recommending it for the manufacture of stamped workpieces with specified properties.

Experimental researches of hydraulic, filtering, mechanical and technological properties of powder porous materials (PPM) from corrosion-resistant steels have shown that the main advantage of this type of permeable blanks is the ability to retain mechanical particles size up to 2 microns by filtering liquids and gases [3, 7]. An important factor limiting the use of PPM for the manufacture of permeable workpieces is bad stampability because the elongation of these materials on fracture under tensile stress does not exceed 1.5%. Made experiments with the EBW of the PPM blanks on standard equipment showed the possibility of synthesis full-strength joints. However, as a result of the uneven structure characteristic of highly-porous PPM, in the abutting joints in the areas with a large pore size, burn-through occurs. For obtaining high-quality seams in the manufacturing of workpieces made from PPM grade H18N15 (0,08% C, 18% Cr, 15% Ni,), work [12] proposed to use the remaining slices made of stainless steel 12H18N10T (0,12% C, 18% Cr, 10% Ni, 1% Ti), the fusing temperature of which is lower than the fusing temperature of the welded blanks.

The analysis of the structures, manufacturing techniques and service conditions of elements for the cleaning of mechanical pollution of liquids and gases [3, 4, 12], thermal protection systems of rocket engines [3, 13], cases of turbine blades for combined cycle gas turbine [14, 15, 16] and gas turbine engines [1, 17, 18], capillary intakes [9], etc. has shown that porous-net materials (PNM) made from steel grade 12H18N10T (0,12% C, 18% Cr, 10% Ni, 1% Ti) have optimal properties. PNM are made by hot reducing in vacuum-processed woven mesh briquette. These materials provide the specified performance characteristics and have high mechanical [19, 20, 21] and technological properties [22], which allows to manufacture a stamped-welded structures with specified properties and the required configuration.

The existing results of theoretical and experimental researches in the area of pressure welding and hot reducing of multilayer sheets [23, 24, 25] were used to create sheet porous-net materials.

Research methodology

Due to the pronounced anisotropy of the PNM properties, the ratio of effective thermal conductivity of edges in the thickness $\lambda_{\perp}$ and in the direction perpendicular to the movement of the welding source λ_w has a significant impact on the seam appearance during fusion welding processes. During the welding of blanks, which are characterized by low thermal conductivity in thickness (for example, highly- porous), depending on the value of λ_w, the origination of defects of the following types is possible:

a) split of the seam during the process of crystallization - typical for blanks with a high value of λ_w as a result of intensive heat removing in the flat surface of sheet the fusion penetration decreases and the temperature gradient across the width of the molten crater increases;

b) incomplete fusion of individual wires in the arrises of the connectable elements with the metal seam - occurs for blanks with a low value of λ_w.

Such defects are experienced from the action of surface tension forces at the interface of the liquid metal of the melted on the surface wires and the molten crater.

Experiments have shown that for blanks with high anisotropy of thermal conductivity in the sheet plane, the width of the impermeable zone (seam) does not depend on the direction of welding, but increases with a decrease of thermal conductivity in the direction of the thickness of the arrises being welded [23].

The study of weldability of PNM from steel grade 12H18N10T (0,12% C, 18% Cr, 10% Ni, 1% Ti) has shown that with decrease in porosity and increase in ratio $\lambda_{\perp} / \lambda_w$ the quality of welded joints increases. For the manufacture of highly-porous welded structures, it is efficient to use PNM with a smaller amount of thermal conductivity anisotropy in the plane of the sheet, for example, based on grids with squarely cells at an angle of 45 °.

The rational values of the EBM conditions of abutting joint are determined by using the coupling equations obtained using the theory of similarity and the regression analysis method. Using the obtained equations and a criteria-based correspondences for the calculation of rate of energy input in welding stainless steel grade 12H18N10T (0,12% C, 18% Cr, 10% Ni, 1% Ti), an equation is obtained for the approximate calculation of the energy required for the origination of abutting joint in EBW:

$$\frac{q}{\sqrt{V_w}} = 20 \cdot h_b \cdot T_f \cdot (1-P) \cdot \sqrt{\lambda_{\perp} \cdot C_m \cdot \rho_m} \quad (1)$$

where q– is the effective power of the welding source; T_f – fusing temperatureof the material; V_w– welding speed; ρ_m – is the density of a compact material; C_m – is the heat capacity of the compact material 12H18N10T (0,12% C, 18% Cr, 10% Ni, 1% Ti) ; h_b – is the thickness of the PNM blanks.

The study of the mechanical properties of welded joints PNM in the conditions of uniaxial and biaxial tensions made it possible to reveal that with an increase in the porosity of the base material, the weldment efficiency of the welded joint decreases, the value of which with porosity $P > 0{,}2$ can be described by the following empirical correspondence:

$$\frac{\sigma_w}{\sigma_{B_{x,\,y}}} = 1{,}3 - 1{,}5 \cdot P. \quad (2)$$

In expression (2), $\sigma_{B_{x,\,y}}$ – is the ultimate stress limit of the PNM under uniaxial tension in the direction of the wire grids; σ_w – seal strength of an abutting joint with the same porosity in the direction of the wire grids.

Using expression (2), we can formulate the following requirement, limiting the amount of porosity of welded blanks:

$$P = 0{,}67 \cdot \left(1{,}3 - \frac{\sigma_w}{\sigma_{B_{x,\,y}}}\right). \quad (3)$$

EBW is the optimal type of welding of porous blanks when creating filtering elements for precision cleaning of liquids and gases from mechanical pollutions. For the development of welding technology and study of the seal strength of welded joints, studies have been carried out where the current of the beam, the welding speed and the porosity of the blanks were varied. The intervals of variation this parameters given in Table 1 were determined due to the analysis of expression (1). The value of the relative durability of welded joints $\frac{\sigma_w}{\sigma_{B_{x,\,y}}}$ was taken as an example of the optimization parameter.

Table1. Main layers and intervals of variation

Parameter	Layers			Intervals of variation	Dimensionality
	lowest-1	zero0	highest+1		
Beam current	45	55	65	10	mA
Welding speed	20	35	50	15	m/h
Porosity	0,15	0,25	0,35	0,1	–

During the welding process, the focal spot diameter (d_s), focal length (l_0) and accelerating voltage (V_v) were not changed and maintained, respectively, $d_s = 0,5\,mm,\ l_0 = 120\,mm,\ V_w = 30\,kV$. The value of $\sigma_{B_{x,\ y}}$ for PNM with 0,35 and 0,15 porosity was 190 and 300 MPa, respectively. As a computational model, a first degree polynomial was defined, the coefficients calculation of which allowed to obtain the following regressional equation:

$$\frac{\sigma_w}{\sigma_{B_{x,\ y}}} = 0,804 + 0,075 \cdot P - 0,023 \cdot P \cdot V_w + 0,0057 \cdot V_w. \tag{4}$$

The manufacture of filter elements was carried out by sheet stamping of PNM in the following sequence (Fig. 1): manufacturing the round blank (a), molding the central part (b), rolling-out (c), percussion and bead forming (d).

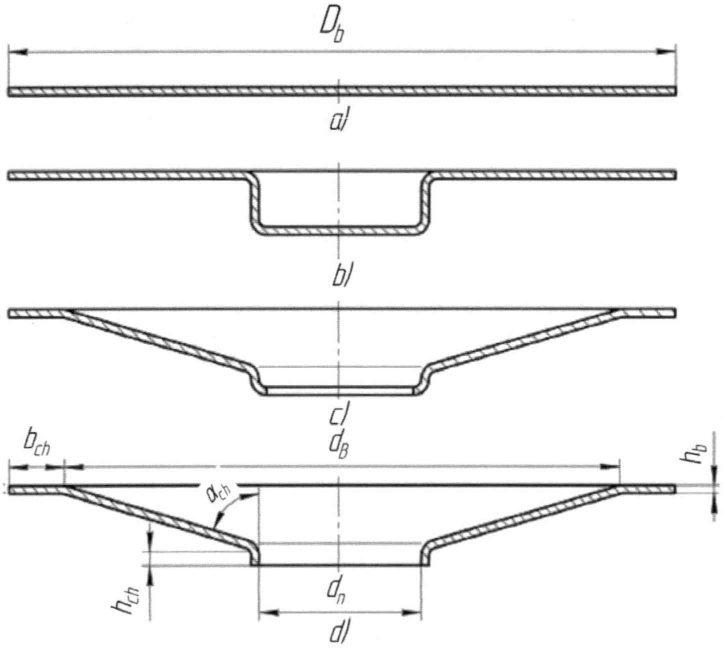

Figure 1. The sequence of sheet stumping operations of filter element

The diameter of the blank required for the sheet stamping manufacture of filter elements can be defined as:

$$D_b = \sqrt{\frac{d_n^2 \cdot (sin\alpha_{ch} - 1) + d_B^2}{sin\alpha_{ch}} + 4 \cdot b_{ch} \cdot (d_B + b_{ch})}. \tag{5}$$

The value of blank elongation during rolling-out of the bell conical section of the shell is defined as:

$$e_s = \frac{d_B - d_n + sin\alpha_{ch} \cdot (2 \cdot b_{ch} - 1)}{(1-P) \cdot (D_b - d_n) \cdot sin\alpha_{ch} + P(d_B - d_n + 2 \cdot b_{ch} \cdot sin\alpha_{ch})}. \tag{6}$$

In formulas (5) and (6) d_B, d_n, b_{ch}, $\alpha_{ch}$are the dimensions of the filter element shown in Figure 1.

The study of the influence of sheet stumping operations on the change in pore size has shown that during the process of rolling out the conical section of the filter element, an increase in the pore size

IOP Conf. Series: Materials Science and Engineering **589** (2019) 012022 doi:10.1088/1757-899X/589/1/012022

was practically not observed, since with $d_n = 15\,mm,\ d_B = 50 \pm 2\,mm,\ b_{ch} = 5 \pm 2\,mm,\ \alpha_{ch} = 75 \pm 5^0$ stretch value was $e_s = 0{,}04\ldots0{,}05$ (stretching degree $k_s = 1{,}06$). In the process of forming the cup as a result of biaxial stretching, the average size increases by 24%, but since this section is short, this increase does not have a significant effect on the maximum filtered substance consumption through the filter element.

The magnitude of the elongation at the arrises of the porous blanks during bead forming the internal diameter of the filter element can be approximately defined as:

$$e_l = \frac{2 \cdot h_{ch}}{\left(\pi \cdot d_n - 2 \cdot h_{ch}\right)}, \tag{7}$$

h_{ch} - the height of the spout diameter d_n (Figure 1).

The analysis of formulas (6) and (7) has allowed to establish that $e_l > e_s$, therefore, in the creating the production process of manufacturing PNM for a filter, the following condition should be envisaged:

$$\delta > e_l \tag{8}$$

δ – is the breaking elongation.

The flow-process chart for the manufacture of a filter intends the performance of ring seams during welding of individual elements. In this case, the quality of welded joints is determined, most commonly, by magnitude of the filter elements porosity.

Research results and discussion

The carried out studies of the weldability of PNM and the developed EBW technology for porous blanks were used in the design and manufacturing of stamped-welded filter elements with cleaning precision of 10...12 μm, the sketch of which is shown in Fig. 2:

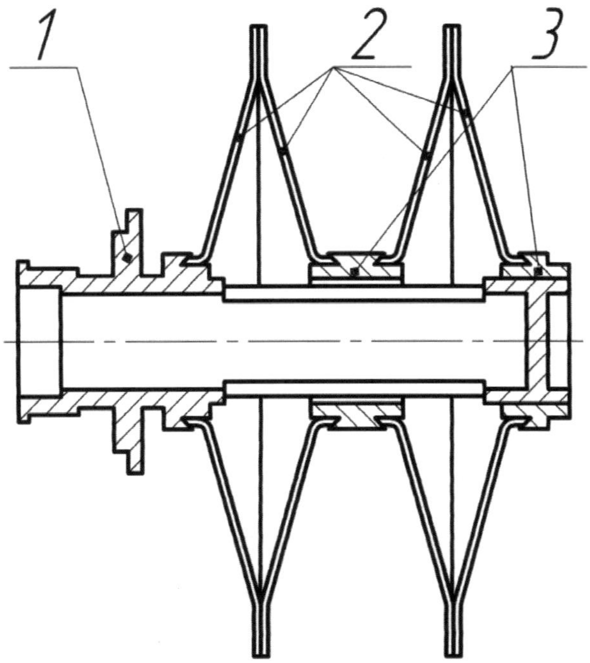

Figure 2. Sketch of the filter, providing the cleaning precision 10 ... 12 microns: 1 - axis; 2 - filter elements; 3 - bushings.

The technological process of filters manufacturing includes EBW of porous elements with an axis, bushings among themselves. Filter elements were made by sheet stamping PNM thickness of 0,5 mm and $P = 0{,}45$, obtained by rolling 13 grids with square cells, the size of which was $a = 0{,}04\,mm$, located at an angle of 45 °. After ultrasonic cleaning in freon gas, porous blanks were set up in special devices guarantying the assembly of the welded elements and their rotation.

Welding was carried out on a current of 5...6 mA with an accelerating voltage of 50 kV with a focal spot diameter of 1 mm. The welding speed was at the connection of the filter elements with the axis and bushings of 6,3 m/h, and among themselves - 22 m/h. Quality control of the welded joints of the filter under hydrostatic test conditions has shown that the probability of obtaining a defect-free welded joint is 0,96.

Analysis of equations (5), (6) and (7) has allowed to establish for obtaining the filter elements with following dimensions: $d_n = 15\,mm,\ d_B = 50\,mm,\ b_{ch} = 5\,mm,\ h_{ch} = 2{,}5\,mm,\ \alpha_{ch} = 75 \pm 5^0$, blank diameter for sheet stamping is 62 mm, the elongation during the rolling-out process of the conical part of the porous element is $e_s = 0{,}04$, and during the bead forming, $e_l = 0{,}118$. As a result of significant deformations ($e_l \approx \delta$) on the arris of the spout, the cracks forming is possible. To eliminate possible defects, the shape of the abutting joints of filter elements with the axis and filter bushings (Fig. 2) is designed in the way that the spout melts during the EBW process.

Analysis of equation (4) showed that with a porosity of the base material less than 0.25, an increase of weldment efficiency can be achieved by increasing of the welding speed, and with a porosity of more than 0,25, a decrease of welding speed results to an increase of $\frac{\sigma_w}{\sigma_{B_{x,\,y}}}$. In this case, the beam current, equal to 0,45...0,65 mA, slightly has an effect on the joint efficiency.

Metallographical researches revealed the undercuts in areas where the wires did not melt with the seam metal, the number of which was increasing with increasing the porosity. The width of the obtained seams on prototypes with a thickness of 2,8…3,8 mm did not change significantly with the variation of welding mode parameters and was 2,5...3,0 mm.

The study of the influence of the welding cycle on the change in the hydraulic and structural characteristics of porous blanks has shown that the decrease of the permeability of welded joints is proportional to the seam area, while the pore sizes in the weld zone do not change.

In the manufacturing of welded structures, the requirement of manufacturability should be taken into account, which is guaranteed by the rational choice of the blank material, the shape of the welded elements, the type of welding and measures to improve the quality of workpieces. To obtain welded structures with high capacity of work, should tend to use PNM with a lower porosity and minimal anisotropy of properties in the flat surface of sheet.

In the case when it is necessary to obtain workpieces from highly porous materials, an improvement in the quality of welded joints can be achieved by increasing the thermal conductivity in the direction of the thickness of the welded arrises, and if possible, by using diffusion welding.

The results of the researches used in the design and manufacture of stamped-welded filters with a cleaning precision of 10-12 microns.

Conclusion

The strength of the seal joints obtained by the EBW is determined by the chemical composition, the shape of the welded elements, the anisotropy of the thermal conductivity and the porosity of the welded edges arrises. With an increase of the ratio of the effective thermal conductivity coefficients in the thickness $\lambda_\perp$ and in the direction perpendicular to the movement of the welding source λ_w and a decrease of the porosity of the arrises, the quality of the welded joints increases.

The manufacturing technology of the filter element by sheet stamping is carried out in the following sequence: manufacturing the round blank, molding the central part, rolling-out, percussion and bead forming. It was established that the value of the workpiece elongation when the conical section of the shell during the rolling-out depends on the geometric parameters of the filter element and its porosity.

List of reference

[1] Zeigarnik U.A., Poliakov A.F., Stratev V.K., Tretiakov A.F., SHekhter U.L. Testing of porous-net materials as the shells of turbine blades, high-temperature gas turbines. Moscow, OIVT RAN, 2010, №2-502, 64 p.

[2] Bunker R.S. Gas turbine cooling. Moving from macro to micro cooling. Proceeding of the ASME TURBO EXPO, 2013. 3p.

[3] Porous-net materials: B.R. /Edited by. Belov S.V. Publishing house.: Metallurgy, 1987. 335 p.

[4] Sparks T., Chase G. Filters and Filtration. Handbook. Elsevier, 2013. 444 p.

[5] Kolesnikov A.G., Safonov A.N. Modern developments of new materials and technological processes for engineering. TyazheloeMashinostroenie, 1998, no.2, pp. 17-21.

[6] Volkov S.S. Ultrasound welding of synthetic fabric for technical purposes. Welding International, 2009, vol. 23, no. 10, pp.193-197.

[7] Belov S.V.,Shmelev L.S., Spiridonov V.S., Ksenofontov A.V., Prikhodko N.G. Structural hydraulic, and filter characteristics of type FNS porous sheets. Soviet Powder Metallurgy and Metal Ceramics, 1986, 25(3), pp. 256-241.

[8] Zubkov N.N., Polykov A.F., ShekhterYu.L. The Hydraulic characteristics of porous materials for a system of transpiration cooling. High Temperature, 2010, 48(2), pp. 231-237.

[9] Kostornov A.G. Capillary transport of low–viscous liquids in porous metallic materials under action of gravitation force. PoroshkovayaMetallurgiya, 2003, no. 9-10, pp.13-26.

[10] Kostornov A.G., Shapoval A.A., Lalor M.J., Mgalobishvili O., Legros J.C. A study of heat transfer in heat pipe evaporators with metal fiber capillary structures. Journal of Enhanced Heat Transfer, 2012, 19(1), pp. 43-51

[11] Kostornov A.G., Krichenko O.V.,Brodnikovskii N.P., Guslienko Y.A., Klimenko V.N. High-porous material made from alloy steel fibers: production, structure, and mechanical properties. Powder Metallurgy and Metal Ceramics, 2008, 47 (5-6), pp. 295-298.

[12] Ovchinnikov V.V., Antonov A.A., Gureev M.A., Magnitov V.S. Special features of welding filtration of the cartridge type produced from porous metallic materials. Welding International, 2002, no.10, pp. 828-831.

[13] Pelevin F.V. Heat transfer in meshed metallic materials with interchanel transpiration and two-dimensional intermesh flow of heat transfer fluid. High Temperature, 2018, 46 (2), pp. 208-216.

[14] VaraksinA.Yu., Polyakov A.F., Reviznikov D.L.,Stratyev V.K., Tretyakov A.F. Investigation of flow in the vicinity of porous materials on a cylindrical body subjected to flow. High Temperature, 2002, 40(6), pp. 843-849.

[15] Polyakov A.F., Stratyev V.K., Tretyakov A.F., ShekhterYu.L. Heat transfer in envelopes made of porous network materials. Thermal Engineering, 2009, 56(3), pp. 227-334.

[16] Polyakov A.F., Stratyev V.K., Tratyakov A.F., ShekhterYu.L. Generalization of experimental data on heat transfer in permeable shells made of porous reticular materials. Thermal Engineering, 2010, 57(6), pp. 516-521.

[17] Wolf J., Moskowits S. Development of the transpiration air-cooled turbine for high – temperature dirty gas stream. Journal of Engineering for Power, 1983, 105(10), pp. 821-825.

[18] Bunker R.S. The role of materials and manufacturing technologies as enablers in gas turbine cooling for high performance engines. Ceramic Transactions, 2009, 205, pp. 3-2.

[19] Tretyakov A.F., Specific strength of porous materials based on metal wire cloths. Strenght of Materials, 1994, 26(2), pp.116-119.

[20] Liu P., He G., Wu L.H. Fabrication of sintered steel wire mesh and its compressive properties. Material Science and Engineering, 2008, vol. 489, no.1-2, pp. 21-28.

[21] Xu G., Liu Y., Luo X., Ma J., Li H. Experimental investigation of traspiration cooling for sintered woven wire mesh structures. International Journal of Heat and Mass Transfer, 2015, vol.91, pp. 898-907.

[22] Tretiakov A.F. Technological heredity in the process of manufacturing products from porous-net materials with desired properties. Report 2. Regularities of the effect of plastical deformation and consolidation of wire mesh on the technological and thermophysical properties of porous-net materials. Roll metal production, 2013, № 6, PP. 29–34.

[23] Kolesnikov A.G., Plokhikh A.I., Komisarchik Y.S., Mikhaltsevich I.V. A study of special features of formation of submicro-and nanosize structure in miltilayer materials by the method of hot rolling. Metal Science and Heat Treatment, 2010, 52 (5-6), pp. 273-278.

[24] Schmiedt A., Luecker L., Kolesnikov A.G, Plokhikh A., Walther F. Production and microstructurebased fatigue multilayer materials produced by hot pack rolling. Materialpruefung/Materials Testing, 2007, 59(2), pp. 123-129.

[25] Vlasova D.N., Plakhikh A.I., Semenov M.V., Polyanskii B.M., Safonov M.D. Effect of diffusion processes during pack rolling on multilayer material stability. Metallurgist, 2018, 62 (5-6), pp. 432-439.

IOP Conf. Series: Materials Science and Engineering **589** (2019) 012023 doi:10.1088/1757-899X/589/1/012023

Methods of dealing with mutual loading of joint hydraulic actuators

P Shcherbachev[1] and D Vdovin[1,2]

[1]Bauman Moscow State Technical University, 5 Second Baumanskaya Street, Moscow, 105005, Russian Federation

[2]E-mail: da_vdovin@mail.ru

Abstract
When developing a hydraulic actuator with several executive engines, it is necessary to take into account their mutual influence on each other, caused by many reasons - technological errors in the manufacture of parts, compressibility of the working fluid, pipeline contamination, etc. In practice, the problem of interference is encountering when using actuators with actuators of translational movements - hydraulic cylinders. In this case, the mutual loading is the greater, the greater the desynchronization of the executive engines. Therefore, we need to achieve synchronization of hydraulic cylinders. This article discusses possible methods for solving this problem.

Introduction

In a hydraulic actuator with several executive engines, it is important to achieve synchronization of the executive engines. In case of non-simultaneous operation of several hydraulic cylinders, their mutual loading occurs [1] - [4]. Mutual influence is causing by many factors: compressibility of the working fluid, different values of overlap in hydraulic distributors, inequality of resistance of hydraulic lines, different loads on hydraulic rams, etc. Since improving the methods of controlling, the amount of overlap does not allow leveling other causes of mutual loading, and errors in forming equality overlaps are inevitable, this method can only be used in combination with other methods. Mechanical synchronization methods also do not eliminate the above factors. Therefore, this article will consider various methods of leveling mutual loading in order to obtain general recommendations for solving this problem.

Methods of dealing with the mutual loading of executive hydraulic motors

Sequential connection of hydraulic cylinders (see Fig. 1) does not solve the problem of mutual loading even under the condition of perfect synchronization of hydraulic cylinders - the pressure in the discharge cavity of the leading hydraulic cylinder is determined by the sum of the pressure drops on both hydraulic cylinders.

Synchronization of hydraulic cylinders with different loads can be ensured by using a two-piece pump (Fig. 2). The disadvantages of the system are limited range of two-piece pumps, relatively low accuracy, due to the manufacturing error and pollution of the hydraulic system, which increases the resistance of hydraulic lines.

IOP Conf. Series: Materials Science and Engineering **589** (2019) 012023 doi:10.1088/1757-899X/589/1/012023

One of the simplest methods for synchronizing the operation of hydraulic cylinders is usingthrottles (Fig. 3) installed in front of the hydraulic cylinders and tuned to the same flow rate of oil passing through the hydraulic lines.

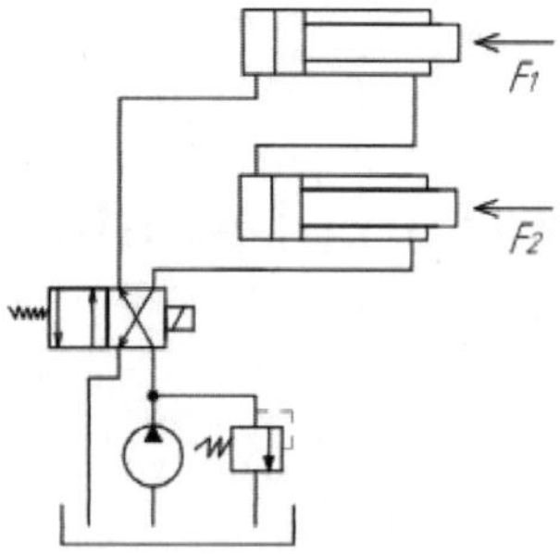

Fig. 1. Series Hydraulic Cylinder Coupling

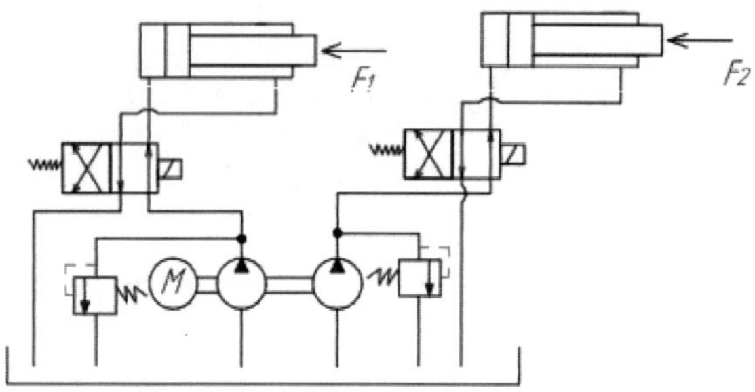

Fig. 2. Hydraulic drive scheme with a two-piece pump

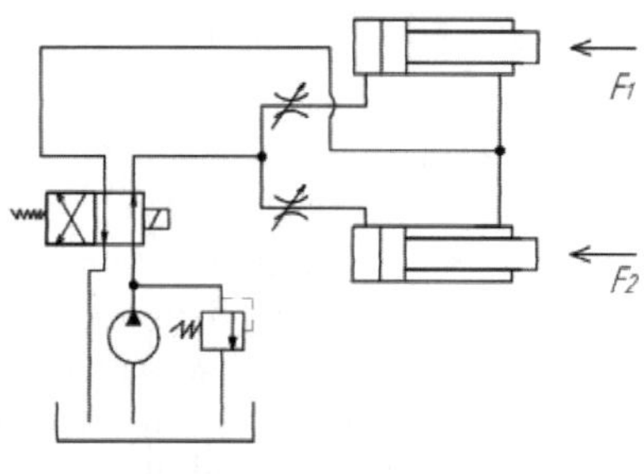

Fig. 3. Synchronization scheme using throttles

The most common method is the use of a flow divider (Fig. 4), the principle of which is based on self-regulating throttling [5]. At point M, the flow is divided into two, each of which passes through a constant throttle 1, and then supplied to sleeve 2 with a floating piston 3 regulating the size of the area of the holes 4 and 5. The piston stops when pressure and flow in the right and left cavities of the sleeve. Therefore, the output speeds of both hydraulic cylinders (Fig. 4) will be equal. Synchronization accuracy is from 1 to 5% and depends on the difference of loads on the rods.

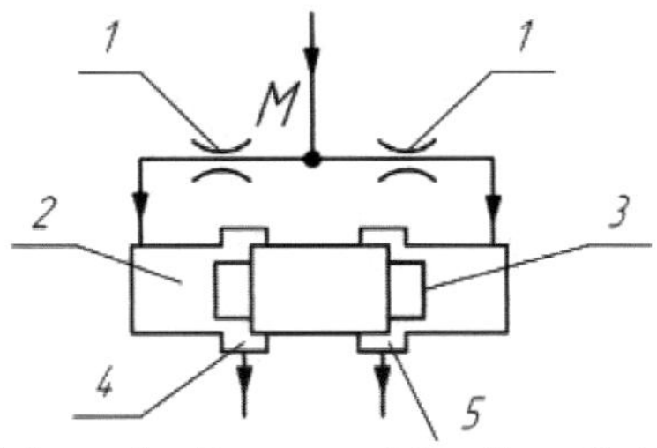

Fig. 4. Schematic diagram of the flow divider.

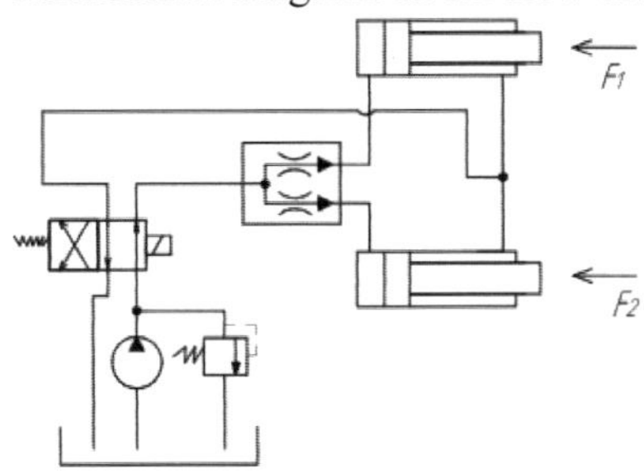

Fig. 5. Synchronization scheme using a flow divider.

Another common method is to use a pair of unregulated hydro motors (see Fig. 6), which will be interconnected by a rigid shaft to ensure equality of costs at the outlet of the hydro motors. In case of inequality of costs, one of the hydro motors goes into pump mode and restores equality of flows in the hydro lines. Synchronization accuracy depends on the type of hydraulic motors and reaches 1% when using axial-piston machines [6].

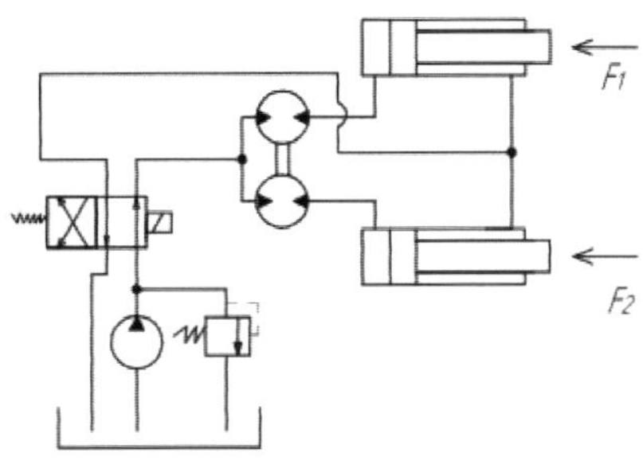

Fig.6. Synchronization scheme using hydraulic motors.

An energetically beneficial method of synchronization is the use of a reciprocating dispenser (Fig. 7). The hydraulic cylinder 1 performs the role of a metering device, providing synchronization of the hydraulic cylinders 2 and 3 with an error value of not more than 0.4 mm when lifting loads by 250 mm. The disadvantage of the method is an increase in the execution time of operations and large dimensions of the system.

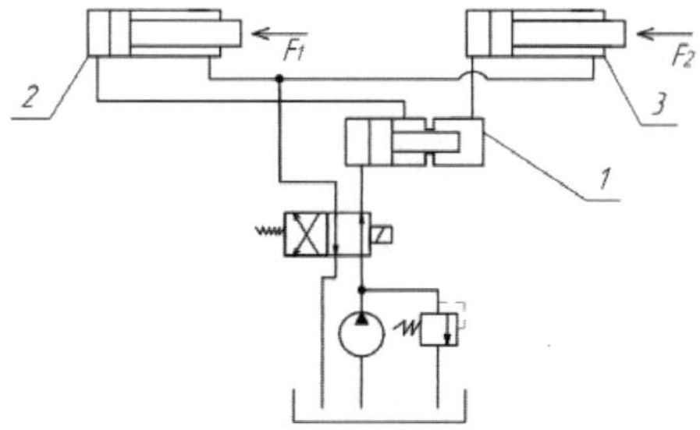

Fig. 7. Synchronization scheme with the use of a linear motion meter.

Very interesting and relatively simple in execution are systems with synchronized backpressure (Fig. 8). The principle of operation is that stopping cylinder 2 causes the oil supply to open valve 5, valve 5 closes line 6 and stops cylinder 7. This method was implemented in the lift system of the weighing machine in a system with sequential cylinders, [6] and was obtained 1 mm error for 250 mm stroke.

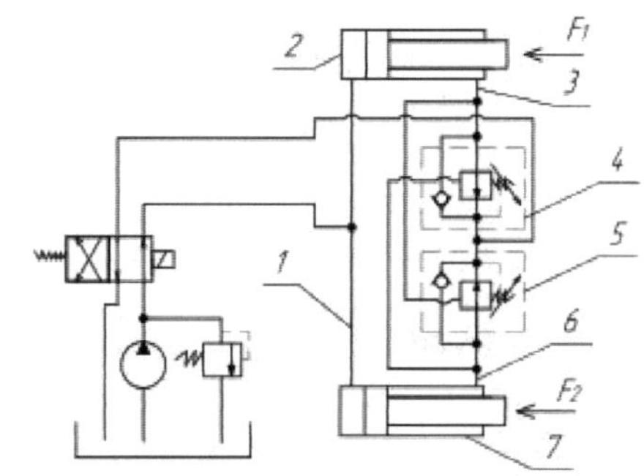

Fig. 8. Synchronization with backpressure

The use of tandem cylinders to synchronize the course has proven itself with asymmetric loads. The synchronism of the stroke is achieved by maintaining the same pressure in the rod cavities (Fig. 9).

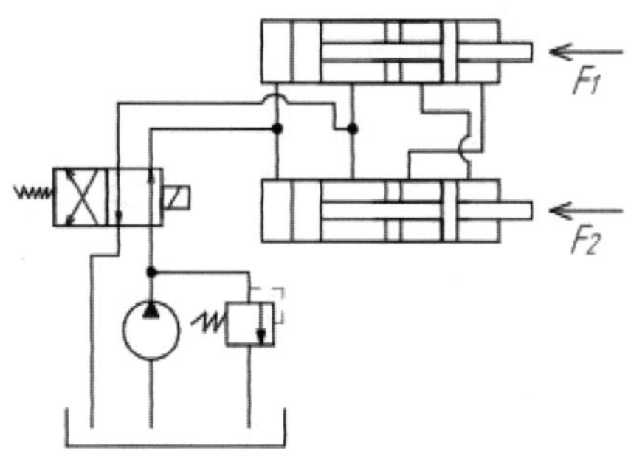

Fig. 9. Schematic diagram of the drive using tandem cylinders

To improve the accuracy of synchronization, they began to try to apply methods using sensors and regulators. For example, a hydraulic circuit using throttles (see Fig. 3) can be supplemented by adding linear sensors to the circuit that measure the extension of the hydraulic cylinder rods (Fig. 10). At the slightest deviation of the parameters of the first sensor from the second, the sensor gives an electrical

IOP Conf. Series: Materials Science and Engineering **589** (2019) 012023 doi:10.1088/1757-899X/589/1/012023

impulse to the throttle, which in turn opens or closes the throttle valve, thereby regulating the flow to each cylinder.

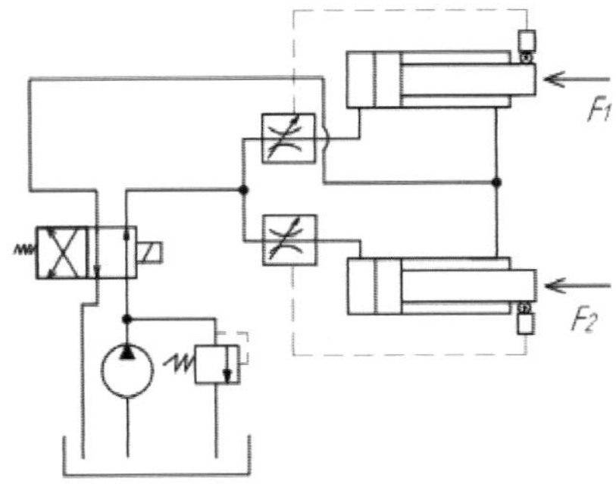

Fig. 10. Scheme of synchronization of hydraulic cylinders using linear sensors

Another method using linear sensors is a system with master and slave hydraulic cylinders. This scheme is applied in a press brake (Fig. 11).

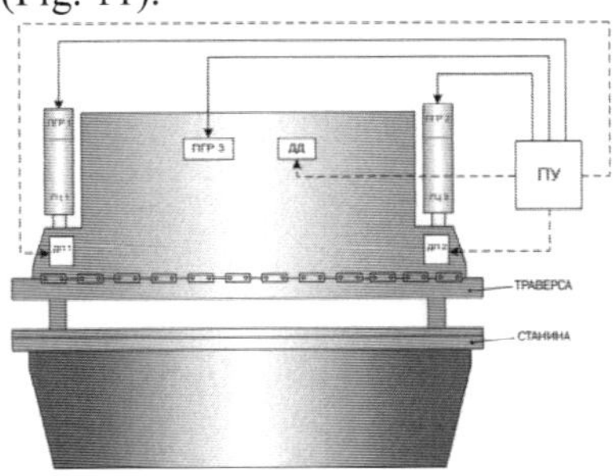

Fig. 11. Diagram of the control system and control hydraulic press

This system consists of proportional directional control valves (DCV), a control panel (CP), position sensors (PosS), a pressure sensor (PS) and hydraulic cylinders (HC). In the machine to control the linear movement of the cylinders used absolute optical displacement sensors, transmitting signals to the microcontroller, which, according to a given algorithm, generates a control signal for the slave cylinder. To reduce the error, the PID controller is used, shown in the block diagram of the control system of hydraulic cylinders (see Fig. 12). It should be noted that this model of the regulator could be changed [7]. The developed control system made it possible to achieve synchronized movement of hydraulic cylinders with an accuracy of 0.05–0.01 mm. The use of multi-coordinate control is required for the systems moving in three-dimensional space.

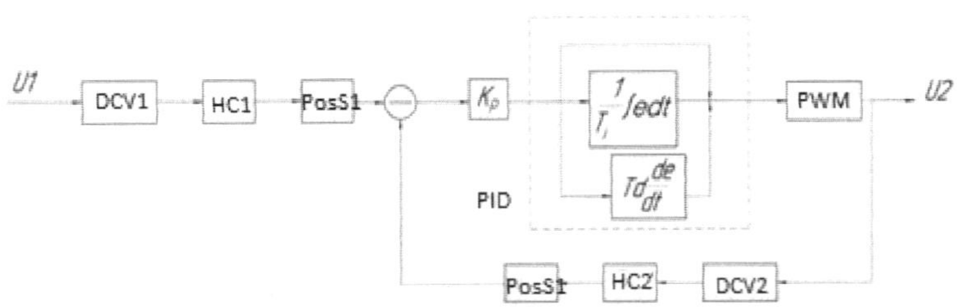

Fig. 12. Hydraulic cylinder control block diagram

Considering the problem of mutual loading, it should be noted systems in which different load on the hydraulic cylinders. It is necessary to distinguish the hydraulic scheme of the system with LS-regulation (Fig. 13), in which the first hydraulic cylinder maintains a constant speed regardless of the load on the second hydraulic cylinder [8]. This system combine's volume and throttle control.

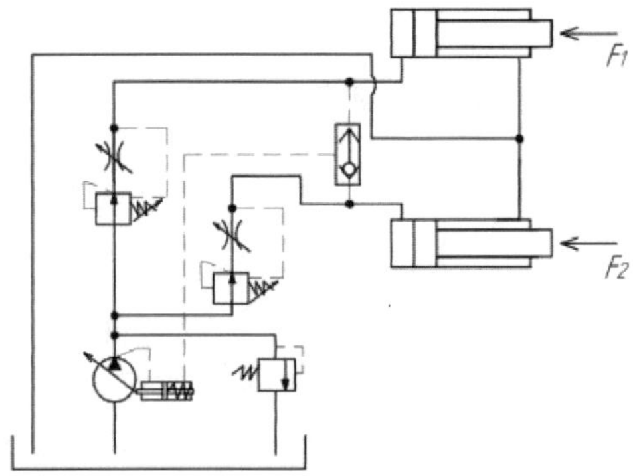

Fig.13. Hydraulic circuit with LS-regulation

Research results

1) For synchronization of hydraulic cylinders with approximately the same load on the rods and synchronization accuracy up to 5%, it is possible to apply circuit methods based on throttling.

2) Synchronization with the help of volumetric regulation allows achieving an order of magnitude greater order of magnitude than with throttle regulation with less power consumption, the asymmetry of the load has less effect.

3) The greatest accuracy is achieved with the help of displacement sensors and various regulators, well proven in other areas of technology [9] - [13]; the asymmetry of the load has the least effect.

Conclusions

The application of the methods directly depends on the specified accuracy and the difference in loads on the hydraulic cylinders. For drives of low accuracy and rare inclusion, methods based on throttle control should be applied. These methods are simple, cheap, but energetically unprofitable. For drives of greater accuracy and higher switching frequency, one should either use volumetric methods for solving the problem, or use various sensors. However, their implementation is more expensive and not always constructively possible. The choice between these two directions should be made because of considerations of the required accuracy (linear sensors are needed for greater accuracy) and energy efficiency (volumetric synchronization methods have the best efficiency). It should be noted that the most common regulator is the PID controller, which has been used in many systems [14] - [21].

References

[1] Motion synchronization control of four multi-stage cylinders electro-hydraulic elevating system. Du, H., Wei, J. 2010 2010 International Conference on Mechanic Automation and Control Engineering, MACE20105536387, p. 5249-5253
DOI: 10.1109/MACE.2010.5536387

[2] Linear extended state Observer-Based motion synchronization control for hybrid actuation system of more electric aircraft. Wang, X., Liao, R., Shi, C., Wang, S. 2017Sensors (Switzerland) 17(11),2444
DOI:10.3390/s17112444

[3] Synchronizing compensation control of electro-hydraulic load simulator using command signal of actuator. Han, S., Jiao, Z., Shang, Y., Wang, C. 2015Beijing

HangkongHangtianDaxueXuebao/Journal of Beijing University of Aeronautics and Astronautics. 41(1), p. 124-132
DOI:10.13700/j.bh.1001-5965.2014.0052

[4] Synchronization modeling and control for two cylinder electro-hydraulic elevating systems. Ni, J., Xiang, Z., Pan, X., Lu, F. 2007Jixie GongchengXuebao/Chinese Journal of Mechanical Engineering.43(2), p. 81-86
DOI: 10.3901/JME.2007.02.081

[5] Control system for electro-hydraulic synchronization on RBPT. Adenuga, O.T., Mpofu, K. 2014 Procedia CIRP17, p. 835-840.
DOI:10.1016/j.procir.2014.01.135

[6] Bogdanovich L.B. "Hydraulic drives" 1980.

[7] Equalization of multi-cylinder electro-hydraulic systems. Sun, Hong, Chiu, George T.-C2000 Proceedings of the American Control Conference 6, p. 4134-4138.

[8] Energy savings in laboratory hydraulic system operation with and without LS regulator. Heřmánek, P., Kviz, Z.2003VDI Berichte(1798), p. 361-366.

[9] Linear Active Disturbance Rejection Controller for VSC-HVDC to Improve System Transient Stability. Huang, Z., Chen, X., Huang, J., Shen, P.2019 2018 International Conference on Power System Technology, POWERCON 2018 – Proceedings 8601725, p. 2888-2895
DOI: 10.1109/POWERCON.2018.8601725

[10] Motion synchronization control of four multi-stage cylinders electro-hydraulic elevating system. Du, H., Wei, J.2010 2010 International Conference on Mechanic Automation and Control Engineering , MACE2010 5536387, p. 5249-5253
DOI: 10.1109/MACE.2010.5536387

[11] Motion synchronization modeling and control for multi-cylinder electro-hydraulic elevating system. Ni, J., Xiang, Z., Pan, X., Lu, F. 2006 JixieGongchengXuebao/Chinese Journal of Mechanical Engineering 42(11), p. 81-87.

[12] Motion synchronization for multi-cylinder electro-hydraulic system. Sun, H., Chiu, G.T.-C.2001IEEE/ASME International Conference on Advanced Intelligent Mechatronics, AIMp. 636-641.
DOI: 10.1109/MACE.2010.5536387

[13] Synchronized motion control for multi-axis electro-hydraulic system of large equipment. Wei, J.-H., Guo, K., Xiong, Y.2013Zhejiang DaxueXuebao(Gongxue Ban)/Journal of Zhejiang University (Engineering Science) 47(5), p. 755-760.
DOI: 10.3785/j.issn.1008-973X.2013.05.003

[14] Nonlinear PID synchro control on pattern drawing machine with four cylinders. Ni, J., Peng, L., Chen, G. 2011 Proceedings of 2011 International Conference on Modelling, Identification and Control, ICMIC 2011 5973725, p. 325-330

[15] Automatic Tuner for PID Controllers with Elements of Artificial Intelligence. Zadorozhnaya, N.M., Lisitsyn, A.N.2018 IOP Conference Series: Earth and Environmental Science194(2),022046
DOI:10.1088/1755-1315/194/2/022046

[16] Synergetic synthesis of control laws for left ventricular assist device rotor on magnetic suspension. Bogdanova, Y., Guskov, A.2016 Proceedings of 2016 International Conference "Stability and Oscillations of Nonlinear Control Systems" (Pyatnitskiy's Conference), STAB 2016 7541168.
DOI: 10.1109/STAB.2016.7541168

[17] Self-learning neural network control system for physical model with one degree of freedom of system of active vibration isolation and pointing of payload spacecraft. Sayapin,S.N.,

Artemenko, Y.N., Panteleev, S.V. 2017Studies in Systems, Decision and Control 95, p. 213-230.
DOI: 10.1007/978-3-319-53327-8_15

[18] Design of Flat Wheel Braking Control System with three Modes of Motion: Rolling, Sliding, Locking. Andrikov, D., Andrikov, D., Mecapeu, C.P.D. 2017 ProcediaComputerScience103, p. 466-469
DOI:10.1016/j.procs.2017.01.024

[19] Minimization of a mismatch time of movement of actuators of a throttle synchronization system. Bushuev, A.Yu., Ivanov, M.Yu., Korotaev, D.V.2018 Journal of Physics: Conference Series1141(1),012090
DOI:10.1088/1742-6596/1141/1/012090

[20] Synchronization control of hydraulic horizontal regulation system. Dou, H., Wang, S. 2011 Proceedings of the 2011 6th IEEE Conference on Industrial Electronics and Applications, ICIEA 2011 5975906, p. 1922-1927.
DOI: 10.1109/ICIEA.2011.5975906

[21] Synchro-control on steel pipe stacking electro-hydraulic servo system with four cylinders. Wang, W., Ni, J., Chen, G. 2011 International Conference on System Science, Engineering Design and Manufacturing Informatization, ICSEM 2011 2,6081327, p. 55-59.
DOI: 10.1109/ICSSEM.2011.6081327

IOP Conf. Series: Materials Science and Engineering **589** (2019) 012024 doi:10.1088/1757-899X/589/1/012024

The multifaceted role of material production in society

A Mukhin[1], G Ganina[1], S Mozgin[1], Yu Ostrovskiy[1] and A Yakovleva[1,2]

[1]Bauman Moscow State Technical University, 5 Second Baumanskaya Street, Moscow, 105005, Russian Federation

[2]Email: yakovleva525@mail.ru

Abstract.
The paper considers a new approach to substantiating the role and place of material production in a person's life. It is shown that material production performs not only the function of satisfying the material needs of a person but is also the source for the development of spiritual production. The evidence base of such allegations is given. The paper may be useful to specialists in the field of production management and development, as well as to employees of educational institutions in the field of technology, production organization, economics, and sociology.

Introduction

It is considered that production can be divided into two areas: material and ideal (spiritual) production. There are a sufficient number of publications on this topic in which attention is paid to the role of spiritual production (science, art, religion) in the life of society and, at the same time, the role of material production is further emphasized [1]. However, attention is drawn to the fact that traditionally material production in the spiritual life of society plays the role of saturating society with material products of such quantity and quality that will allow society in the spiritual sphere not to care about existence [2– 9]. At the same time, some newest studies indicate the erroneousness of such a "one-sided" judgment about the role of material production and its effect on the spiritual life [10,12–16].

The main feature of this situation is that it does not notice that in the "nature" of material production there are factors that determine the character of the spiritual development of man and society at the ontological level.

If one does not notice this, relegating to material production the role of "product" provision of the spiritual life of society, it is not surprising that in some communities where any means of material support are located (besides production), production is gradually "curtailed" and this is given to quite convincing practical arguments [7, 10, 11]. For example, one explains this phenomenon and argues its attractiveness by the fact that the release of people from industrial activity increases the possibility of their spiritual development.

Such arguments are not only theoretical in nature - some communities or countries in practice implement such a doctrine, moving production to where labor is cheaper and operating profit from production becomes higher [8,13]. However, after some time, such communities "unexpectedly" face negative phenomena in their lives, for example, such as a "moral decline" among the non-working population, and the degradation of a part of society gradually goes on [14,15,16].

All this taken together forces ones to return production facilities back, which has some examples [9]. At the same time, one focuses primarily on the prospects for profit growth; wherein spiritual and moral aspects are “forgotten”, although there is a positive trend of moral “cleansing” of society when production is resumed.

Such phenomena, sometimes occurring in a latent form, make us more attentive to the scientific and practical problem of the mutual influence of material and spiritual production. It requires new approaches to the study of such a phenomenon.

Materials and methods

This paper presents an approach based on the principles of ergodynamics - a new direction in science and practice, based on the study of energy conversion in various systems. In terms of the development of society, which are dynamic processes based on human labor, ergodynamics concerns, for example, ergonomics, social development [16].

For science and manufacturing practice, one applies an approach based on the regularities of the transformation of labor activity by examining the role of different forms of thinking of the decision maker. The essence of this approach is that a person with the help of his intuitive thinking mechanisms is able to form models of his activity at its various stages. In accordance with the goal, the structural model of relations between the elements of the problem situation is transformed, and this is where the creative activity of a person consists.

We can separate all possible actions of a person into two types:

- carried out on the basis of decisions made in a formal-logical way;
- performed on the basis of decisions made in an intuitive way.

The combination of purposeful actions, each of which is based on decisions made in an intuitive way, is called creative work. In contrast to the creative work, the routine work will be called a set of targeted actions, each of which is based on decisions taken in a formal-logical way.

Obviously, when a new situation appears, we are forced to make an intuitive decision, therefore, at the first stage of solving any task, by definition, work is creative. When the situation is repeated, successful solutions are remembered, which to some extent can be described in a formal-logical way. Now, in a certain situation, the availability of knowledge allows a person to use formal logical thinking, and his subsequent work, in this case, becomes routine, including by automating routine operations.

Naturally, being freed from routine work in some situations, a person can engage in creative work in another situation, which in turn can lead to the appearance of formal description and a new transition to routine work, see, for example, [6,18]. The set of transitions indicated here is called the ergatic transformation of human labor activity. The presence of two forms of thinking leads to a constant ergatic transformation of labor activity, regardless of the type of activity itself and the period of human development.

Material production for a long time develops as the main type of production that satisfies human needs through the manufacture of material products and the direct consumption of various means of production. In parallel with material production, humanity was engaged in the "manufacture" of products of spiritual content in such areas of spiritual production as science, art, and religion.

In Russian literature, the term “spiritual production” became popular thanks to the works of the philosopher V. P. Tugarinov [19]. He argued that the production of things and the production of ideas such as the difference between two different spheres of social life. A close approach to this is declared, for example, in the work [20] devoted to the improvement of man.

A comparison of material and spiritual production is carried out through the identification of their differences. The difficulty here lies in the fact that material and spiritual production in real conditions is inseparable. Thus, material production is carried out in accordance with the intentions of people, and the products of spiritual activity enter the public consciousness only when they acquire a material shell.

In social reality, matter and spirit do not exist separately, in a “pure” form, independently of each other. The ratio between material and spiritual production in different spheres of public life is different.

In economics, for example, material production has a relative share, and spiritual freedom is limited by material conditions. In culture, on the contrary, “the kingdom of the spirit reigns supreme,” and the material form adapts itself to the spiritual content. In politics, material and spiritual production exchange roles, and in different eras, material, spiritual production becomes decisive.

The mutual influence of material and spiritual production is carried out through the products of activity. Nevertheless, the "product" approach when comparing both productions does not imply that they even "speculatively" can "merge" with each other, because the products of activity are fundamentally different.

The prospect of a "merger" of material and spiritual production through a producer approach may look differently, based on the general scheme of a producer system suitable for both material and spiritual production (Fig. 1).

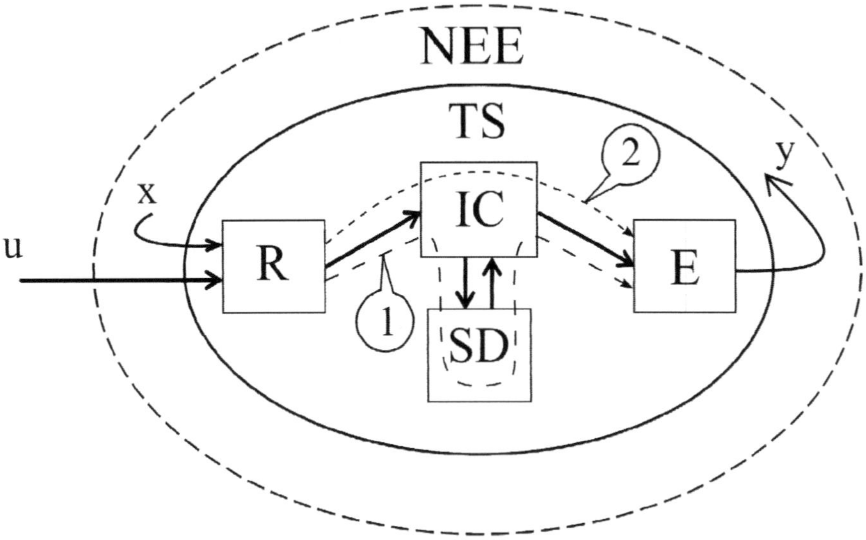

Fig. 1. Production system for any kind of production:

Here: TS is the transforming system, NEE is the near external environment, IC is the information converter, SD – is the storage device, R is the receptor, E is the effector, x is the effect on the NEE, y is the result of the NEE conversion, u is the control influence.

The production system can be a person in the workplace, the product of which functioning is a product with certain specified requirements or a creative team whose product of activity is design documentation. Here, as the 2u" can be variable performance characteristics of the designed product.

For material production, the decisive factor is that NEE and TS are material, including all other attributes. For spiritual production, all the listed structural elements are ideal.

The unifying feature for the two types of production from the point of view of the production system (Fig. 1) is the presence of two "trajectories" of decision-making. Trajectory 1 means "movement of information" in the case when the decision on the impact of y is made formally and logically when the M based on past experience contains the logical statements «x $\wedge$u$\rightarrow$y». Trajectory 2 is typical for the case when you have to make a decision in a deliberately unknown situation (new u). In this case, the decision is made by intuitive judgment. Labor, based on trajectory 2 is creative, and labor based on trajectory 1 is routine.

Based on the fact that, in spiritual production, the predominant form of labor is creativity, and in material production, creativity is one of the main attributes of labor, it follows that creativity is an attribute property of the fusion of material and spiritual production.

The role of material production in the spiritual life of society is reduced to the fact that the process of transformation of labor activity, which releases the creative potential of production, at the same time nourishes society with creative individuals capable of developing the spiritual life of society.

Studies have shown that the fusion of material and spiritual production through creativity, as an attribute property of a merger, does not boil down only to proving that material production plays an important role in the spiritual life of society. It is necessary to understand how significant this feature is in human life and society; is it possible, by ignoring material production, to develop spiritual production and what is the danger for the society of such ignoring.

It is necessary to approach the problem of merging material and spiritual production comprehensively, due to expanding the area of interrelation and mutual influence of material and spiritual production.

Results and discussion

In order to assess the relationship of material and spiritual production, we will consider only the "engineering" aspects of spiritual production - we will be interested in such creative specialists as designers, technologists and other specialists who create new methods and production in general.

We summarize the known interrelationships and interdependencies in Table 1.

Table 1.

№	Kindofrelationship	Predicateform
1	If there is material production, then you can produce a material product that meets the needs	$MS1(y) \rightarrow MK1(z)$
2	If there is material production, then there is an increase in creative potential in spiritual production.	$MS1(y) \rightarrow DT1(x)$
3	If there is an increase in creative potential in spiritual production, then it is possible to organize work on the creation of advanced material production.	$DT1(x) \rightarrow MS2(y)$
4	If the required material product is not produced in the sphere of production, then natural resources are spent on its acquisition.	$\overline{MK1}(z) \rightarrow MR1(u)$
5	If we continue to waste natural resources, there is no perspective for the creation of the "next" material production	$MR1(u) \rightarrow \overline{MS2}(y)$
6	Material production at the next stage is based on material production at the previous stage.	$MS1(y) \rightarrow MS2(y)$

We open the informative form of the predicate symbols presented here:

x - people (specialists, creative professions), y - production, z - products of activity, u - resources, MS1 - material producer of the present time, MS2 - material producer of the subsequent time, MK1 - material product of the present time, MK2 - material product of the subsequent time, MR1 is the material resources of the present time, MR2 is the material resources of the subsequent time, DT1 is the creative potential of the spiritual producer of the present time.

For example, consider the situation around the existing standing with material production. In some communities, the possibility of comfortable existence without own material production is theoretically justified and practically realized, when the activities of the working-age population are focused on the management of financial flows from the sale or exchange of resources.

Opponents of this approach believe that the lack of their own material production in the future will lead not only to a decrease in the efficiency of spiritual production but also to a drop in the level of consumption of material products.

Without going into a philosophical discussion about the correctness of those and other judgments, we will try to describe the situation formally, using the predicate apparatus, using the symbols and relations given in Table 1.

We describe at a substantive level one of the possible tasks:

1. Having material production, one can create a material product that meets the needs of people;
2. If there is no material production in the community, then the community spends natural resources to purchase a tangible product;

IOP Conf. Series: Materials Science and Engineering **589** (2019) 012024 doi:10.1088/1757-899X/589/1/012024

3. Without material production, the community cannot count on the replenishment of creative potential;
4. Having no creative potential, the community stops working further on the creation of material production;
5. Spent resources do not allow in the future to invest in the creation of new material production;
6. If the strategy to create your own product is chosen, then the following production is created after the existing production;
7. The absence in the future of new material production cuts off the possibility of creating a new material production capable of meeting the needs of people.

To prove the verity, we apply an effective method of derivation, which is based on the proof of the contrariety of the negation of the original formula. It is important that such procedures are applicable to the formulas of the predicate logic, presented in a standard form when any matrix of a formula can be represented as a conjunction of a finite number of disjunction of atoms or their negations [21].

In this regard, statements (1 ÷ 7) are given in a form convenient for formal proof (table 2).

Table 2.

№	Formalizationofstatements	Standardform
1	$MS1(y) \rightarrow MK1(z)$	$\overline{MS1} \vee MK1$
2	$\overline{MS1}(y) \rightarrow MR1(u)$	$MS1 \vee MR1$
3	$\overline{MS1}(y) \rightarrow \overline{DT1}(x)$	$MS1 \vee \overline{DT1}$
4	$\overline{DT1}(x) \rightarrow \overline{MS2}(y)$	$DT1 \vee \overline{MS2}$
5	$MR1(u) \rightarrow \overline{MS2}(y)$	$\overline{MR1} \vee \overline{MS2}$
6	$MS1(y) \rightarrow MS2(y)$	$\overline{MS1} \vee MS2$
7	$\overline{MS2}(y) \rightarrow \overline{MK2}(z)$	$MS2 \vee \overline{MK2}$
8 Conclusion		$MK1 \wedge \overline{MK2}$

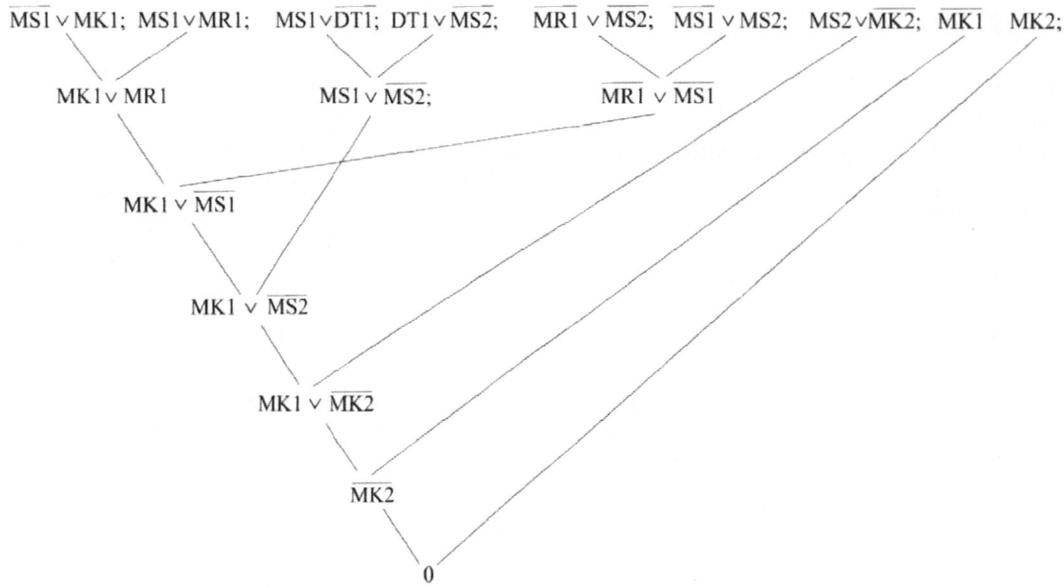

Fig.2. Output tree

The diagram (Fig. 2) shows how new pairs (resolvents) are formed from the initial set of disjuncts sequentially by crossing out contradictory ones. For example, in pairs ($\overline{MS1} \vee MK1$) and ($MS1 \vee MR1$), MS1 and $\overline{MS1}$ are contradictory, and the new disjunct is obtained from the remaining i.e. $MK1 \vee MR1$.

Getting an empty disjunct at the end of the procedure, as shown in fig. 2, indicates that the withdrawal process is completed and a refutation of the formula ($\overline{MK1}$; MK2) is found. Consequently, it can be concluded that under given conditions there is MK1 and there is no MK2.

At a substantive level, the conclusion is as follows: if there is no material production in the community, then the material needs can be satisfied for a certain time due to the consumption of natural resources. However, after a while, such satisfaction will cease both due to its exhaustion, and because the community will not remain in a sufficient amount of creative potential for the organization of new material production.

This conclusion may seem obvious to many experts in the field of economy, but the formal proof held here is necessary for the following reasons:

- the verity of the above statement is not obvious to everyone, otherwise, it is difficult to explain the adherence of some economists to dogmas that lead to the curtailment of industrial production; wherein the formal proof is the highest degree of credibility;
- acquaintance of practitioners with the methods of formal proof of production statements will help them to solve practical production problems more confidently.

Findings

The relationship of material production with other facets of human life has long been recognized but is not always taken into account in the strategy and tactics of managing community development.

In particular, the lack of understanding of the close relationship between material production and spiritual production leads to the fact that decisions on material production are sometimes made on the basis of the sole goal - providing society with the necessary material products. Since the satisfaction of material, needs can be ensured and at the expense of the consumption of natural resources, in practice, such a policy is also built.

If even in the community the potential of natural resources is great, it turns out that the absence of own material production leads to a fall in the creative potential that can start creating new production in the future.

Arguments and considerations in favor of the development of material production are not always convincing for decision makers. In this regard, it is useful to formalize task solving as the highest form of persuasiveness.

The findings of this paper, supported by formal evidence, indicate that material production not only affects all aspects of life but is also a determining principal link.

References

[1] H. Braun und M. Riedel (Hg.). Natur und Geschichte. Karl Lowithzum 70. Geburtstag. Stuttgart, 1967. pp. 132—155.

[2] How Automation And Key Performance Indicators (KPI's) Contribute To Sustainable Development In The Mining Industry Juliana Parreira, Zoe Mullard, John Meech, Magaly Garcia Vasquez Second International Conference on Multinational Enterprises and Sustainable Development, Nancy-Mets, France 2009

[3] David H. Autor Why Are There Still So Many Jobs? The History and Future of Workplace Automation. – Journal of Economic Perspectives. –Volume 29, Number 3. – Summer 2015. – pp. 3-30

[4] Robots at Work Georg Graetz, Uppsala University, Guy Michaels, London School of Economics - February 27, 2015

[5] Robots and industrialization in developing countries. – UNCTAD policy brief, 2016. – UNCTAD/PRESS/PB/2016/6 (No. 50)

[6] Osipkov, V., Ksenevich, T., Belousov, B., Karasev, O. et al., "Intelligent Transport Systems: Revolutionary Threats and Evolutionary Solutions," SAE Technical Paper 2016-01-0157, 2016, https://doi.org/10.4271/2016-01-0157.

[7] Chernikov, B.V., Antonchikov, S.N. Modeling of adaptive organization (2017) Proceedings of 2017 10th International Conference Management of Large-Scale System Development, MLSD 2017, paper № 8109605.

[8] Cf. Gerald Epstein and Dorothy Power, "The Return of Finance and Finance's Returns: Recent Trends in Rentier Incomes in OECD Countries, 1960-2000, Research Brief, Political Economy Research Institute (University of Massachusetts Amherst). – no. 2, November 2002.

[9] Vida Vanchan Rachel Mulhall John Bryson. Repatriation or Reshoring of Manufacturing to the U.S. and UK: Dynamics and Global Production Networks or from Here to There and Back Again. – Growth and Change, Vol. 49, Issue 1, March 2018, pp. 97-121. Accessmode: https://doi.org/10.1111/grow.12224 (dateofappeal 26.10.2018).

[10] K. Lynette James, Dave Barlow, Anne Bithell, Sarah Hiom, Sue Lord, Pat Oakley, Mike Pollard, Dave Roberts, Cheryl Way, Cate Whittlesea The impact of automation on pharmacy staff experience of workplace stressors. – International Journal of Pharmacy Practice, Volume 21, Issue 2, pages 105–116, April 2013. – DOI: 10.1111/j.2042-7174.2012.00231.x

[11] Akberdina Victoria V., Tretyakova Oksana V., Vlasov Andrey I. A Methodological Approach To Forecasting Spatial Distribution Of Workplaces In An Industrial Metropolis. Problems and Perspectives in Management, 2017, 15(4), pp. 50-61.

[12] Vertakova, Y., Larionov, V., Kondratova, E. Accounting of environmental imperatives in the public regulation of sustainable energetics development (2017) Proceedings of the 29th International Business Information Management Association Conference - Education Excellence and Innovation Management through Vision 2020: From Regional Development Sustainability to Global Economic Growth, pp. 1982-1994.

[13] Daron Acemoglu, PascualRestrepo Robots and Jobs: Evidence from US Labor Markets. – Working Paper Series (National Bureau of Economic Research). – no. w23285. Access mode: http://www.nber.org/papers/w23285 (date of appeal 26.10.2018).

[14] Mike Davis, Prisoners of the American Dream. Politics and Economy in the History of the US Working Class. - London: Verso, 1986

[15] Naomi Klein, The Rise of Disaster Capitalism. Rebuilding is no longer the primary purpose of the reconstruction industry. – The Nation, 14 April 2005, Access mode: https://www.thenation.com/paper/rise-disaster-capitalism (date of appeal 26.10.2018).

[16] Tim Di Muzio, “The Art of Colonisation: Capitalising Sovereign Power and the Ongoing Nature of Primitive Accumulation” – New Political Economy, 12(4). – 2007, pp.519, 531-2

[17] Venda Valery F., Venda Yuri V. Dynamics in ergonomics, psychology, and decisions: Introduction to Ergodynamics. – Norwood, NJ.:Ablex Publishing Corporation, 1995. – 503 p.

[18] Larionova A.A., Zaitseva N.A., Fadeev A.S., Zhenzhebir V.N., Filatov V.V., Pshava T.S. The Use of Organizational and Technological Innovations in the Process of Managerial and Engineering Personnel’s Training. Eurasian Journal of Analytical Chemistry, 2017. - Vol. 12. - N 7b. - pp.1573-1580

[19] James J. O'Rourke The Value Theory Of V. P. Tugarinov in Studies in Soviet Thought 28 (1984) 109-116. 0039-3797/84/0282-0109 © 1984 by D. Reidel Publishing Company.

[20] Merkur. Heft 243. Juli 1968. S. 591-610; Merkur. Heft 244. August 1968. pp. 682-693

[21] J. A. Robinson. A machine-oriented logic based on the resolution principle. Journal of the Association for Computing Machinery, vol. 12 (1965), pp. 23-41.

IOP Conf. Series: Materials Science and Engineering **589** (2019) 012025 doi:10.1088/1757-899X/589/1/012025

Quad bike frame dynamic load evaluation using full vehicle simulation model

D Vdovin[1,2], I Chichekin[1] and O Ryakhovsky[1]

[1]Bauman Moscow State Technical University, 5 Second Baumanskaya Street, Moscow, 105005, Russian Federation

[2]E-mail: vdovin@bmstu.com

Abstract. The paper describes a quad bike multi-body model intended to extract maximum dynamic forces acting on the vehicle frame. This complex full-vehicle model includes all essential systems such assuspension, steering, frame, power train, brakes, payload, as well as tire-road interaction model. A number of the quad bike operation events has been simulated. Time histories of the forces acting on the vehicle frame and tire contact patch have been analyzed and load cases with maximum forces have been extracted for subsequent finite-element analysis of the frame. In case the sufficient number of the events is simulated, the load cases can be used in the design optimization process, such as topology optimization, at early stages of the frame development.

1. Introduction

A quad bike frame is a load-bearing structure intended to withstand numerous load cases during vehicle ride. Strength and reliability of this particular structure could be predicted by finite element methods [1]-[3]. These strength calculations should be performed for all possible loads, which are usually dynamic. The purpose of this study is to evaluate these dynamic loads acting on a quad bike frame by using full-vehicle simulation model capable of capturing full range of environment conditions in which the quad bike is operated. The more ride situations are simulated in virtual tests, the higher the reliability of the quad bike and of its frame can be guaranteed.

Another purpose of dynamic load evaluation models is to use the mat the early stage of the frame design process. Modern methods, like topology optimization, provide calculation of the beneficial and reliable load paths for the frame structure only when all major loads are known in advance.

2. Quad bike multi-body dynamic model description

The quad bike and its dynamic model are shown in Figure 1. The model consists of front and rear suspensions, steering, power train, driver and passenger, front and rear payloads, load-bearing frame. The model has been built under the following assumptions: all the parts of the model are rigid; joints are frictionless; roadbed is rigid; driver, passenger and payload are not moving relative to the quad bike during ride. The quad bike model was built in an all-purpose multi-body dynamics software [4]-[9].

The models of the suspensions and steering system are shown in figure 2. The front and rear suspensions are double wishbone suspensions with telescopic hydraulic shock absorbers and springs. The rear suspension is equipped with an anti-roll bar [10]-[12].

IOP Conf. Series: Materials Science and Engineering **589** (2019) 012025 doi:10.1088/1757-899X/589/1/012025

Figure 1. Quad bike (left) and its multi-body dynamic model with driver, passenger and payloads (right)

Power train model shown in figure 3 consists of a power unit and trans mission. The power unit includes an engine, a CVT (continuously variable transmission) and a transfer case. The CVT, transfer case gear ratios and engine's torque-speed curve are defined in the model by means of an external DLL file.

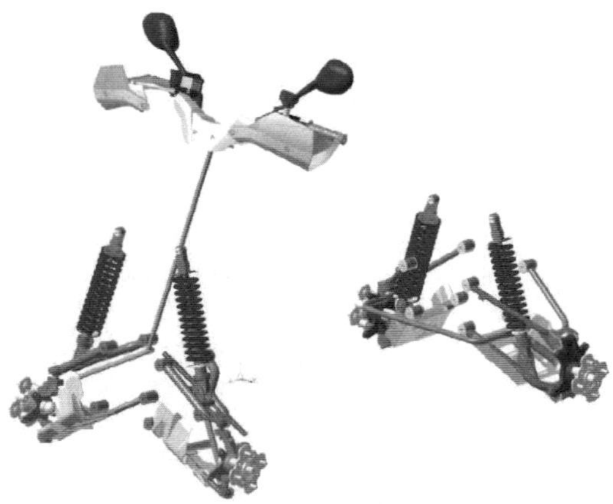

Figure 2. Front and rear suspensions model with steering model

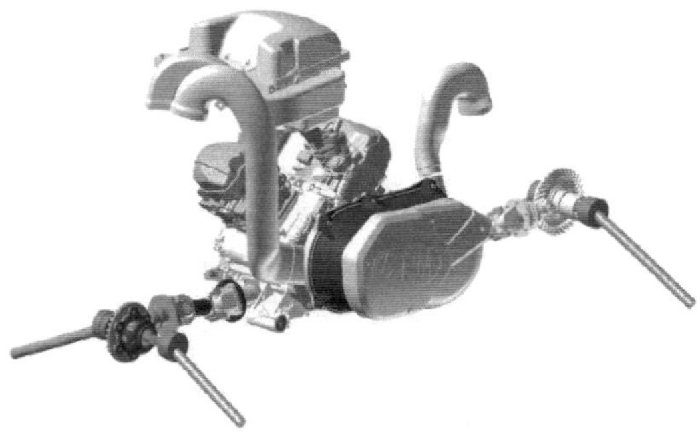

Figure 3. Power unit and transmission model

The engine torque is calculated as a function of the transmission rotation speed and accelerator pedal rate and is applied to the front and rear final drives, modelled as rigid bodies with gear coupling ratios. Final drives models also include a locking differential. Torque-speed curves for the power unit and CVT unit are shown in figure 4.

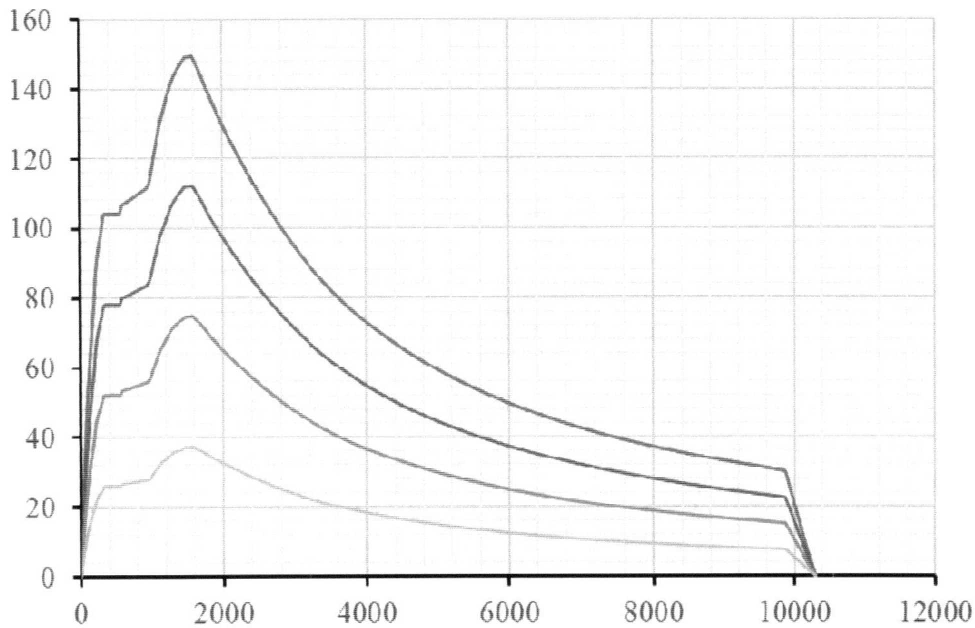

Figure 4. Torque vs rotation speed curvesat different accelerator rates for power unit coupled with CVT

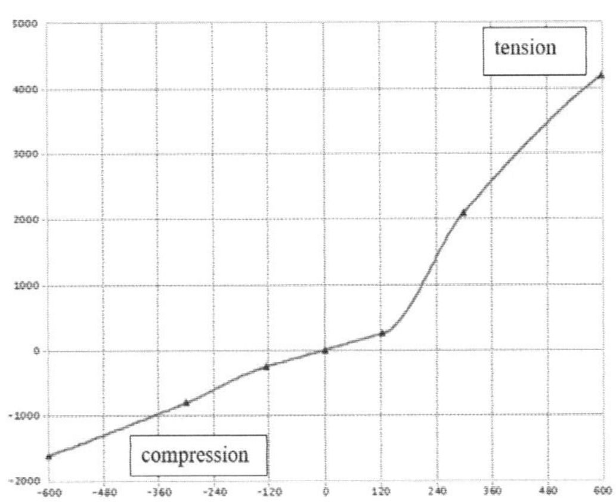

Figure 5. Damping cuvre, [N-sec/mm vs mm/s] for suspension

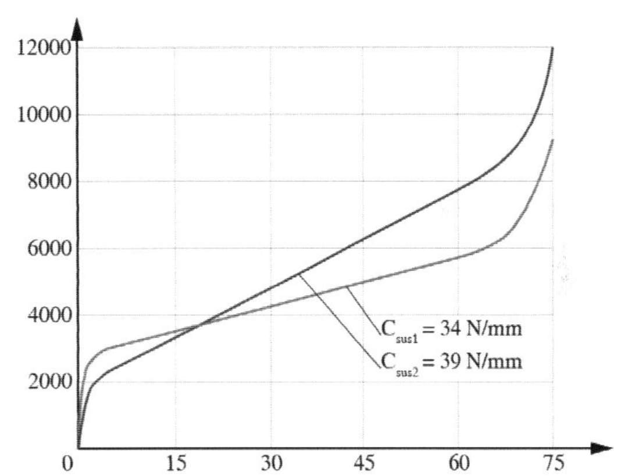

Figure 6. Spring stiffness curves, [N/mm vs mm] for front (Csus1)and rear (Csus2) suspensions

Suspension-frame joints are modelled using standard bushing elements with no friction or damping. Stiffnesses of the bushings areshownin table 1.

Table 1. Suspension-frame joints bushings stiffness

Linearstiffness, N/mm		Angular stiffness, N*mm/rad	
Radial	Axial	Bending	Torsion
10000	1000	1000	0

Tire-road interaction model choice is very important. After analysis of numerous types of these models [13]-[19]we have chosenPacejka's MF-Tire interaction model described in [20], [21], [22]. Its reaction forces are shown in figure 4.

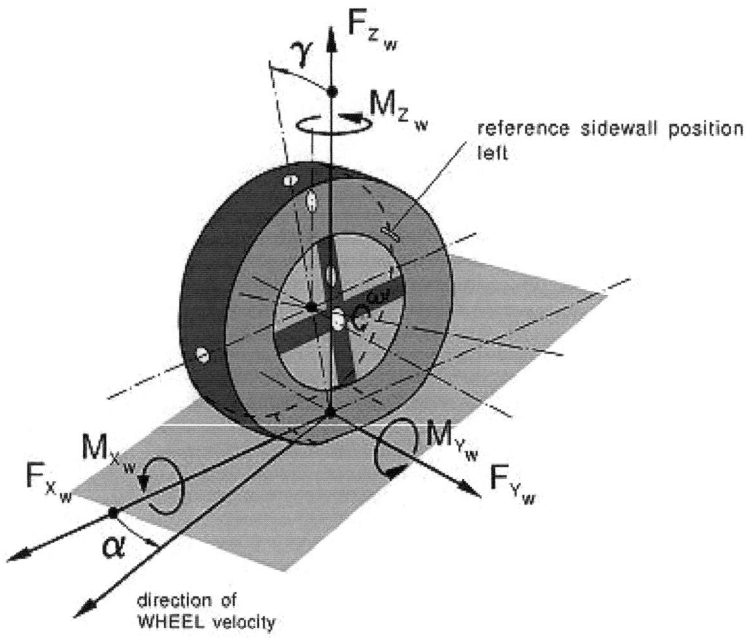

Figure 7. Tire-road interaction model

Dynamic loads acting on the quad bike frame are evaluated for the following events:

1. Static equilibriumon a flat road (a reference loadcase);
2. Maximum straight-line accelerationforward and subsequent emergency braking;
3. Moving over a 30 deg ramp;
4. Moving on a road with a side slope;
5. Cross-axling;
6. Vertical loading with acceleration 3g (jump simulation);
7. Winching;
8. Towing;
9. Bumping intoan obstacle;
10. Loadingthe quad bike frame sidestep by the combined weight of the driver and passenger.

The list of events considered as loadcases could be extended asthe experience ofthequadbike operation in different conditions is being gained.

3. Simulation results

Every simulated event listed above produces dynamic force time history in the contact tire patch, suspension, frame, steering joints, etc [22]. For instance, cross-axlingshown in figure 8 results in the tire contact patch force time histories presented in figure 9. These graphs can be used to determine maximum reactions acting on the wheels of the vehicle during the event. Corresponding force acting on the frame in the frame-strut joint of the front left suspension at the same timeframe as in figure 9 is shown in figure 10. Figure 11 zooms this graphs for the time period from 29 to 31 secduringwhichthe cross-axlingtakes place. As one can see, at a time t=29.84 sec, the front left and rear right wheels are lifted (their contact patch forces are set to zero), and the other two wheels are under increased load.

Figure 8. Cross-axling simulation at a **time = 29.84 s**. Blue arrows showreaction forces in the tire contact patch, red arrows show drive torques apllied to the wheels.

This combination of the contact patch reactions corresponds to a quite heavy loadcase for the vehicle, at which its frame is simultaneously being bended in longitudinal direction and twisted. Analysis of the force time histories depicted in figure 8 and 9 shows that one event can produce several loadcases: loads acting on the frame and at different jointshave multiple peaks at different time moments. Therefore, the whole cross-axling event should be represented as several loadcases, at whichthe worst combination of the rear and front wheel vertical, longitudinal and transversal forces with the corresponding gravity and inertia forces acting on the whole vehicle takes place.

Besides the maximum loads that determine ultimate strength of the quadbike frame, it is also possible to simulate regular vehicle rides. Figures 11 and 12 presentan uneven road cruising event and the time history of the contact patch and strut-frame joint forces during this ride. These time histories can be used for further fatigue calculations of the vehicle frame.

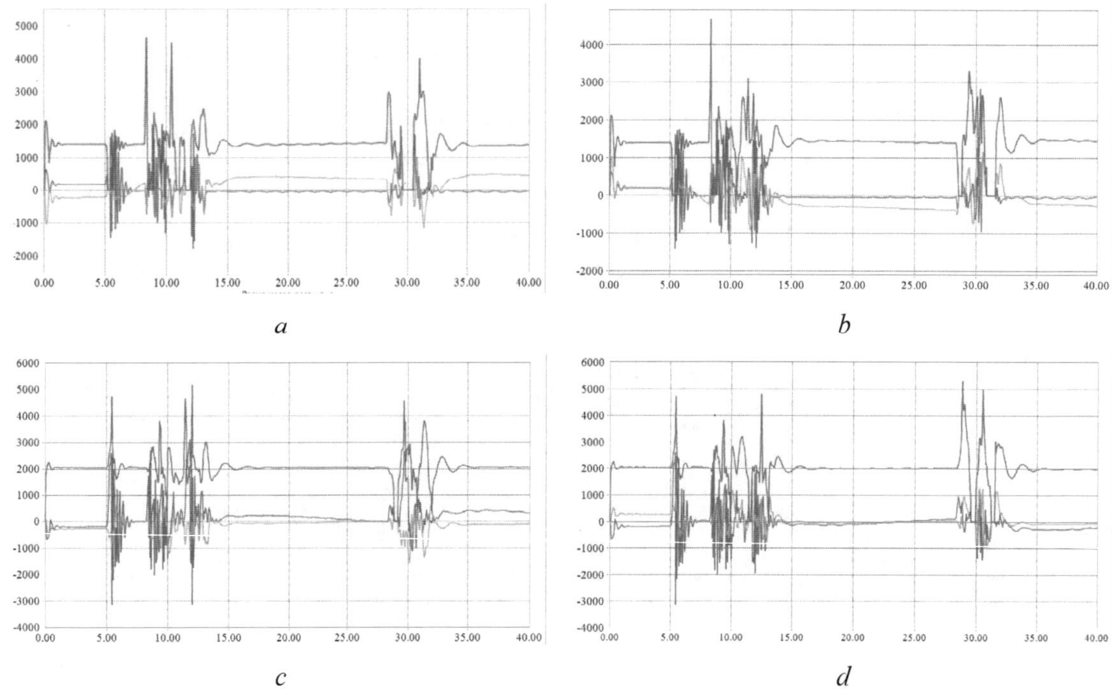

a *b*

c *d*

Figure 9. Tire contact patch forces:
a – front left wheel, b – front right wheel, c – rear left wheel, d – rear right wheel
– Fx – traction force, – Fy – lateral force, –Fz – vertical reaction

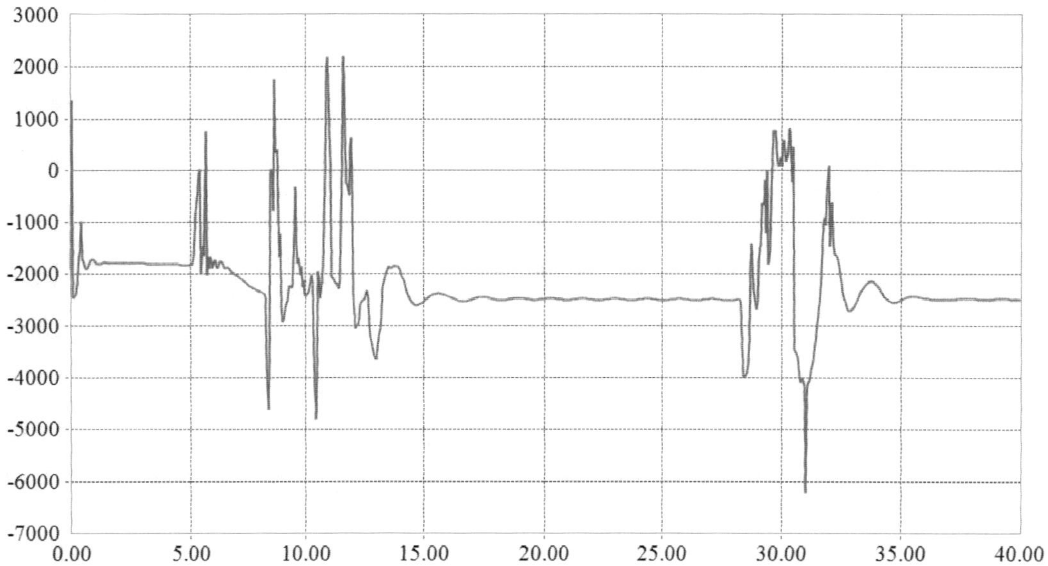

Figure 10. Force magnitude vs time in the frame-strut joint of the front left suspension

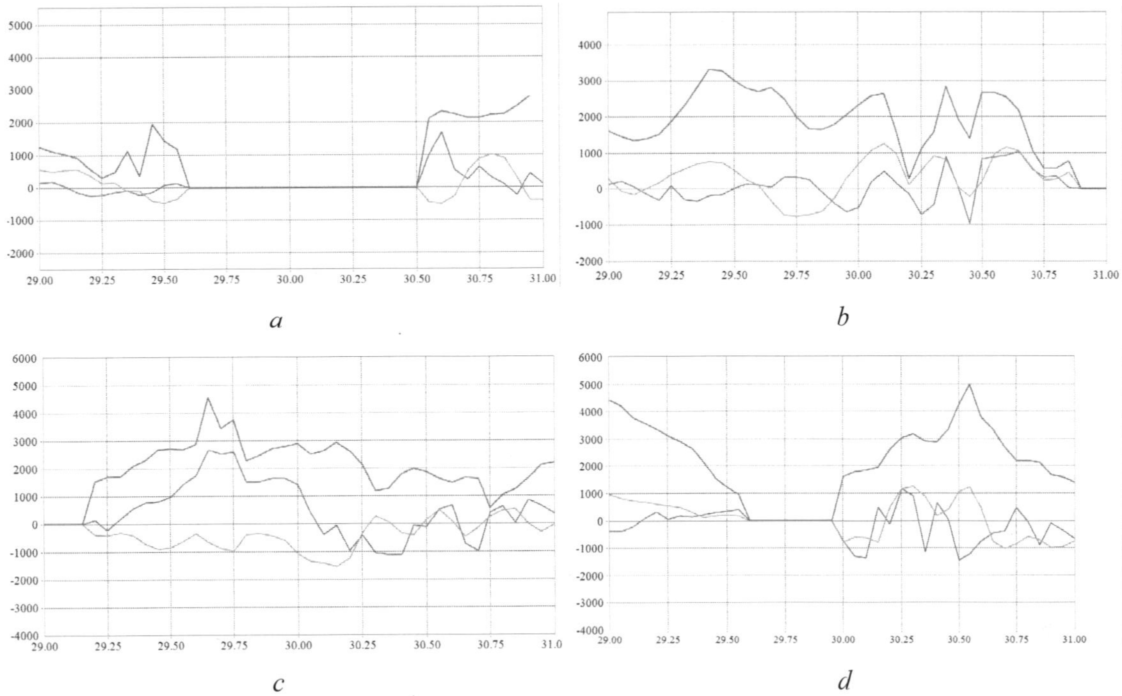

Figure 11. Tire contact patch forces during cross-axling:
a – front left wheel, b – front right wheel, c – rear left wheel, d – rear right wheel – Fx – traction force,– Fy – lateral force, –Fz – vertical reaction

Figure 12. Uneven road cruising event at a time = 81.5 sec

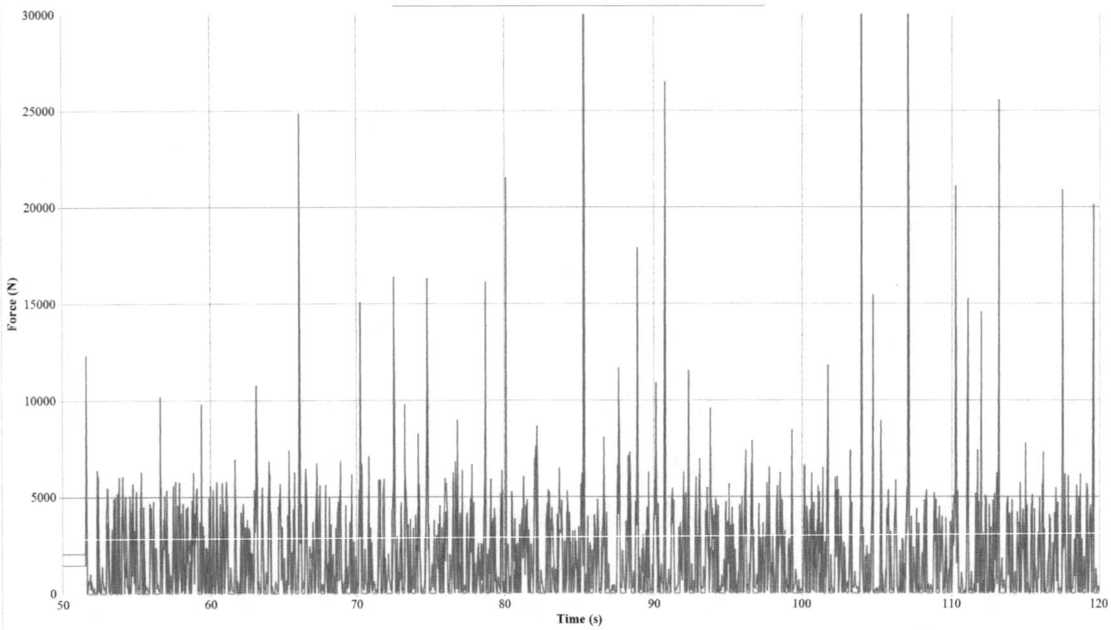

Figure 13. Forces duringuneven road cruising event:
–Fz – vertical reaction at wheel contact patch, –Fz – strut-frame joint force

4. Conclusions and discussion

Analysis of the time historiesof the forces in the tire contact patch and suspension-frame joints allows detection of the quasi-static loadcases with maximum forces during the event. This maximum loads can be transferred to finite element analysis software for the frame strength estimation. In practice, it is convenient to use inertia relief technique and special scripts to transfer loads from multi-body model to finite element model, as described in [3]. This script-based automation allows performing stress analysis of many loadcases without tremendous time-consuming handwork. Calculated loads can also be used in design optimization process, like topology optimization, at early stages of the frame development.

It is also handy to use full vehicle dynamic model to determine the time histories for the forces acting on the vehicle and its frame during regular ride for further fatigue analysis.

The simulation technique described in this paper produces vast amount of data containing time histories of the forces acting on the vehicle frame in every joint during an event. Moreover, the number of events can be also quite big consequently multiplying the amount of data. Automated analysis of this data for extracting maximum loads acting on the vehicle frame presents a problem which has not been fully solved by now.

Acknowledgments

It should be noted that this work was carried out at the Bauman Moscow State Technical University, with financial support from the government in the face of the Russian Ministry of Education under the project: №14.577.21.0272. (Identification number: RFMEFI57717X0272).

References

[1] KotievG O, Padalkin B V, Kartashov A B, Diakov A S 2017. ARPN J. of Eng. and Appl. Sci. **12** 1064-71

[2] Vdovin DS 2014 Society of Automotive Engineers, Moscow State University of Mechanical Engineering (MAMI), p. 2–6

[3] Lin Liao A 2011 *IOP Conf*Denver, Colorado***52*** *1-10.*

[4] Ryan RR 1993**22** 144–152.
[5] Thomas D Gillespie February 1992.
[6] Gorobtsov AS, Shurygin VA, Serov VA, Dyakov AS, Lapteva VO, Makarov AA 2014 Gruzovik **11** 2–5
[7] Gorelov VA, Komissarov AI, Miroshnichenko AV 2015. Procedia Engineering.**129**. 300-307
[8] Keller AV, Gorelov VA, Vdovin DS, 2015 Universitat Politecnica de Catalunya Barcelona; Spain, 1285-1296.
[9] Kong YS, Abdullah S, Omar MZ, Haris SM 2016. **78**. 6 85-90.
[10] Levenkov YaYu, Chichekin IV. 2016 Inzhenernyy vestnik **12** 4.
[11] Boby George, Akhil T Benny, Albert John, Aswin Jose, Denny Francis. 2016 **7**
[12] VdovinD S, Chichekin I V. 2016 Procedia Engineering**150** 1276-1279.
[13] Skotnikov G I, Jileykin M M and Komissarov A.I 2018. IOP Conf. Series: Materials Science and Engineering **315** 012027
[14] Levenkov YaYu, Vol'skaya NS 2015 Technology of Wheeled and Tracked Machines**1** 20–26
[15] Mike Dempsey, Garron Fish, Juan Gabriel Delgado Beltran, 2015 Proceedings of the 11th International Modelica Conference Versailles, France*IOP Conf. Serie*. 273 – 280.
[16] Volskaya N S, Zhileykin M M and Zakharov A Y 2018. IOP Conf. Series: Materials Science and Engineering 315 012028
[17] Vol'skaya NS, Levenkov YaYu, Rusanov OA,2013 Nauka i obrazovanie: nauchnoe izdanie MGTU im. N.E. Bamumana **5** 107–124
[18] Gladov GI, Popov SD, Sarkisov PI, 2015, Proceedings of the 24th Symposium of the International Association for Vehicle System Dynamics (IAVSD 2015) 749-758.
[19] Mike Dempsey, Garron Fish, Juan Gabriel Delgado Beltran, 2015 Proceedings of the 11th International Modelica Conference 273 – 280.
[20] Pacejka, H.B.: "Tyre and Vehicle Dynamics", Second Edition, Butterworth-Heinemann, Oxford, 2006.
[21] Pacejka, H.B., I.J.M. Besselink: "Magic Formula Tyre model with Transient Properties", Supplement to Vehicle System Dynamics, Vol. 27, pp. 234-249, 1997.
[22] TNO Automotive: "MF-Tool 6.1 User Manual", TNO Automotive, The Netherlands, 2008.

IOP Conf. Series: Materials Science and Engineering **589** (2019) 012026 doi:10.1088/1757-899X/589/1/012026

Light frame design for quad bike using topology optimization

D Vdovin[1,2], Y Levenkov[1] and V Chichekin[1]

[1]Bauman Moscow State Technical University, 5 Second Baumanskaya Street, Moscow, 105005, Russian Federation

[2]E- mail: vdovin@bmstu.ru

Abstract. The paper describes a quad bike frame design procedure with the use of topology optimization technique. The design space specific properties, quad bike specific load cases and problem formulation for topology optimization are presented. Topology optimization results – beneficial load paths of the frame structure – have been interpreted into a new frame design providing sufficient space for placement of all systems of the vehicle. The new frame strength has been proved using finite element method. Comparing with the previous steel frame design, the new frame of the quad bike has become 2 times as stiff and 17% lighter while having low stress levels due to the substitution of aluminum material for steel and a new optimized topology of the frame load-bearing structure.

1. Introduction

Weight reduction is an actual problem in mobile transport. Lower vehicle weight leads to lower fuel consumption, hence lower air pollution and operating costs, reduced material and energy amount needed for production, and, possibly, lower production and utilizing costs. Lightweight vehicles also have better off road capabilities and floatability on snow and swamps, amphibious vehicles gain better buoyancy and less drag coefficient on water.

Topology optimization is a promising technique, that could be used to reduce weight of the load-bearing automotive structures, like frames and bodies, powertrain cases and housings, suspension parts, etc. [1-3]. Theory of topology optimization and solutions of the basic simple problems are presented in [4-5].

This paper describes a quad bike lightweight frame design process with the use of topology optimization, see Figure 1. The frame is the main load-bearing structure of the quad bike intended to withstand all loads acting on the vehicle during its ride. All major systems like the powertrain, the fuel tank, the suspension with wheels, payload sub-frames, driver and passenger seats are connected to the frame by means of joints and brackets.

There are a number of similar works, dedicated to topology optimization of regular and electric bike supporting frames [5-6], where authors mostly use 2d design space to determine beneficial load paths of the bike frames or its separate parts. In work [7] a 3d design space was used to calculate the load-bearing structure of an electric bike with respect to the non-design space needed for the placement of the electric battery inside the frame. Polaris, a manufacturer of snowmobiles, used topology optimization to reduce weight of snowmobile chassis components up to 40%.

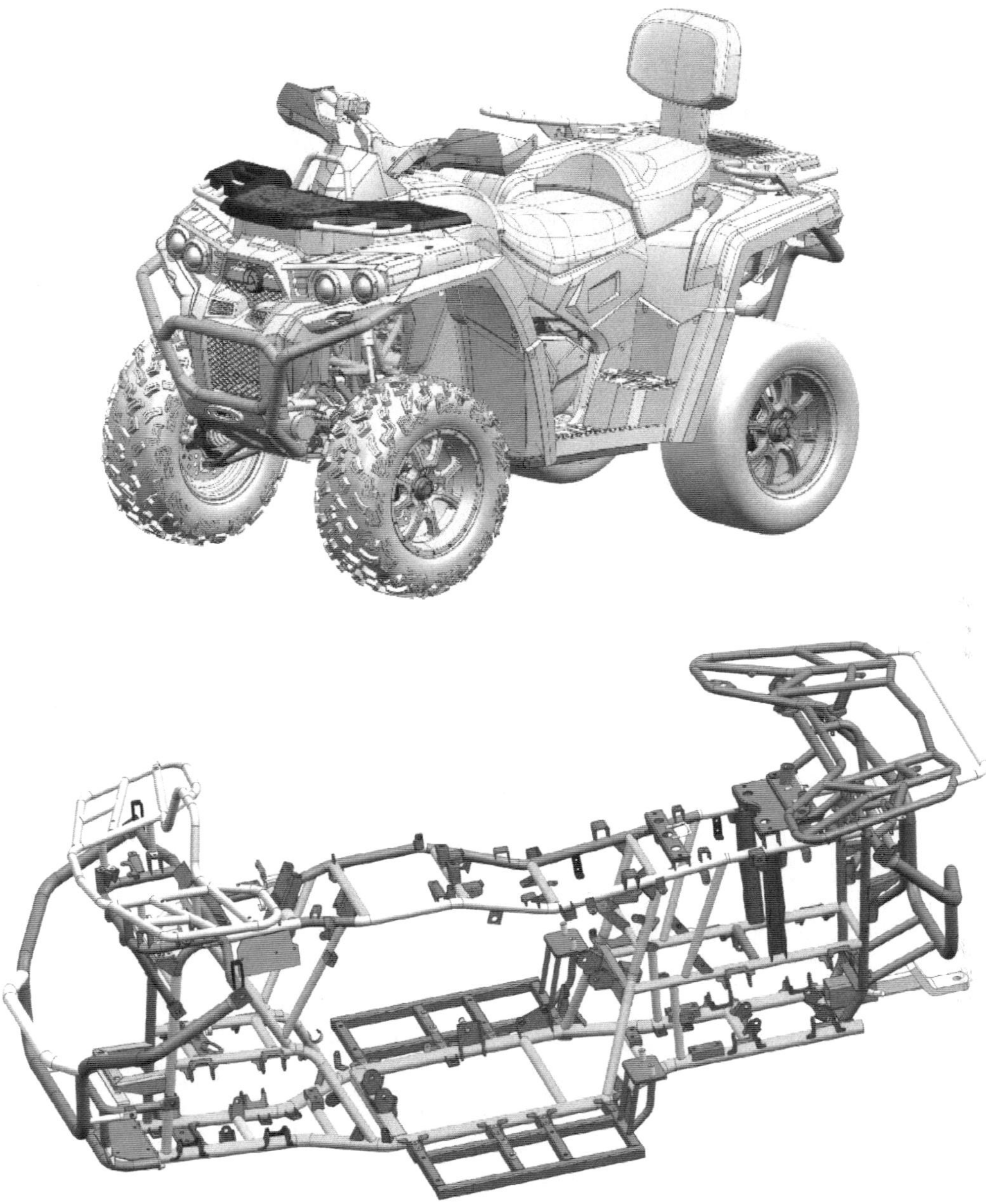

Figure 1. Quad bike CAD-model and its steel frame

There are three main issues of the topology optimization process: 1) building the appropriate design space; 2) define reliable set of loads which would cover all possible events during the ride of the vehicle; 3) results interpretation into a technologically feasible structural design of the frame.

Generation of the design space for the frame of the quad bike is a problem closely related to the major packaging task of the vehicles systems. A designer should place all the systems, like the powertrain, the suspension, the braking system, the payload and the driver/passenger, etc., according to all their requirements in one 3d-space. Since all this major parts of the vehicle are placed in the 3d-space without intersections, it is possible to define a 3d-figure of the design space for the frame.

The set of the defined load cases for the vehicle frame will determine the reliability of the frame. If some of the vehicle ride events during load calculations are not taken into account, its frame built with

IOP Conf. Series: Materials Science and Engineering **589** (2019) 012026 doi:10.1088/1757-899X/589/1/012026

topology optimization probably will fail. In this work, the appropriate set of load cases is taken from the previous work, where loads are calculated using a full vehicle multi-body dynamic model. It is convenient to use inertia relief technique to transfer loads from the multi-body dynamic model to the finite element analysis model, as described in [8-9]. Dynamic models for the full vehicle calculations must be suitable for reliable loads estimation and can be quite complex, including dozen of sub-models like tire-road interaction, analytical and semi-analytical powertrain and braking systems and so on [10-30].

2. Model description

Figure 2shows the design space built for the quad bike. The design space has been built only for the major central load-bearing structure without the forward and rear sub-frame for the payload. The sidesteps for the driver and the passenger have also been excluded from the design process. The joints and brackets where the forces from other systems of the vehicle are acting on the frame have been defined as a non-design space. The standard 8-node hexagonal finite element having a size of 4 mm was used to mesh the 3d-figure of the design and non-design space. The overall number of the nodes was 221003, the number of the elements was 187179. The design space figure has been defined with the internal hollows for the powertrain system (power unit with cooling, fuel injection system, exhaust system, CVT transmission, transfer case, final gears and driveshafts), fuel tank, steering, winch, etc. External bounds of the design space have been defined to avoid intersection of the frame with moving suspension parts and steerable wheels, and provide ergonomics of the driver and passenger.

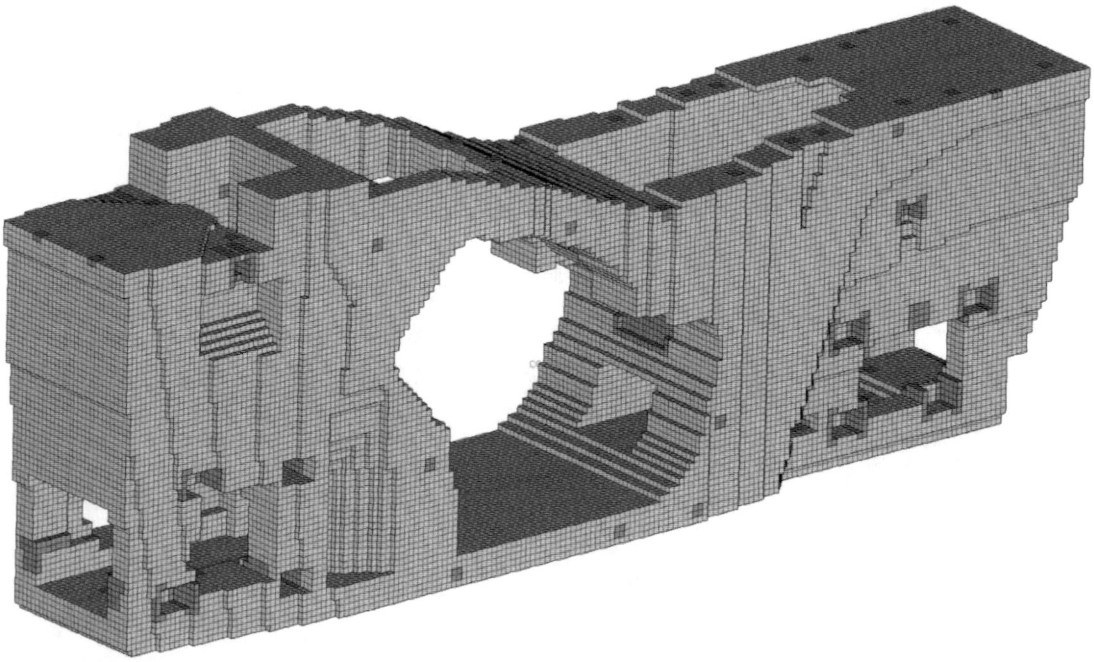

Figure 2. Design space for the quad bike load-bearing frame with hollows for internal systems of the quad bike

1. Load cases taken from the full vehicle multi-body dynamic model represent a number of balanced concentrated loads and moments acting on the frame in the non-design space joints and brackets. In this work, the following list of load cases has been defined for topology optimization:
2. Static equilibrium on a flat road (a reference load case);
3. Maximum straight-line acceleration forward and subsequent emergency braking;
4. Moving over a 30 deg ramp;

5. Moving on a road with a side slope;
6. Cross-axling;
7. Vertical loading with acceleration 3g (jump simulation);
8. Winching;
9. Towing;
10. Bumping into an obstacle;
11. Loading the quad bike frame sidestep by the combined weight of the driver and passenger.

Figure 3shows the system of balanced forces and moments acting on the design space during vertical loading with acceleration 3g. Inertial loads acting at the gravity centers of massive parts have been redistributed between the joints and brackets of the non-design space using special RBE3 elements.

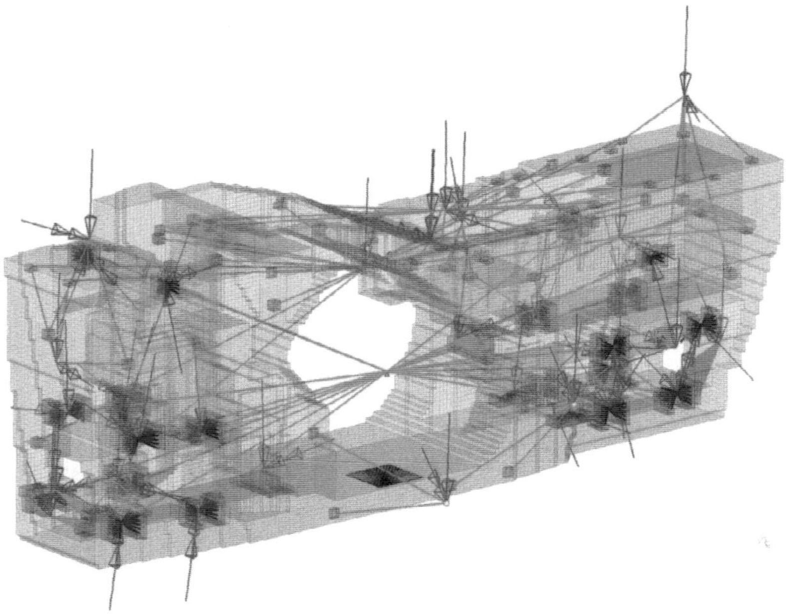

Figure 3. Load case "vertical loading with acceleration 3g". The arrows represent a balanced system of concentrated inertial forces, reactions, and moments acting on the design space

The objective function during topology optimization was defined as a structure minimum compliance for all linear static load cases. Maximum amount of the design space volume fraction was constrained by 15%. Mechanical properties of the design space were defined as linear elastic material with elastic modulus
E = 72000 MPa and Poisson's ratio nu=0,33 (aluminum).

3. Optimization results and its interpretation

Load paths calculated by the topology optimization were visualized as a 0.3 density isosurface dividing the design space into needed and unnecessary material, see **Figure 4.**

Next step in the design process of the frame is technology selection and appropriate interpretation of the found load paths into technologically feasible structure. Aluminum 1560 alloy 30x2 mm tubes and their welding have been chosen as a primary technology. Highly loaded parts of the frame near the suspension joints could be manufactured as solid brackets using milling process at the stage of prototyping or using aluminum casting in serial production. Interpretation also includes a detailed design of the joint and bracket zones and packaging check of the frame in the full vehicle 3d-model presented in figureFigure5.

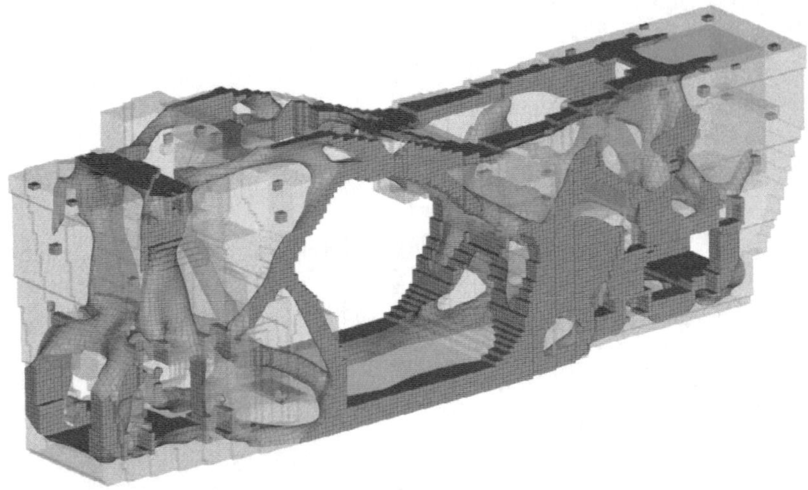

Figure 4. Topology optimization results for quad bike frame: isosurface for the density with the value of 0,3

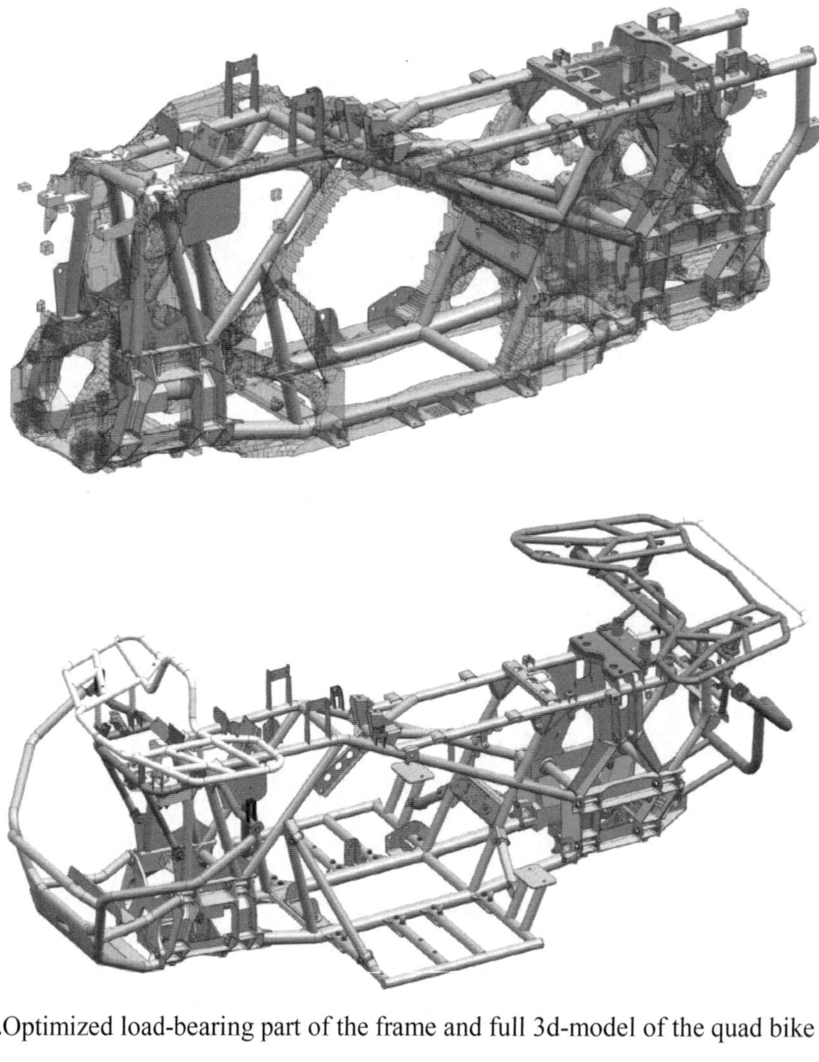

Figure 5. Optimized load-bearing part of the frame and full 3d-model of the quad bike frame

IOP Conf. Series: Materials Science and Engineering **589** (2019) 012026 doi:10.1088/1757-899X/589/1/012026

4. Strength check analysis for optimized quad bike frame

Mechanical properties of the aluminum 1560 tubes and brackets are presented inTable 1.

Table 1. Aluminum 1560 mechanical properties

Alloy	Elastic modulus, MPa	Poisson's ratio	Yield stress, MPa	Stress limit, MPa	Elongation at brake, %	Near welding zone stress limit, MPa
Aluminum 1560	72000	0,3	160	320	16	224

Welding of aluminum tubes was tested on tension samples, shown in figure 6, left. In figure 6 (on the right), corresponding finite element analysis results of the same sample presented. Test results have shown that the strength of the welded tubes is at a level of 70% from their initial strength, so the allowed stress level in the near-welding zones has become lower accordingly.

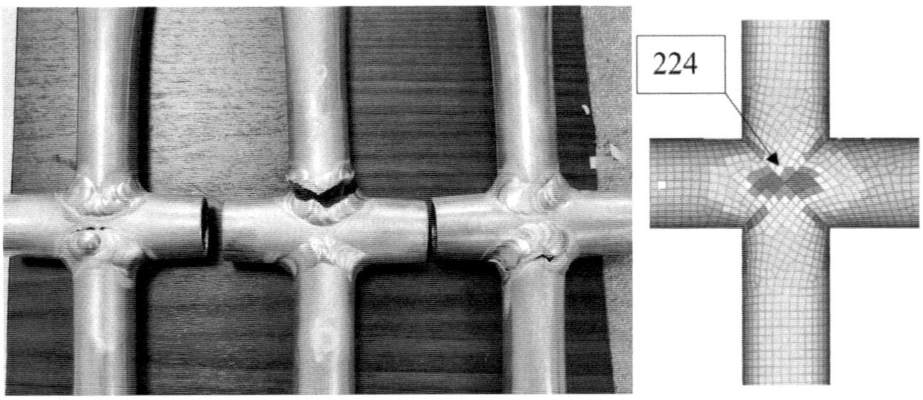

Figure 6. Samples of the welded Aluminum 1560 alloy tubes after tension tests (left) and corresponding von Mises stress distribution, MPa, calculated by the finite element method (right)

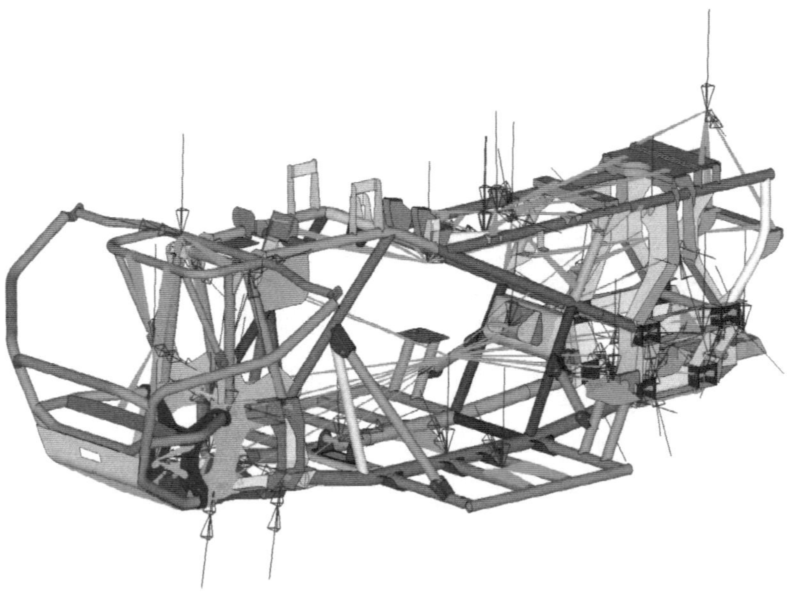

Figure 7. Shell finite element model of the optimized frame and balanced system of loads for the vertical loading with acceleration 3g

The standard finite element method was used for the stress and strain calculation of the new quad bike frame. The finite element model built from the shell-type elements has been elaborated as shown inFigure 7. The total number of the elements is 558572, the number of the nodes is 781318.

For the all previously defined load cases, linear static analysis was performed, and maximum stresses were checked by the allowed stress level for the material. Von Mises stresses at cross-axling event are presented inFigure 8. Von Mises stress envelope for all load cases presented in Figure 9. Maximum stresses, mostly concentrated in the tubes intersection zones (near-welding zones) are not exceeding allowed stresses, so the strength of these zones is guaranteed. All the load cases were also checked for linear buckling modes with the resultant minimum safety factor of 3,7.

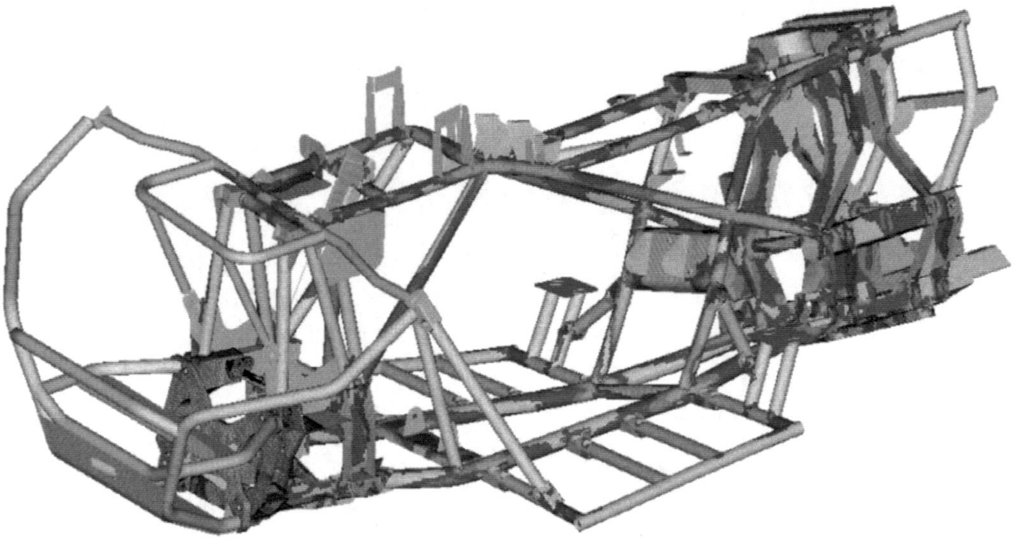

Figure 8. Von Mises stress distribution and deformed shape of the frame at cross-axling event

Figure 9. Envelope of von Mises stresses, MPa, for all loadcases

5. Results and discussion

Overall bending and torsional stiffness of the optimized aluminum quad bike frame is 2.5 times as much than the one of the initial steel frame. Weight reduction of the optimized load-bearing structure (central part of the frame) is 15% (the weight has dropped from 40 kg to 34 kg). Overall weight reduction of the full quad bike frame is 17%, due to additional substitute of aluminum for steel in the bulbar and winch brackets.

Stress state analysis of the optimized frame made of regular aluminum tubes shows that additional weight reduction is possible. Due to low stress level at tubes zones located far from welding, butted tubes[31]could provide significant weight reduction. Reengineering of the payload sub-frame, possibly with the use of plastic or composite material, can also save mass. Tubes made of carbon fiber is also a promising solution, but it requires careful technologically reasonable design of the joints in the tube intersection zones.

6. Acknowledgments

It should be noted that this work was carried out at the Bauman Moscow State Technical University, with financial support from the government in the face of the Russian Ministry of Education under the project: №14.577.21.0272. (Identification number: RFMEFI57717X0272).

References

[1] Vdovin D S, Kotiev G O 2014. *Tractors and agricultural machinery* **8** 33
[2] Vdovin D S, Shabolin M L 2016. *Izvestiya MGTU "MAMI"* **4** 90
[3] Bendsoe M P, Kikuchi N 1988. Generating optimal topologies in structural design using a homogenization method.*Computer Methods in Applied Mechanics and Engineering*. 1988**71** 197
[4] Bendsoe M P, Sigmund O 2003.*Topology Optimization: Theory, Methods, and Applications*(Springer Science & Business Media) p 370
[5] Zeleny P, Cadek M 2015.*MM Science Journal* **October** 696-700
[6] Denghong Xiao, Xiandong Liu, Wenhua Du, Junyuan Wang, Tian He 2012. *Struct. Multidisc.Optim.* **46** 913
[7] ManiosS E, LagarosN D, NassiopoulosE 2019.*Frontiers in Built Environment* **5** Art 34
[8] Vdovin D, Chichekin I 2016.*Procedia Engineering***150** 1276-79.
[9] Vdovin D S, Chichekin I V,Levenkov Y Y 2018.*Trudy NAMI***1** 36
[10] Hashemi E, Pirani M, Khajepour A, Kasaiezadeh A2016. *Vehicle System Dynamics* **54** 1736-61
[11] Wei Y, Liu Y, Li X, Oertel C2016.*Vehicle System Dynamics***54** 463-73
[12] Taheri M, Ahmadian M2016.*Vehicle System Dynamics***54** 653-66
[13] Rodriguez J, Freeman PT, Wagner J, Pidgeon P, Alexander K2016.*Int. J. of Automotive Technol.***17** 71-81
[14] Xia X, Xiong L, Sun K, Yu ZP2016.*Int. J. of Automotive Technol.***17**991-1002
[15] Diakov AS, Kotiev GO 2018. *MATEC Web of Conf.***224**02096
[16] Evseev K B, Kartashov A B, Dashtiev I Z and Pozdeev A V 2018. *MATEC Web of Conf.***224**02039
[17] Kotiev GO, Padalkin BV, KartashovAB,Diakov AS 2017.*ARPN J. of Eng. and Appl. Sci.***12** 1064-71.
[18] Sarach EB, Kotiev GO, Beketov SA 2018. *MATEC Web of Conf.***224** 04009
[19] KlubnichkinVE,Diakov AS, Klubnichkin EE, Zakharov AY, Vakhidov U Sh, Suchenina AS and Basmanov IV 2019. J. of Phys.: Conf. Series**1177** 012048
[20] Volskaya NS, Zhileykin MM and Zakharov AY 2018.*IOP Conf. Series: Materials Science and Engineering* **315** 012028
[21] Klubnichkin EE, Klubnichkin VE, Kotiev GO 2018. *IOP Conf. Series: Materials Science and Engineering* **386** 012025

[22] Zhileykin M M, Kotiev G O, Nagatsev M V 2018. *IOP Conf. Series: Materials Science and Engineering* **315** 012031
[23] Kotiev GO, Butarovich DO, Kositsyn BB 2018. *IOP Conf. Series: Materials Science and Engineering* **315** 012014
[24] Skotnikov GI, JileykinMM and Komissarov A.I 2018 *IOP Conf. Series: Materials Science and Engineering* **315** 012027
[25] GorelovVA, Komissarov AI 2016 *Procedia Engineering* **150** 1322-28.
[26] Gorelov VA, Komissarov AI, Miroshnichenko AV 2015.*Procedia Engineering* **129** 300-7
[27] Keller A V, Gorelov V A, Anchukov V V 2015. *Procedia Engineering***129** 280-7.
[28] Ejsmont J, Taryma S, Ronowski G, Swieczko-Zurek B2016. *Int. J. of Automotive Technol.* **17**237
[29] Pacejka, HB 2006.*Tyre and Vehicle Dynamics. Second Edition*(Oxford: Butterworth-Heinemann)
[30] Pacejka, HB, BesselinkIJM 1997.*Supplement to Vehicle System Dynamics* **27** 234
[31] CovillD, DrouetJ.-M 2018.*Proceedings* **2** 216

IOP Conf. Series: Materials Science and Engineering **589** (2019) 012027 doi:10.1088/1757-899X/589/1/012027

Mathematical model of interaction of a rolling elastic wheel with elastic-visco-plastic support base taking into account the history of loading

M Zhileykin[1], A Zakharov[1] and M Panshin[1,2]

[1]Bauman Moscow State Technical University, 5 Second Baumanskaya Street, Moscow, 105005, Russian Federation

[2]E-mail: panshin.m.v@bmstu.ru

Abstract. One of the main methods to predict permeability is a simulation of the motion of wheeled vehicles in different operating conditions, which are based on mathematical models of interaction between elastic tires with deformable irregularities of the support base. Currently in the theoretical study of the interaction of propellers with the ground formed two main directions: analytical method involving a mathematical description of the process under study and the finite element method based on computer simulation. The aim of this study is to develop a mathematical model of rolling elastic wheels on deformable rough reference base, taking into account the deformation of the contact patch at each end of the elementary playground. A mathematical model of the curvilinear rolling of the elastic wheel on an uneven elastic-visco-plastic support base is developed, taking into account the deformation of the contact spot in each finite elementary site; the influence of ground hooks on the parameters of the wheel movement; coupling properties of the support base in the horizontal direction under the action of a vertical load; change in the direction of radial and tangential reactions; elastic-visco-plastic properties of the support base;" history " of loading of each elementary platform of the support base; condition of the support base on its physical and mechanical characteristics.

1. Introduction

One of the main direction of economical growth in Russia remains to be a speedy development of north and northeast regions that are the constituents of the 60 percent of the country territory.

The further development of these territories requires new methods and technologies for solving transport and technological problems when off-road transportation of cargoes and people is conducting. In this way the movement of vehicle is difficult and impossible in some cases.

The existing models of wheeled, tracked and rotary screw vehicles don't correspond to present functional, efficiency, reliability and ecological requirements for wheeled propulsion devices when operating in north regions of the country where supporting surfaces are weak-bearing. In this way there are technical, economic and social need in creating and using vehicles with pneumatic wheeled propulsion devices including ultra-low pressure devices that satisfy the requirements.

It should be noted that our country doesn't have necessary collection of energy efficient off-road vehicles currently. The existing off-road machines performed according to old traditional schemes and produced serially by industry don't meet the requirements defining the efficiency and environmental

IOP Conf. Series: Materials Science and Engineering **589** (2019) 012027 doi:10.1088/1757-899X/589/1/012027

friendliness of wheeled propulsion devices in difficult climatic operating conditions. In current situation the problem of wheeled vehicles patency prediction of movement over supporting surface with weak-bearing properties is relevant [1,2].

One of the fundamental methods of patency prediction is simulation method of wheeled vehicle movement in different operational conditions that is based on mathematical models of interaction the elastic tire with deformable support base unevenness.

Established [3, 4] that the traction patency wheeled vehicle mutually affect both the deformation properties of tire and physical and mechanical characteristics of soil: normal deflection of the tire and the depth gauge; the change in the reference area of the contact patch depending on load and air pressure in the tire; the presence of hysteresis losses in the material of the tire that affect rolling resistance caused by friction in the contact patch on the ground; creating a tangent reactions of the soil over the entire area of contact.

Along with the definition of the dependencies to describe "load-deformation of the ground " and "load-deformation mover", not less important task is the selection of modelreset vertical deformations of soil in contact with the pneumatic tire. On the correct choice of the type and kind of approximating dependence is largely determined by both qualitative and quantitative aspects of General solution of the problem of contact interaction of elastic mover with deformable ground [5].

Currently in the theoretical study of the interaction of propellers with the ground formed two main directions: analytical method involving a mathematical description of the process under consideration [6 – 8] and finite element method (FEM), based on computer simulation [9 – 13]. In the study of the interaction of wheel elements with the soil, an analytical method has found wide application. Developed mathematical models of the interaction of the tyre with the ground allow to solve various tasks. These models are used as in the study of processes of interaction of a single wheel mover with the soil array and at research of dynamic models of mobile machines operating in specific road-soil (RSC) and climatic conditions. One of the most significant shortcomings of these models is the description of the interaction of the wheel with a smooth deformable base, while the real profile of the support surface has a significant height of irregularities is comparable with the radius of the wheel.

The study of the processes occurring in the soil mass under the influence of the wheel mover, the finite element method are relatively new, currently the largest application. This method is better than others provided numerical procedures for the study of mathematical models of objects. The most important advantage is the availability of implicit methods of integration of systems of differential equations. The application of this method is the most accurate to describe the process of interaction of wheel elements with the soil, to determine the stresses in the soil mass deformation of tire and soil compaction. In contrast to analytical methods form the contact patch of the tyre with an elastic support surface is the result of the simulation taking into account the independent soil characteristics and the constructive and operational parameters of the engine. However, a significant disadvantage of this method is its high computational complexity, therefore, with currently available computers to use these models as part of the overall model of the motion of multi-wheeled vehicles difficult.

The aim of this study is to develop a mathematical model of rolling elastic wheels on deformable rough reference base, taking into account the deformation of the contact patch.

2. Mathematical model of deformable support base

The solution of problems associated with the action of short-term loads on the deformable support base is based on model representations of the properties of the continuous medium on which the load acts [14]. One of the approaches in solving this problem is to replace the real spatial system "load-deformable half-space" with an idealized mechanical system, the parameters of which correspond to the physical and mechanical characteristics of the support base [15,16]. Analysis of various computational schemes and mathematical models of support bases showed that to determine their complete and residual deformations under the action of short-term load it is advisable to use a mechanical system with one degree of freedom, the movement of which is determined by the model of a visco-elastic medium devoid of inertia (figure 1).

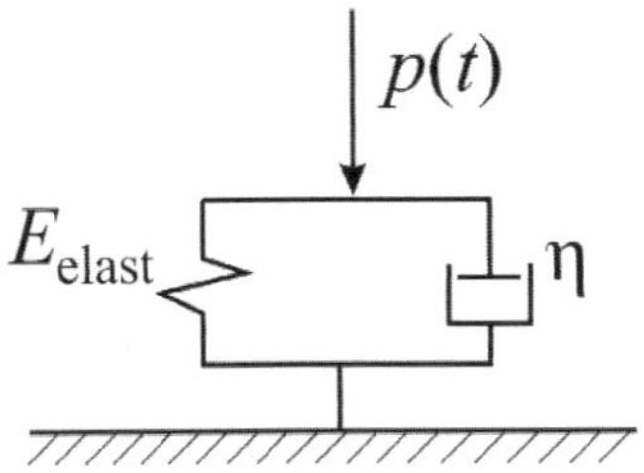

Figure 1. A design scheme of a deformable support base: E_{elast} – elastic modulus, η – the coefficient of viscous friction, $p(t)$ – pressure.

As presented in figure 1, the design scheme in parallel with a spring with elasticity modulus E_{elast} connected to the damper with viscosity η.

The differential equation that determines the behavior of the support base under uniaxial compression, according to the Kelvin-Voigt model, has the form

$$\dot{h}_{gr} = \frac{h_{gr}}{\eta} + \frac{p(t)}{\eta E_{\text{elast}}}, \tag{1}$$

where h_{gr} – current deformation (draft) of the support base.

The diagram of the ultimate deformation h_{gri} of the *i*-th platform of the support base is established on the basis of the Bernstein-Letoshnev formula [9]]

$$p_{0zi} = c h_{gri}^{\mu}, \text{ MPa}, \tag{2}$$

Where p_{0zi} – limiting pressure of the i-th platform of the support base at the draft h_{gri}; c – the coefficient of deformation of the soil; μ– the density of the soil.

The meaning of the strain limit diagram is as follows: at $p_i(t) \le p_{0zi}$ of soil sediment does not occur.

When re-increasing the load, when the pressure on the i-th support platform of the support base $p_i(t) > p_{0zi}$, the soil sediment h_{gr} begins to grow according to equation (1).

In connection with the non-linear character of the dependence (2) the modulus of elasticity $E_{\text{elast}i}$ of the *i*-th platform of the support base is a variable value $E_{\text{elast}i} = \frac{dp_{0zi}}{dh_{gri}} = \mu c h_{gri}^{\mu-1}$. Increasing the modulus of elasticity $E_{\text{elast}i}$ with increasing precipitation h_{gri} allows to take into account the effect of soil compaction.

3. Mathematical model of interaction between elastic tires with the irregularities of deformed support base

The proposed model uses two different coordinate systems (figure 2) due to the structure and form of the equations of motion of the object.

First, the stationary coordinate system (SCS)$O_2X_2Y_2Z_2$, is used to simulate the specified ground traffic conditions. The origin of the system of point O_2 coincides with the beginning of the simulated track.

To determine the forces acting on the car from the ground we introduce a micro mobile coordinate system (MMCS) under which we understand the system $O_TX_TY_TZ_T$, the center of which O_T coincides with the geometric center of rotation, the axis of O_TX_T coincides with the projection of the longitudinal axis of symmetry of the wheel on the support surface.

Key assumptions:

1) The normal pressure in the contact spot is distributed proportionally to the deflection of the tire on each elementary section of the contact platform.

2) The force of interaction of the wheel with its support base directed in the opposite direction from the slip speed.

Consider the scheme of movement of the wheel in figure 2.

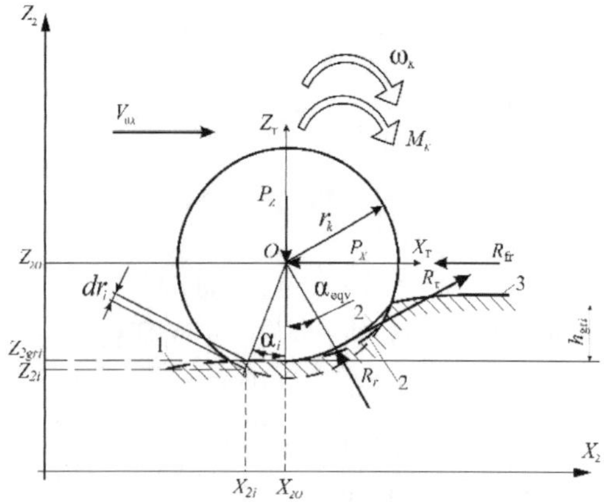

Figure 2. A design scheme of wheel rolling over uneven sub-base:
1 – strain profile of soil; 2 – deformable soil profile; 3 – the undeformed profile of the soil; O – the center of the wheel; Z_{2O}, X_{2O} –the coordinates of the center wheel SCS; $\omega_к$ – the angular velocity of rotation of the wheel; $M_к$ – torque applied to the wheel; V_{0x} – longitudinal velocity of the wheel center; P_Z, P_X– vertical and longitudinal forces applied to the centre of the wheel from the wheeled vehicle axis; R_r, R_τ – radial and tangential projection of the wheel interaction reaction with the support base; R_{fr}– wheel rolling drag force; $r_к$ – free radius of the wheel; Z_{2gri}, X_{2i} – the coordinates of a point under the support base of the i-th point of the undeformed profile wheels; α_i – the angle between the vertical axis and the i-th point of the undeformed profile wheels; α_{eqv} – the equivalent angle of the application point of the total reaction; dr_i – radial deflection of the tyre for the i-th point of the undeformed wheel profile; h_{gri} – the rut depth under the i-th point of the undeformed profile wheels.

For model development we use the results of [9, 17-20]. On the lower semicircle of the undeformed profile of the wheel will select a certain number of points, n, the position of which will determine the angle α_i between the vertical line dropped from the center of the wheel on the axis X_2, and a ray connecting a point of the undeformed profile of the wheel with its center (figure 2). The number of points is chosen based on a compromise between accuracy of the model and its performance. Define the coordinates X_{2i} and Y_{2i} of selected points of the profile at the SCS (figure 3).

$$X_{2i} = X_{20} + r_k \sin\alpha_i \cos(\theta + \beta);$$
$$Y_{2i} = Y_{20} + r_k \sin\alpha_i \sin(\theta + \beta);$$
$$-\frac{\pi}{2} \le \alpha_i \le \frac{\pi}{2};$$

whereX_{20}, Y_{20} are the coordinates of the center of the wheel O SCS.

The vertical coordinate $Z_{2i}i$-th point of the undeformed profile of the wheel in the SCS, we define by the formula

$$Z_{2i} = Z_{20} - r_k \cos\alpha_i,$$

whereZ_{20} is the vertical coordinate of the wheel center at the SCS.

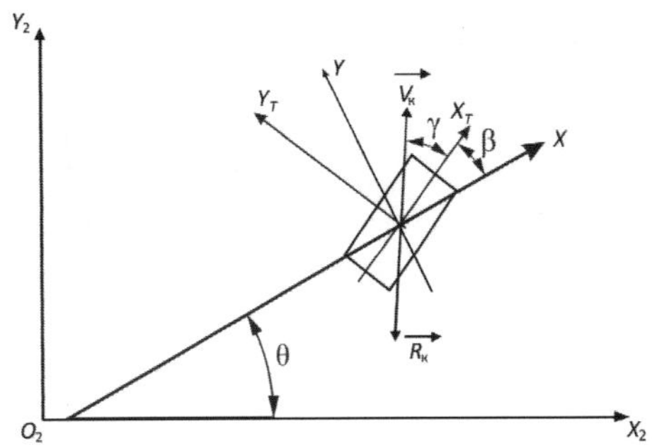

Figure 3. Design scheme for determining the force of interaction between the wheel and the support surface: $\vec{V}_K$ – the vector of linear speed of the wheel center; $\vec{R}_K$ – reaction vector of the wheel interaction with the support base; θ – the angle between the longitudinal axis of the machine and the axis X_2 SCS; β – the rotation angle of the steering wheel; γ – the angle between the plane of symmetry of the wheel and the vector $\vec{V}_K$

The formation of the longitudinal profile Z_{2gri} is carried out according to the following algorithm

$$Z_{2gri} = Z_{2gri}^{undef} - h_{gri},$$

where h_{gri} — vertical draft of soil under the i-th elementary platform of the contact spot; Z_{2gri}^{undef} – the vertical coordinate of the undeformed profile of the gauge (simulated in advance by a known method described, for example, in [19]).

The d_{ri} deflection of the tire in the radial direction for the$i$-th point of the undeformed profile is determined from the following ratios

$$dr_i = \begin{cases} 0, \ Z_{2gri} \le Z_i \\ (Z_{2gri} - Z_{2i})\cos\alpha_i, \ Z_{2gri} > Z_{2i} \end{cases},$$

whereZ_{2gri} – vertical coordinate of the profile of the support base for the i-th point of the wheel.

Thus, to determine the reactions of interaction of wheels with the supporting surface Rx and R_z in MMCS in the presence of several areas of "overlap" profile substructure undeformed contour of the wheel it is necessary to determine the equivalent angle α_{eqv} the point of application of the total radial reaction R_r in the radial direction and the tangential reaction R_τ (figure 2).

Define α_{eqv} as a weighted average value

$$\alpha_{eqv} = \frac{\sum_{i=1}^{n} \alpha_i dr_i}{\sum_{i=1}^{n} dr_i}.$$

Radial reaction R_r is the sum of two components: elastic R_{ry} and damping R_{rd}: $R_r = R_{ry} + R_{rd}$.R_{ry} depends on the equivalent deflection of the tire

$$dr_{eqv} = \frac{\sum_{i=1}^{n} dr_i}{n_k},$$

wheren_k is the number of points in the undeformed profile in contact with the support surface.

R_{rd} depends on the speed of the deflection of the tire in the radial direction. We define the projection of velocity of points of the contour of the wheel on the axis X_T and Z_T:

IOP Conf. Series: Materials Science and Engineering **589** (2019) 012027 doi:10.1088/1757-899X/589/1/012027

$$V_{iX_T} = \omega_k (r_k - dr_i)\cos\alpha_i + V_{0X_T};$$
$$V_{iZ_T} = \omega_k (r_k - dr_i)\sin\alpha_i + V_{0Z_T},$$

whereω_k – angular velocity of rotation of the wheel; $V_{0XT}$and V_{0ZT} – projection of the velocity vector of the wheel center (point O) on the axis X_T and Z_T respectively.

The vector of linear speed of the i-th point of the undeformed profile of the wheel in the radial direction

$$V_{ri} = V_{iX_T}\sin\alpha_i + V_{iZ_T}\cos\alpha_i.$$

The speed of profile deformation of the i-th point in the radial direction

$$\frac{d}{dt}(dr_i) = \frac{dZ_{2gri}}{dt}\cos\alpha_i - V_{ri}.$$

Equivalent speed of deflection

$$\frac{dr_{\text{э}kb}}{dt} = \frac{\sum_{i=1}^{n}\left(\frac{d}{dt}(dr_i)\right)}{n_k}.$$

Further, knowing the elastic and damping characteristics of the tire in the radial direction, we find R_r.

Determination of the parameters of the contact spot and the magnitude of ground subsidence h_{gr} given the "history" of its loading

The area of the contact patch F_{ti} (figure 4) define by the formula

$$F_{ti} = 2\int_{\alpha_{i-1}}^{\alpha_i} b(\alpha)\left[r_k - dr(\alpha)\right]d\alpha,$$

$$b(\alpha) = b_t\left(1 + \frac{dr(\alpha)}{r_k}\right) \qquad (3)$$

where b_t – the width of the tyre.

The discrete analogue of the formula (3) can be written as

$$F_{ti} = b_t\left(2 + \frac{dr_{i+1} + dr_i}{r_K}\right)\left(r_K - \frac{dr_i + dr_{i-1}}{2}\right)(\alpha_{i+1} - \alpha_i)$$

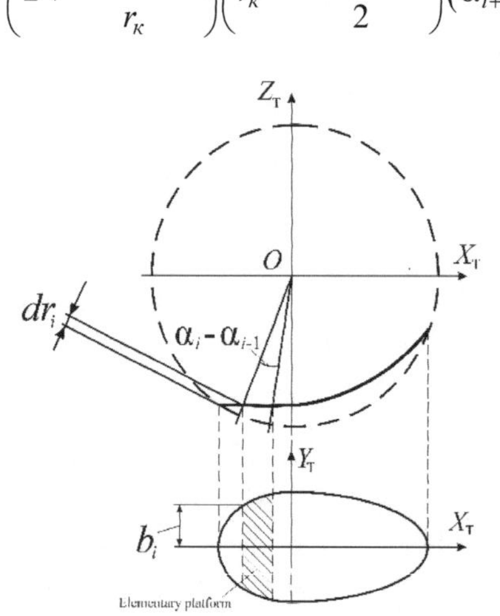

Figure 4. Design scheme for determining the contact spot area.

Determine the normal pressure p_{zi} in the i-th elementary area of the contact spot. Each such platform (figure 5) is formed by two planes passing through the axis of rotation of the wheel and rotated relative to each other at an angle $\alpha_i - \alpha_{i-1}$.

$$p_{zi} = \frac{P_z}{F_{ti} \cdot 10^6} \frac{dr_i}{\sum_{i=1}^{n} dr_i} cos\alpha_i, \text{ MPa}, \qquad (4)$$

where P_z – vertical wheel load.

Then the draught of each elementary i-th site is calculated using equation (1).

The contact conditions of the tyre with the support base must be met:

a) if $h_{gri} > dr_i$, then $h_{gri} = dr_i$, which means that the punching of the soil in each elementary area of the contact spot can not be more than the elastic deflection of the tire in this area;

b) if the cumulative sediment of the soil on the i-th elementary site of the contact spot $h_{gr\Sigma i}$, calculated by formula (1), allows to provide such its bearing capacity, which can withstand the pressure (4), the condition for further precipitation will be $\frac{P_z}{F_{ti}} \frac{dr_i}{\sum_{i=1}^{n} dr_i} cos\alpha_i > ch_{gr\Sigma i}^{\mu} 10^6$;

c) if $Z_{2gri} < Z_{20} - r_\kappa cos\alpha_i$, then $h_{\text{гpi}} = 0$, which means that the soil sediment inthe i-th support area can not go beyond the contour of the undeformed tire.

The total length of the midline contact spot L can be defined as the sum of the lengths of the elementary pads L_i

$$L = \sum_{i=1}^{n_k} L_i, \; L_i = \left(r_\kappa - \frac{dr_i + dr_{i-1}}{2} \right) (\alpha_{i+1} - \alpha_i)$$

Influence of lugs on the parameters of the wheel movement

In the presence of the cleats must also calculate the shear forces in the areas of the projections and depressions, as well as the removal of soil from the zone of contact in case of intense slipping of the wheel.

Additional vertical penetration dh_{vp} the center of the wheel in the ground caused by the excavation of soil from the area of contact is calculated by the formula [21]

$$dh_{vp} = \frac{t_{lt} h_{gro} S_k}{t_{gro} (1 - S_k)}, \; t_{gro} = \frac{2\pi r_k}{n_{gro}},$$

where t_{lt} – length trench lug; h_{gro} — grouser height; t_{gro} — step lug; n_{gro} — the total number of lugs on the tyre circumference.

Condition of removal of soil from the contact zone of the wheel with the support base

$$r_\kappa \int_{t_{j-1}}^{t_j} \omega_\kappa dt < L, \qquad (5)$$

where t_j, t_{j-1} – the current and previous time points.

If the condition (5) is not met, then $dh_{vp} = 0$.

Then the total vertical depth h_Σ wheel center

$$h_\Sigma = h_r + dh_{vp}.$$

We assume that if $h_\Sigma \geq r_k$, you lose the mobility of the machine due to the hanging of the machine body on the ground.

To determine the additional shear forces $R_{\tau j}^{gro}$ in the areas of the projections and depressions of the cleats (figure 5) use the results obtained in [20]:

$$R_{\tau j}^{gro} = F_{gro} c_s 10^6 exp\left[-\frac{\left(\left|e_x\right| - e_{xm}\right)^2}{0,05 e_{xm}} tg\varphi_s^*\right],$$

$$e_x = S(\Delta t) - \int_{t_1}^{t_2} \omega_k(t)\left(r_k - dr_{eqv}\right)dt,$$

$$\Delta t = t_2 - t_1,$$

where F_{gro} – frontal area of the projection of the lug; e_{xm} – the maximum shift of the soil in which the cohesion of soil particles is not broken; e_x – current shift of the soil; $S(\Delta t)$ is the path traversed by the center of the wheel during the time Δt; $\omega_k(t)$ is the current angular speed of rotation of the wheel.

Total tangent force $R_{\tau X_T}^{gro}$ grsv projection on the axis X_T MMSC for all for all m grouser in the zone of contact of the wheel with its support base

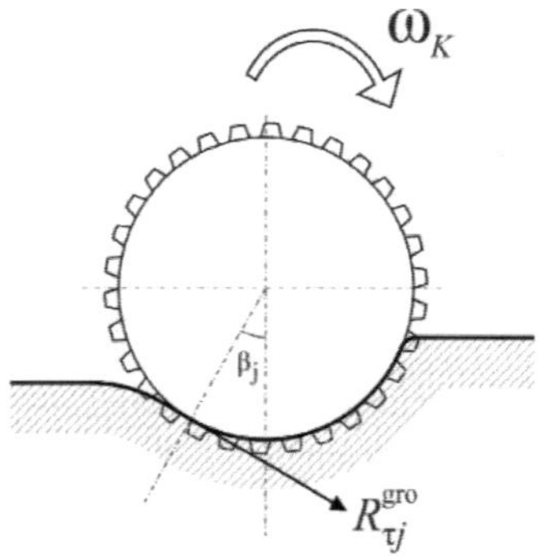

Figure5. Calculation scheme for determination of the tangential forces $R_{\tau j}^{gro}$ in the areas of the projections and depressions of the cleats.

$$R_{\tau X_T}^{gro} = \sum_{j=1}^{m} R_{\tau j}^{gro} cos\beta_j,\ m = \frac{L_1 + L_2}{2\pi r_K} n_{gro}.$$

Then the expression for determining the longitudinal reaction R_{KX_T} wheels with a support base in a projection on the axis X_T MMSC

$$R_{KX_T} = R_\tau cos\alpha_{eqv} - R_r sin\alpha_{eqv} - R_{fr} + R_{\tau X_T}^{gro}.$$

Account the bearing capacity of the Foundation soil in horizontal direction under the action of vertical loads

The magnitude of the horizontal reaction R_{KX_T} may be limited by two factors [21]: a wheel slip on the surface of the compacted soil after overcoming the forces of adhesion and loss of bearing capacity from the shear weight of the soil in the direction of action of the horizontal reaction. And if the coupling properties are accounted for by the factor μ_S, the loss of bearing capacity can be assessed by comparing the existing shear stresses τ with the maximum allowable stresses τ_{max}.

The maximum tangential stress in the contact spot is calculated using the Coulomb expression

$$\tau_{max} = p_z tg\varphi_s^* + c_s 10^6,$$

where c_s – the coefficient of cohesion (adhesion) of the soil, MPa.

The current tangent stress τdefine by the formula [20]:

$$\tau = p_{0z} tg\varphi_s^* \left[1 - exp\left(-\frac{|e_x|}{0{,}1e_{xm}}\right)\right] + c_s 10^6 exp\left[-\frac{\left(|e_x| - e_{xm}\right)^2}{0{,}05e_{xm}} tg\varphi_s^*\right],$$

Finally, the expression for R_{KX_T} takes the form

$$R_{X_T} = k_\tau \left(R_\tau cos\alpha_{eqv} - R_r sin\alpha_{eqv} - R_{fr} + R_{\tau X_T}^{gro}\right),$$

$$k_\tau = \begin{cases} 1, if\ \tau \le \tau_{max} \\ \frac{\tau_{max}}{\tau} if\ \tau > \tau_{max} \end{cases}.$$

Conclusions

A mathematical model of the curvilinear rolling of an elastic wheel on an uneven elastic-visco-plastic support base is developed:

- deformation of the contact spot in each finite element area;
- influence of lugs on the parameters of the wheel movement;
- coupling properties of the support base in the horizontal direction under vertical load;
- change of direction of radial and tangential reactions;
- elastic-visco-plastic properties of the support base;
- "history" of loading of each elementary platform of the base;
- the state of the support base on its physical and mechanical characteristics.

References

[1] Rybansky, M. The impact of terrain on cross-country mobility geographic factors and their characteristics / M. Rybansky [et. al] // 18th International Conference of the International Society for TerrainVehicle Systems, ISTVS 2014. – Seoul, 2014.

[2] Kotiev, G.O., Padalkin, B.V., Kartashov, A.B., Diakov, A.S. Designs and development of Russian scientific schools in the field of cross-country ground vehicles building (2017) ARPN Journal of Engineering and Applied Sciences, 12 (4), pp. 1064-1071.

[3] Belousov, B., Ksenevich, T.I., Vantsevich, V., Komissarov, D. 8×8 platform for studing terrain mobility and traction performance of unmanned articulated ground vehicles with steered wheels (2013) SAE Technical Papers, 9.

[4] Belousov, B.N., Shelomkov, S.A., Ksenevich, T.I., Kupreyanov, A.A. Experimental verification of a mathematical model of the wheel-supporting surface interaction during nonstationary rolling motion (2009) Journal of Machinery Manufacture and Reliability, 38 (5), pp. 501-505.

[5] Keller, A.V., Gorelov, V.A., Vdovin, D.S., Taranenko, P.A., Anchukov, V.V. Mathematical model of all-terrain truck (2015) Proceedings of the ECCOMAS Thematic Conference on Multibody Dynamics 2015, Multibody Dynamics 2015, pp. 1285-1296.

[6] Gorelov, V.A., Komissarov, A.I. Mathematical model of the straight-line rolling tire - Rigid terrain irregularities interaction (2016) Procedia Engineering, 150, pp. 1322-1328. DOI: 10.1016/j.proeng.2016.07.309

[7] Gorelov, V.A., Komissarov, A.I., Miroshnichenko, A.V. 8×8 wheeled vehicle modeling in a multibody dynamics simulation software (2015) Procedia Engineering, 129, pp. 300-307. DOI: 10.1016/j.proeng.2015.12.066.

[8] Keller, A.V., Gorelov, V.A., Anchukov, V.V. Modeling truck driveline dynamic loads at differential locking unit engagement (2015) Procedia Engineering, 129, pp. 280-287. DOI: 10.1016/j.proeng.2015.12.063.

[9] Volskaya, M.M. Zhileykin and A.Y. Zakharov. Mathematical model of rolling an elastic wheel over deformable support base. IASF-2017. IOP Conf. Series: Materials Science and Engineering 315 (2018) 012028 doi: 10.1088/1757-899X/315/1/012028.

[10] Kupreyanov, A.A., Morozov, M.V., Belousov, B.N., Ksenevich, T.I., Vantsevich, V.V. Experimental research of tire elastomer-surface tribological properties(2014) Proceedings of the ASME Design Engineering Technical Conference, 3.

[11] Klubnichkin, V.E., Klubnichkin, E.E., Kotiev, G.O., Beketov, S.A., Makarov, V.S. Interaction between elements of the track ground contacting area with the soil at curvilinear motion of the timber harvesting machine (2018) IOP Conference Series: Materials Science and Engineering, 386 (1), article № 012016.

[12] Wong, J.Y. Theory of Ground Vehicles / J.Y. Wong. – New York: Wiley IEEE, 2001. – 560 p.

[13] Sarach, E., Kotiev, G., Beketov, S. Methods for road microprofile statistical data transformation (2018) MATEC Web of Conferences, 224, article № 04009.

[14] Bekker M.G. Introduction to terrain-vehicle systems. – Ann Arbor: University of Michigan Press, 1969. – 520 p.

[15] Benes L., Hermanek P., Novak P. Tensile resistance of wheeled combine harvester. Engineering for Rural Development // MM Science Journal, Volume. - 2018, pp. 2481-2483; doi:10.17973/mmsj.2018_10_201848.

[16] Shaheb M.R., Grift T., Godwin R., Dickin E., White D., Misiewicz P. Effect of tire inflation pressure on soil properties and yield in a corn - soybean rotation for three tillage systems in the Midwestern United States. -2018. doi: 10.13031/aim.201801834.

[17] Špokas L., Adamčuk V., Bulgakov V., Nozdrovický L. The experimental research of combine harvesters. Res. Agr. Eng., 62: (2016): p. 106-112.

[18] Kaneko, T. and Kageyama, I. 2003 A study on the braking stability of articulated heavy vehicles, JSAE Review, Vol. 24, pp.157–164

[19] Shoop S., Kestler K., Haehnel R. Finite element modeling of tires on snow // Tire Science and Technology. 2006. Vol. 34, no. 1. P. 2–37.

[20] Botero J., Gobbi M., Mastinu G. A new mathematical model of the traction force in pneumatic tire snow chain systems // AssociazioneItalianaPerL'AnalisiDelleSollecitazioni / Dipartimento di Meccanica, Politecnico di Milano. 2005. 10p. www.aiasonline.org/AIAS2005/Articoli/art084.pdf.

[21] V VLarin Theory of motion of four-wheel drive wheeled vehicles: the textbook / – M.: Izd-vo MSTU im. N. Uh. Bauman 2010. – 391 c.

IOP Conf. Series: Materials Science and Engineering **589** (2019) 012028 doi:10.1088/1757-899X/589/1/012028

Comparison of various simulation methods of a two-phase flow in a multiphase pump

A Boyarshinova[1,2], V Lomakin[1] and A Petrov[1]

[1]Bauman Moscow State Technical University, 5 Second Baumanskaya Street, Moscow, 105005, Russian Federation

[2]E-mail: a.boyarsh@yandex.ru

Annotation The article presents a mathematical model of a two-phase flow in a multiphase helico-axial pump. In this paper, the model of a multiphase incompressible fluid flow (ρ = const) were used. Numerical simulation is based on solving discrete analogues of the basic hydrodynamic equations. The hydrodynamic simulation of the described mathematical model was performed in two ways: in a simplified form, using a multiphase volume of fluid model (VOF), and the segregated flow method. The results of the calculation of two-phase flows for various values of the gas content are given. The influence of gas content on the pump performance and its pressure is considered. A comparison of calculation results of two methods of hydrodynamic simulation of two-phase flow is shown.

Introduction

Multiphase flow is a flow of a mixture of substances in various aggregative states - phases. A continuous or discrete phase can be a solid, liquid or gaseous state of matter. In terms of simulation, a phase can be defined as the amount of a substance in a system that has its own physical properties in order to distinguish it from other phases in the system [1]-[2]. Two-phase flow is a special case of multiphase flow, containing only two different components. These are liquids with solid or gas inclusions, gases with liquid drops or solid particles. The structure of two-phase flows can be very diverse. It is determined by the size, distribution of the elements of the dispersed phase in a continuous medium and their direct interaction. Multiphase media in nature are smoke, smog, fog, rain, in turn, in the technique typical examples of such media are emulsion, sludge, boiling liquid, etc. [3]-[4].

The problem of numerical simulation of the hydrodynamics of two-phase flows is relevant, since for many years one of the main methods for studying such media has been experiment. However, the development of numerical methods and computing hardware made it possible to calculate the fields of velocities, pressures and other parameters of two-phase flows in different engineering tasks[5]-[7].

The aim of the work is to describe a mathematical model of a two-phase flow and develop methods for hydrodynamic simulation of water flow with different gas content in a multiphase pump, as well as to determine the best way for calculating two-phase flow.

One of the possible solutions to the task is to use a helico-axial multiphase pump for the calculation [8]-[9]. It was designed for pumping a media with gas content up to 99%, during continuous operation[10]. Such pumps were designed primarily for oil production with associated gas without the use of additional units [11]-[12]. This two-phase pumping technology reduces pollutants and CO_2 emissions, minimizes environmental impact, avoids gas or crude oil leaks, and is also more cost-effective than oil extraction using separation plants [13]. The design of one stage of the pump is shown in Fig. one.

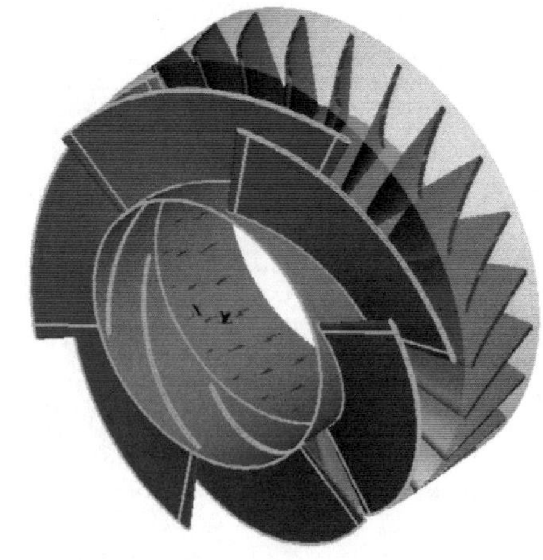

Fig. 1. The design of the stage of the helico-axial multiphase pump

The stage of the helico-axial multiphase pump is an axial impeller followed by a guide vanes [14].The idea is that the impeller disperses the multiphase flow, and due to expansion of cross section of guide vane, the volume of gas bubbles decreases [15]-[16]. Thus, the gas is not shut off the flow.

In this paper hydrodynamic simulation was carried out using several methods of two-phase flow. The simulation was performed in the STAR CCM + software package.

Methodology

Hydrodynamic simulation of a two-phase flow can be calculated by following methods:

1) homogeneous multiphase volume of fluid model (VOF), is a simplified model that simulates a multiphase flow in the form of a suspension. It is used when two phases of the same substance (for example, water and steam) are involved, which are in thermodynamic equilibrium and have common fields of speed, pressure and temperature[17]-[18].

Basic assumptions of the two-phase mixture model:

- The mixture of phases behaves like a single fluid. A set of following equations: transport equations for mass, momentum, and energy, is solved for the mixture as a whole.
- Phases are in thermal equilibrium. The two phases have the same temperature, and the phase distribution is calculated on the basis of the heat balance.
- There is no relative (sliding) speed between phases.

2) segregatedmultiphase flow model uses the same set of equations as the two-phase thermodynamic equilibrium model (mass and momentum transfer equations), but solves them separately for each phase (except for the general pressure field).

The mathematical model of segregated multiphase flow consists of a set of differential and algebraic equations [19]-[20]:

1. The volume of the i-th phase in each cell is calculated as:

$$V_i = \int_V \alpha_i dV$$

where α_i – is the concentration of the i-th phase in the cell.

The sum of the concentrations of all phases in the cell is equal the one.

$$\sum_{i6}^{n} \alpha_i = 1$$

2. The equation of mass conservation (continuity equation):

$$\frac{\partial}{\partial t}\int_V \alpha_i \rho_i dV + \int_A \alpha_i \rho_i \vec{V_i} d\vec{a} = 0$$

where ρ_i – i-th phase density;

$\vec{V_i}$ – i-th phase velocity (in the case of turbulent flow simulation by a RANS-type model, velocity averaged over time [21]-[23]).

3. The quantity of motion changing:

$$\frac{\partial}{\partial t}\int_V \alpha_i \rho_i \vec{V_i} dV + \int_A \alpha_i \rho_i \left(\vec{V_i}\vec{V_i}\right) d\vec{a} = -\int_V \alpha_i \nabla p dV$$

$$+\int_V \alpha_i \rho_i \vec{g} dV + \int_A \left[\alpha_i \left(T_i + T_i^t\right)\right] d\vec{a} + \int_V \overrightarrow{M_i} dV$$

where $\left(\vec{V_i}\vec{V_i}\right)$ – tensormultiplication of the velocity vector of the i-th phase;

p – pressure;

$\vec{g}$ – mass intensity vector (in this case, the gravity force is 9.81 m/s^2 and the inertial force due to the rotation of the computational domain);

T_i - molecular viscosity stress tensor;

T_i^t – turbulent stress tensor;

$\vec{M}$ – the vector of the total intensity of phase interaction forces per unit volume, for the vector $\overrightarrow{M_i}$ the equality is true:

$$\sum_i \overrightarrow{M_i} = 0$$

The vector $\overrightarrow{M_i}$ characterizes all the forces that separate phases interact with each other.

$$\overrightarrow{M_i} = \sum_{i \neq j}\left(\overrightarrow{F_{ij}^D} + \overrightarrow{F_{ij}^{VM}} + \overrightarrow{F_{ij}^L} + \overrightarrow{F_{ij}^{TD}} + \overrightarrow{F_{ij}^{WL}}\right)$$

где $\overrightarrow{F_{ij}^D}$ – resistance force;

$\overrightarrow{F_{ij}^{VM}}$ – virtual mass force;

$\overrightarrow{F_{ij}^L}$ – lifting force;

$\overrightarrow{F_{ij}^{TD}}$ – force caused by turbulent dispersion;

$\overrightarrow{F_{ij}^{WL}}$ – force caused by wall effect.

4. Resistance force

In flow simulation of continuous and dispersed media, the resistance force acting on the dispersed medium I from the side of the continuous medium j is equal to:

$$\overrightarrow{F_{ij}^D} = A_D \overrightarrow{V_r}$$

Where A_D – linearized resistant coefficient;

$\overrightarrow{V_r} = \overrightarrow{V_j} - \overrightarrow{V_i}$ – relative velocity of mediums;

$$A_D = C_D \frac{1}{2} \rho_C \left|\overrightarrow{V_r}\right| \left(\frac{a_{cd}}{4}\right)$$

Where C_D – resistant coefficient;

ρ_C – continuous phase density;

a_{cd} – interaction area of the phases (in this case, the term $\frac{a_{cd}}{4}$ is the area of the projection of the spherical particle on the plane).

Resistant coefficient is based on the relation:

$$C_D = f_D C_{D\infty}$$

Where $C_{D\infty}$ - coefficient of resistance of a single spherical particle moving in an infinite flow;

f_D – coefficient taking into account the concentration of particles.

$C_{D\infty}$ can be found from relation:

$C_{D\infty} = \frac{24}{Re_d}\left(1 + 0.15 Re_d^{0.687}\right)$ when $Re_d < 1000$

or

$C_{D\infty} = 0{,}44$ when $Re_d > 1000$

$$Re_d = \frac{\rho_C \left|V_r\right| l}{\mu_C}$$

l – characteristic interaction length or bubble size;

μ_C – dynamic viscosity coefficient of continuous medium;

Coefficient f_D can be found from relation:

$$f_D = \alpha_C{}^{nD}$$

α_C – continuous phase concentration;

$n_D = -8.3$ for spherical particle.

5. Virtual mass

The inertia of the surrounding fluid affects the acceleration of the particle immersed in the fluid. This effect is simulated by adding mass to the dispersed particle.

The force from the virtual mass acting on phase i moving accelerated relative to phase j is found as:

$$\overrightarrow{F_{ij}^{VM}} = C_{VM} \rho_C \alpha_C \left(\overrightarrow{a_j} - \overrightarrow{a_i}\right)$$

$\overrightarrow{a_{j,i}}$ – acceleration of j-th and i-th phase;

$C_{VM} = 0{,}5$ – virtual mass coefficient for a spherical particle.

6. Lifting force

In the case when the flow surrounding the dispersed particle is inhomogeneous or the particle is swirling, the force is perpendicular to the relative velocity..

$$\overrightarrow{F_{ij}^{L}} = C_{Leff} \rho_C \alpha_d \left(\overrightarrow{V_r} \times \left(\nabla \times \overrightarrow{V_c}\right)\right)$$

IOP Conf. Series: Materials Science and Engineering **589** (2019) 012028 doi:10.1088/1757-899X/589/1/012028

$\overrightarrow{V_r}$ – relative phase velocity;

$\overrightarrow{V_c}$ – continuous phase speed;

α_d – dispersed phase concentration;

$C_L = 0,25$ –lifting force coefficient.

7. Turbulent dispersion force

Additional change in phase concentrations caused by flow turbulence is modeled as the force of turbulent dispersion

$$\overrightarrow{F_{ij}^{TD}} = A_D \overrightarrow{V_{TD}}$$

– resistant force coefficient;

$\overrightarrow{V_{TD}}$ – relative slip speed;

$$\overrightarrow{V_{TD}} = D_{TD}\left(\frac{\nabla\alpha_d}{\alpha_d} - \frac{\nabla\alpha_c}{\alpha_c}\right)$$

$D_{TD} = C_0 \dfrac{\nu_c^t}{\sigma_\alpha} I$ – turbulent diffusion tensor; $C_0 = 1$;

ν_c^t –kinematic coefficient of turbulent viscosity;

σ_α – turbulentPrandtl number;

I – unit tensor;

$$\sigma_\alpha = \sigma_0 \sqrt{1 + C_\beta \xi^2}\,\frac{1+\eta}{b+\eta}$$

$\sigma_0 = 1$ – unmodified turbulent Prandtl number;

$C_\beta = 1{,}8$ – correction factor;

ξ – article sliding velocity related to the speed of turbulent fluctuations;

η – particle interaction time related to relaxation time;

b – the ratio of the accelerations of the continuous / dispersed phase;

$$b = \frac{1 + C_{VM}}{\dfrac{\rho_d}{\rho_c} + C_{VM}}$$

$$\eta = \frac{\tau_I}{\tau_R}$$

τ_I – characteristic time of interaction of a particle and a turbulent vortex;

τ_R – particle relaxation time;

$$\tau_I = \frac{\tau_T}{\sigma_0}\frac{1}{\sqrt{1 + C_\beta \xi^2}}$$

η_c - continuous phase turbulence energy dissipation rate;

$$\tau_R = \tau_D\left(1 + \frac{\rho_c}{\rho_d} C_{VM}\right)$$

$\tau_D = \frac{\rho_d d^2}{18\mu_c}$ – characteristic time scale for a dispersed particle;

$$\xi = \frac{\left|\overrightarrow{V_r}\right|}{\sqrt{\frac{2}{3}k_c}}$$

k_c – kinetic turbulence energy of the continuous phase.

8. Wall influence

Whengasbubble located near the solid wall, it undergoes an asymmetrical action from the liquid. The force per unit volume that a gas bubble experiences is equal to:

$$\overrightarrow{F_{ij}^{WL}} = -C_{WL} y_w \alpha_D \rho_C \frac{\left|\overrightarrow{V_{r,\tau}}\right|^2}{d_p} \vec{n}$$

где y_w – distance from the wall;

α_D – dispersed phase concentration;

ρ_C – dispersed phase density;

d_p – bubble diameter;

C_{WL} – coefficient, which is the function of the distance from the wall and decreasing with increasing distance;

$\vec{n}$ – normally oriented to the wall unit vector at the point closest to the bubble;

$\overrightarrow{V_{r,\tau}}$ – tangent to the wall component of the relative velocity;

Coefficient C_{WL} is as:

$$C_{WL} = max\left(C_{W1} + \left(\frac{C_{W2}}{y_w}\right)d_p, 0\right)$$

Coefficientsequal to: $C_{W1} = -0,01$, C_{W2}=0,05. Thus the force caused by the influence of the wall disappears at a distance from the wall equal to the five diameters of the bubble.

The 3D model of the helico-axial multiphase pump for the calculation is shown in Fig. 2.

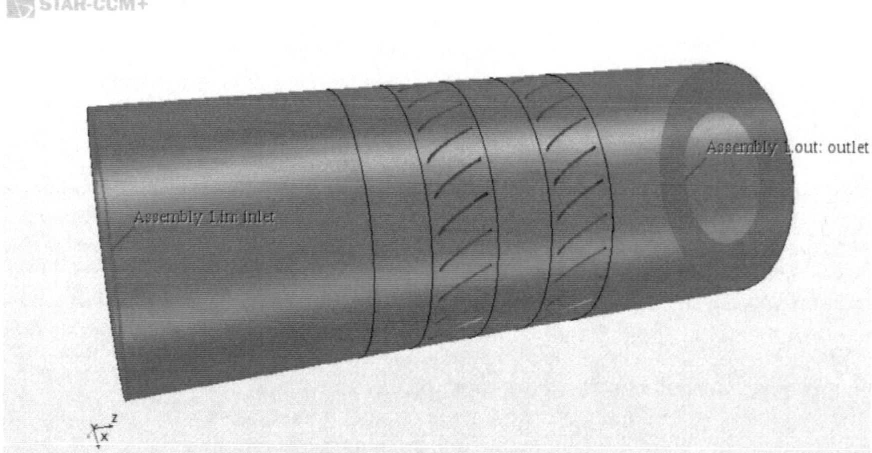

Fig. 2. 3D model of the flow part of the pump

For the calculation two pump stages were chosen for a more visual representation of the two-phase flow.

Boundary conditions: inlet velocity and outlet pressure, and for a split flow model, it is necessary to indicate the velocity of both the liquid and the gas. The calculation was carried out by gradually increasing the gas content value at the inlet boundary of the pump.

The grid in longitudinal and cross-section, presented in Fig. 3, consists of 508,000 cells. Near the solid walls, the cells elongated in the direction perpendicular to the wall (boundary layer), in the core of the flow the grid is not structured, consists of polyhedra of various shapes and sizes.

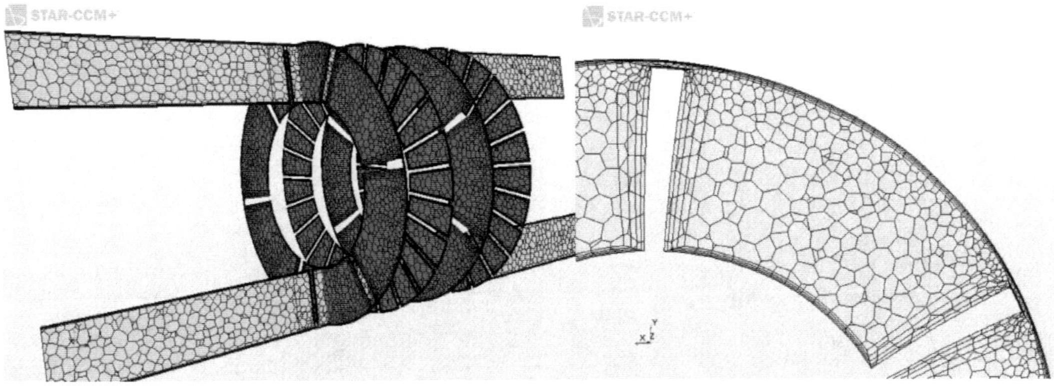

Fig. 3. The grid in the cross section of the pump

Although the simulation was performed in different ways, the grid was constructed identically.

Results

The flow of the two-phase medium in the flow part of the pump, obtained as a result of the simulation, is shown in the illustrations below.

For a visual representation of the two-phase flow, the scalar distribution of the volume fraction of water is showed on Fig.4 (the liquid phase is red, the gas phase is blue).

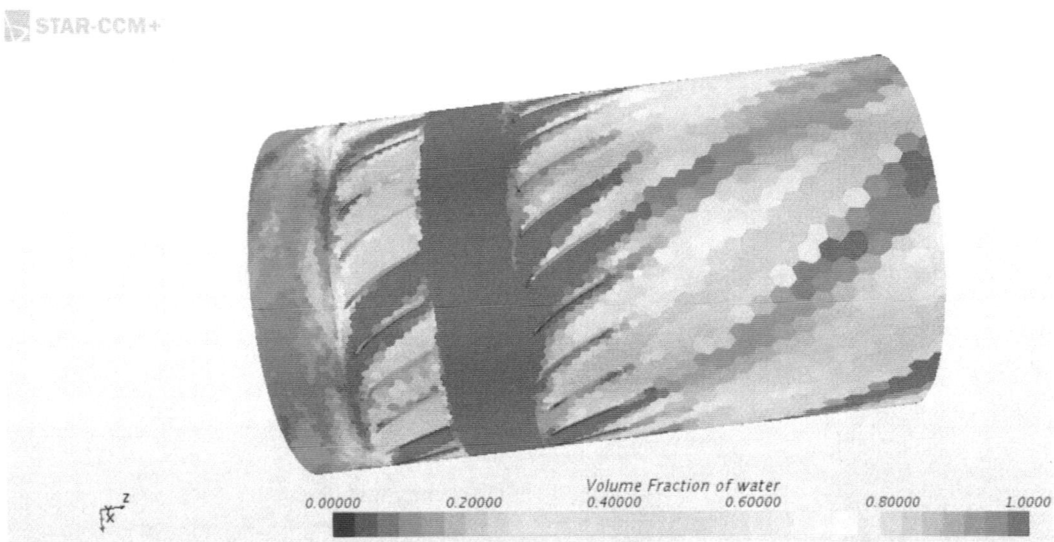

Fig. 4. The volume fraction of water in the pump at 50% of the gas at the inlet (simplified model)

At Fig. 4 it can be seen that the pump push through the liquid-gas mixture in a 50/50 ratio, and after the first stage the amount of gas has decreased.

At Fig. 5 showed the current flow lines in first stage of a multiphase pump.

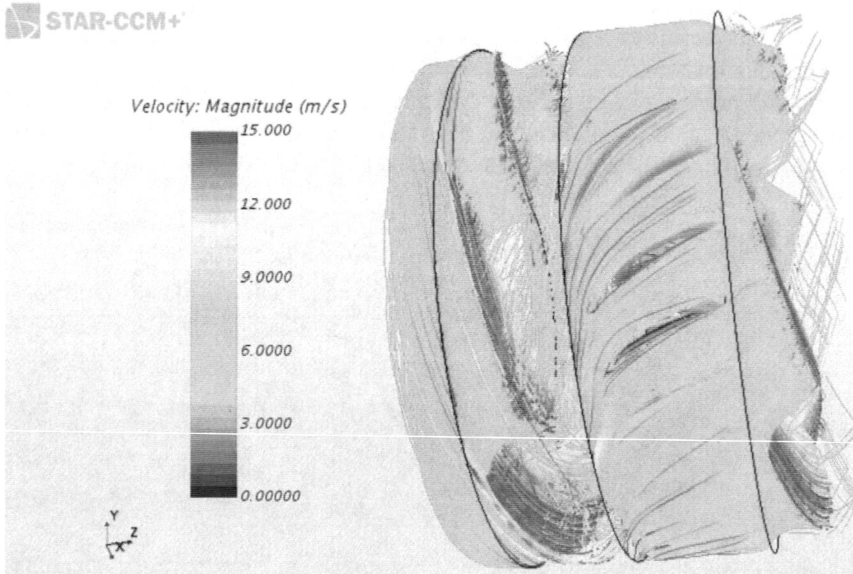

Fig. 5. Current flow lines in the pump stage

The following are the scalar scenes of the distribution of the volume fraction of water in the cross section of the pump so as to get a visual representation of what is happening in impeller and guide vanes. In order to compare the two calculation methods, all the results are presented for the simplified model, then for the segregate flow model. In Fig. 6-11 are presented from left to right: impeller of the first stage, guide vanes of first stage, then impeller of second stage and the guide vanes of second stage (the liquid phase is red, the gas phase is blue).

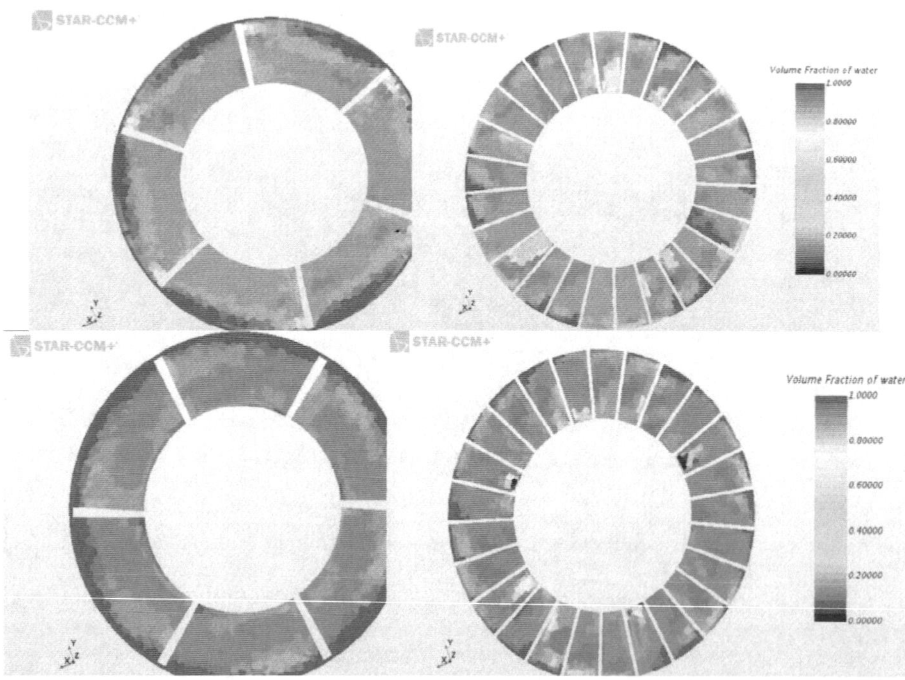

Fig. 6. The volume fraction of water in the cross sections of the pump with 20% of the gas at the inlet (simplified model)

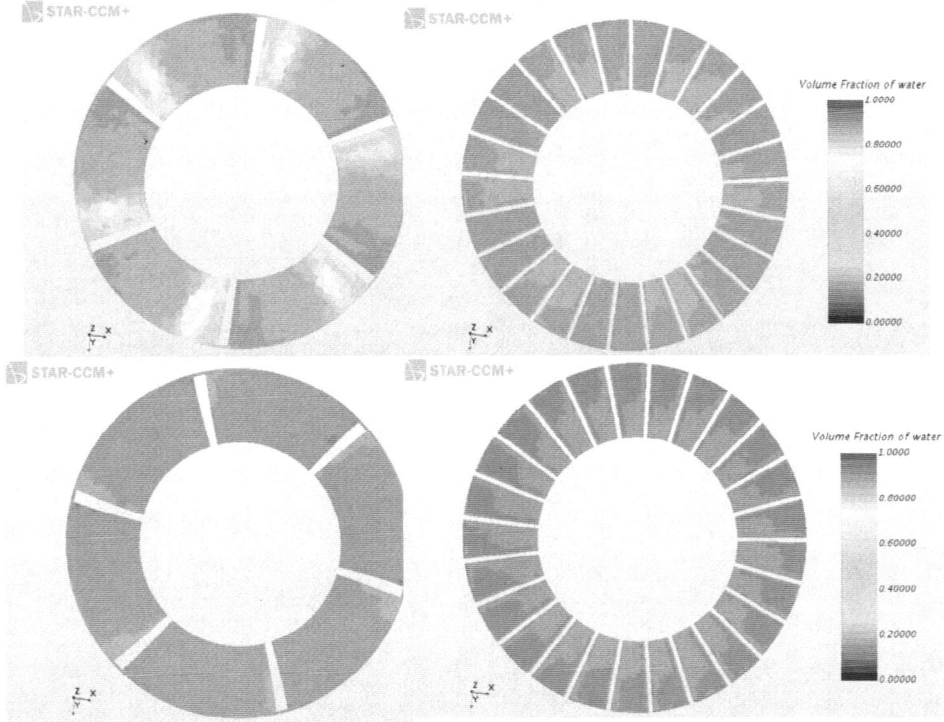

Fig. 7. The volume fraction of water in the cross sections of the pump at 20% of the gas at the inlet (segregated flow model)

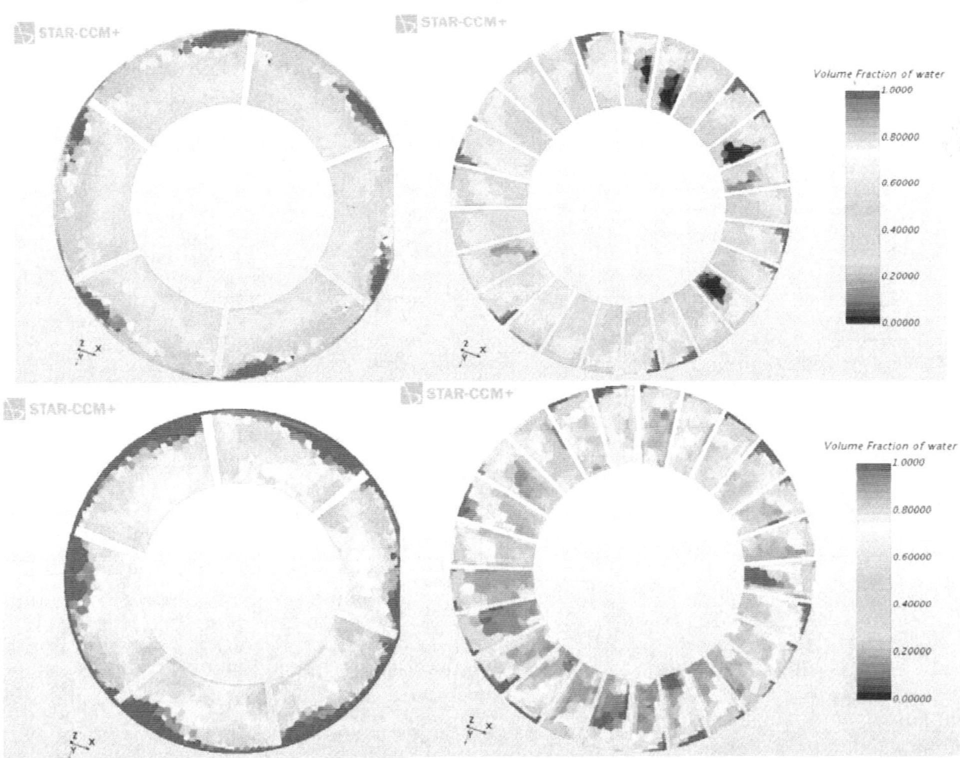

Fig. 8. The volume fraction of water in the cross sections of the pump with 50% of the gas at the inlet (simplified model)

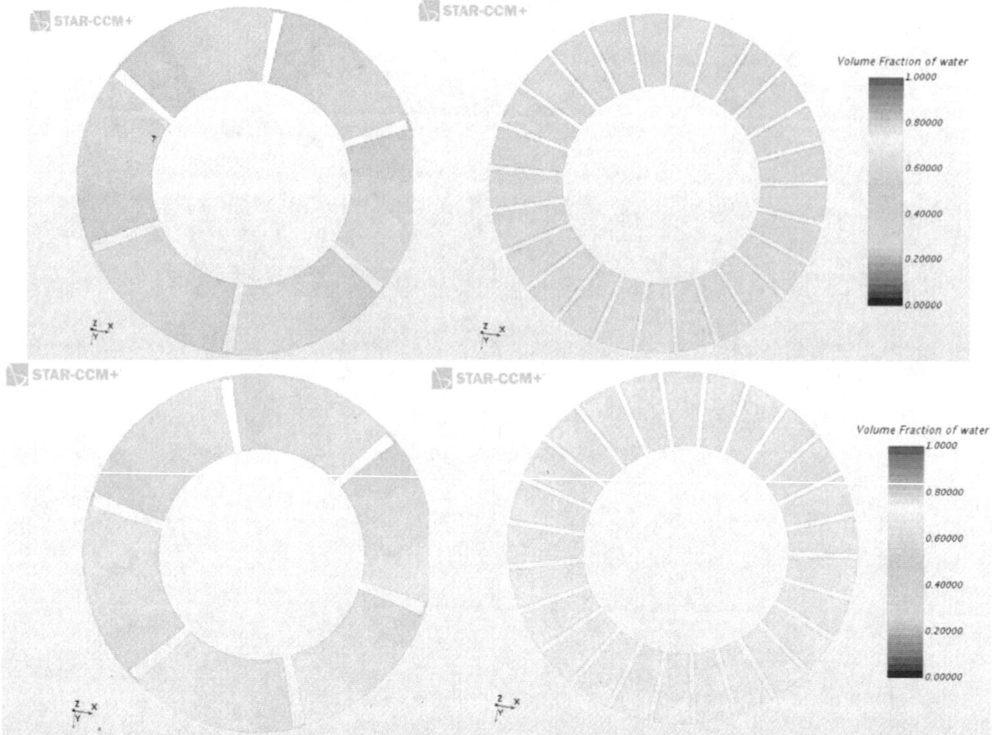

Fig. 9. The volume fraction of water in the cross sections of the pump with 50% of the gas at the inlet (segregated flow model)

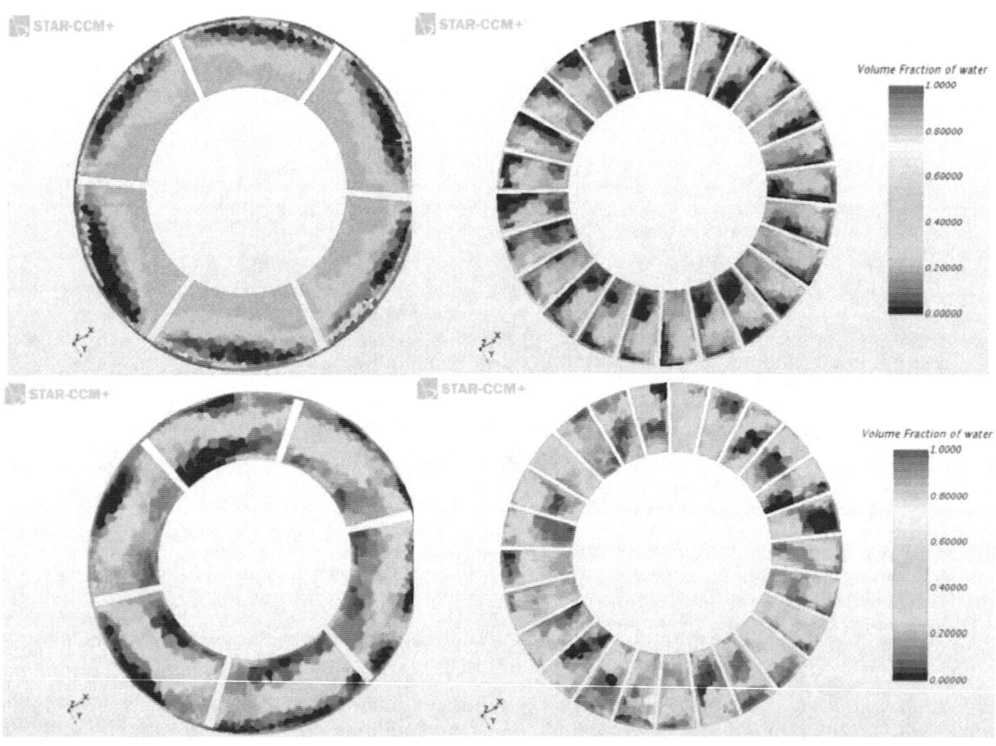

IOP Conf. Series: Materials Science and Engineering **589** (2019) 012028 doi:10.1088/1757-899X/589/1/012028

Fig. 10. The volume fraction of water in the cross sections of the pump with 80% of the gas at the inlet (simplified model)

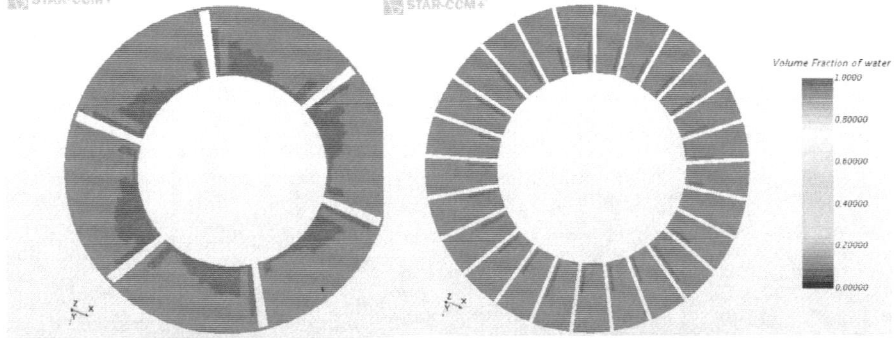

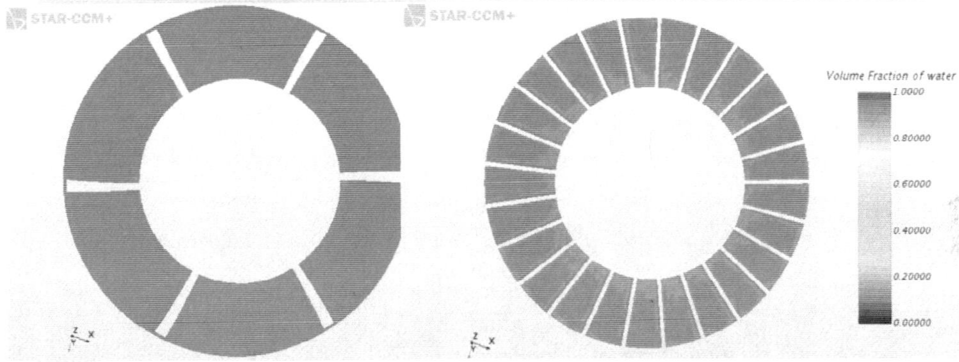

Fig. 11. The volume fraction of water in the cross sections of the pump with 90% of the gas at the inlet (segregated flow model)

Comparing Fig. 6, 8, 10 with fig. 7, 9, 11, it can be seen that the flow distribution significantly differs in the simplified model and in the segregated flow model. This can be explained by the fact that in full model appear diffusion of gas and liquid, which corresponds to reality. However, in the simplified model, the physics of the process is clearly visible: the liquid in impeller is thrown away to the periphery by centrifugal force and gas spreads throughout the flow, but accumulates closer to the center. Also in the guide vanes gas accumulates on the back side of the blades. But in reality, most likely, such an obvious phase separation in the pump does not occur. In both models, the amount of gas in the second stage is less than in the first.

The simulation results are summarized in table 1.

Table 1. SimulationResults

Gas content α, %	Homogeneous multiphase modelVOF			Segregated multiphase flow model		
	H, м	Q, м^3/ч	η, %	H, м	Q, м^3/ч	η, %
0	6,9	80,2	58,6	6,53	81,05	56,2
10	6,93	72	79,6	6,5	72,3	53,1
20	7	64,6	84,2	6,43	64,24	48
30	6,9	55,9	73,5	6,2	56,7	43,1
40	6,7	48,52	38,7	5,7	48,2	36,2
50	6	40	32,4	4,9	39,9	30,9
60	4,78	32,3	25,8	3,7	33	25,7
70	2,8	24,3	18,8	2,6	24,3	18,5
80	1,6	15,8	11,5	1,58	16,4	12,5

90	-	-	-	0,6	7,2	6,2

The dependences of pump head, efficiency and productivity on gas content are presented in Fig. 12-14 (dependencies obtained using a homogeneous multiphase VOF model are shown in blue, red shows a multiphase segregated flow model).

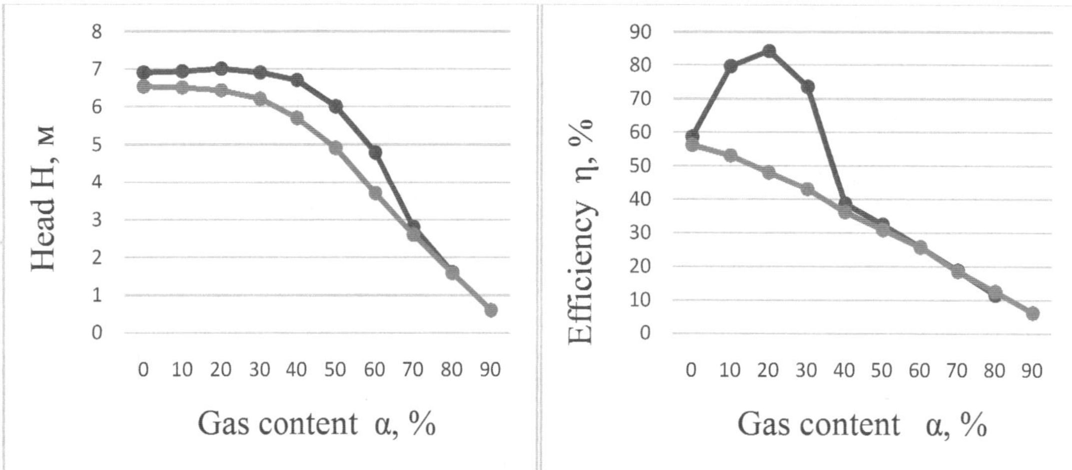

Fig.12. Dependencies of head and pump efficiency on gas content

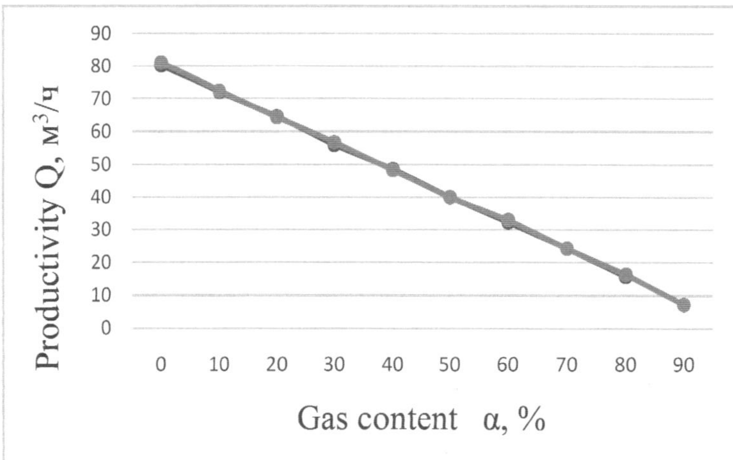

Fig.13. Dependence of pump flow on gas content

From Fig.12-13, it can be concluded that pump head calculated by the VOF method is overestimated relative to the other model, and the efficiency with gas content from 10% to 40% is unreasonably high.

Conclusion

The paper describes a mathematical model of a multi-phase flow, which was implemented by two methods of calculation in the STAR CCM + software package.

The performed simulation showed physicaly results in both cases. As a first approximation the homogeneous multi-phase VOF model can be used. Its advantage is that the calculation requires significantly less computing power. But still, this is a simplified model and errors in it are no exception, for example, too high efficiency in the calculated flow section of the pump. Using a multiphase segregated flow model, it was possible to calculate fluid flow with gas content up to 95%; in turn, the simplified model showed the result only up to 80% of the gas phase.

In fact, a more complex two-phase flow structure in a multiphase pump is possible. The multiphase flow models contains many empirical coefficients, so an experiment is required to validation of this model.

Bibliography

[1] Gläser, D., Flemisch, B., Helmig, R., Class, H. A hybrid-dimensional discrete fracture model for non-isothermal two-phase flow in fractured porous media (2019) GEM - International Journal on Geomathematics, 10 (1), № 5.

[2] Mikelić, A., Wheeler, M.F., Wick, T. Phase-field modeling through iterative splitting of hydraulic fractures in a poroelastic medium (2019) GEM - International Journal on Geomathematics, 10 (1), № 2.

[3] Pinilla, J.A., Guerrero, E., Pineda, H., Posada, R., Pereyra, E., Ratkovich, N. CFD modeling and validation for two-phase medium viscosity oil-air flow in horizontal pipes (2019) Chemical Engineering Communications, 206 (5), pp. 654-671.

[4] Feng, T., Wang, M., Song, P., Liu, L., Tian, W., Su, G.H., Qiu, S. Numerical research on thermal mixing characteristics in a 45-degree T-junction for two-phase stratified flow during the emergency core cooling safety injection (2019) Progress in Nuclear Energy, 114, pp. 91-104.

[5] Lomakin, V.O. Investigation of two-phase flow in axial-centrifugal impeller by hydrodynamic modeling methods (2015) Proceedings of 2015 International Conference on Fluid Power and Mechatronics, FPM 2015, № 7337302, pp. 1204-1206.

[6] Putra, R.A., Schäfer, T., Neumann, M., Lucas, D. CFD studies on the gas-liquid flow in the swirl generating device (2018) Nuclear Engineering and Design, 332, pp. 213-225.

[7] Cao, S., Peng, G., Yu, Z. Hydrodynamic design of rotodynamic pump impeller for multiphase pumping by combined approach of inverse design and CFD analysis (2005) Journal of Fluids Engineering, Transactions of the ASME, 127 (2), pp. 330-338.

[8] Shi, G., Wang, Z., Wang, Z., Li, Z., Liu, X. Research on the pressurization performance of an impeller in a multi-phase pump under different working conditions (2019) Advances in Mechanical Engineering, 11 (3), .

[9] El Hajem, M., Morel, R., Champagne, J.Y., Vilagines, R., Pagnier, P. The study of the flow in anhelico-axial pump using laser Doppler velocimetry [Article@Exploration par anémométrie laser Doppler de l'écoulementdansunepompehélico-axiale] (2001) Houille Blanche, (3-4), pp. 34-39.

[10] Shi, G.-T., Wang, Z.-W., Luo, K. Analysis of the Turbulent Flow Intensity and Dissipation Characteristics of an Oil-gas Multiphase Pump in Its Compression Stages [Article@油气混输泵压缩级内湍流强度及湍流耗散特性分析] (2018) RenengDongliGongcheng/Journal of Engineering for Thermal Energy and Power, 33 (6), pp. 115-121.

[11] Falcimaigne, J., Brac, J., Charron, Y., Pagnier, P., Vilagines, R. Multiphase pumping: Achievements and perspectives [Article@Pompagepolyphasique: Réalisationsetperpectives] (2002) Oil and Gas Science and Technology, 57 (1), pp. 99-107.

[12] ANIN Framo Engineering Multiphase pumping - a technology and field testing status report (1994) World Pumps, 339, pp. 29-30.

[13] Jiang, J., Rui, Z., Hazlett, R., Lu, J. An integrated technical-economic model for evaluating CO 2 enhanced oil recovery development (2019) Applied Energy, 247, pp. 190-211.

[14] Shi, Y., Zhu, H., Zhang, J., Yin, B., Xu, R., Zhao, J. Investigation of condition parameters in each stage of a three-stage helico-axial multiphase pump via numerical simulation (2017) Proceedings of the International Offshore and Polar Engineering Conference, pp. 156-162.

[15] Zhang, J.Y., Li, Y.J., Cai, S.J., Zhu, H.W., Zhang, Y.X. Investigation on the gas pockets in a rotodynamic multiphase pump (2016) IOP Conference Series: Materials Science and Engineering, 129 (1), № 012007.

[16] Zhang, Y., Di, Y., Liu, P., Li, W. Simulation of tight fluid flow with the consideration of capillarity and stress-change effect (2019) Scientific Reports, 9 (1), № 5324.

[17] Lomakin, V.O., Kuleshovav, M.S., Bozh'eva, S.M. Numerical Modeling of Liquid Flow in a Pump Station (2016) Power Technology and Engineering, 49 (5), pp. 324-327.

[18] Lomakin, V.O., Kuleshova, M.S., Kraeva, E.A. Fluid Flow in the Throttle Channel in the Presence of Cavitation (2015) Procedia Engineering, 106, pp. 27-35.

[19] Brennen, C.E. Hydrodynamics of pumps (2011) Hydrodynamics of Pumps, 9781107002371, pp. 1-270.

[20] Brennen, C.E. Fundamentals of Multiphase Flows (2005) Cambridge:Cambridge University Press, pp. 1-410.

[21] Gouskov, A.M., Lomakin, V.O., Banin, E.P., Kuleshova, M.S. Minimization of Hemolysis and Improvement of the Hydrodynamic Efficiency of a Circulatory Support Pump by Optimizing the Pump Flowpath (2017) Biomedical Engineering, 51 (4), pp. 229-233.

[22] Lomakin, V., Cheremushkin, V., Petrov, A. The development of the theory of calculation of the hydrodynamic coupling (2019) IOP Conference Series: Materials Science and Engineering, 492 (1), № 012012.

[23] Lomakin, V.O., Chaburko, P.S., Kuleshova, M.S. Multi-criteria Optimization of the Flow of a Centrifugal Pump on Energy and Vibroacoustic Characteristics (2017) Procedia Engineering, 176, pp. 476-482.

IOP Conf. Series: Materials Science and Engineering **589** (2019) 012029 doi:10.1088/1757-899X/589/1/012029

Comparative analysis of the use of the spiral and bladed stators in a compact high-speed centrifugal pump with hydrodynamic modeling methods

A Martynyuk[1], **A Petrov**[1] **and M Kuleshova**[1,2]

[1]Bauman Moscow State Technical University, 5 Second Baumanskaya Street, Moscow, 105005, Russian Federation

[2]E-mail: kuleshova@bmstu.ru

Abstract. The relevance of improving pumps used in thermal stabilization systems and increasing their service life by reducing the radial forces on the rotor of the impeller is considered. The influence of the guide vane on the characteristics of a compact high-speed centrifugal pump is investigated. With the help of hydrodynamic modeling in STAR CCM + software package, the values of head, efficiency, radial force and power for two pumps were obtained. A comparison of the obtained characteristics of a pump with a spiral and a bladed stator was carried. The advantages and disadvantages of the guide vane are determined and recommendations for improving its characteristics are given.

Introduction

Compact high-speed centrifugal pumps are used in thermal stabilization systems. They are used in aviation, radio-locator stations, military industry, as part of satellite equipment and manned space stations, for cooling lasers, etc. All of them have great prospects for development, therefore, nowadays the improvement of pumps for thermal stabilization systems is very important. Their service life is especially essential. Radial force on a rotor of a centrifugal pump has a great influence on it: in case of deviation from the optimal operation modes, it increases significantly. As a result, pump wear increases and equipment life cycle shortens. To reduce radial force load on the rotor a guide vane is used. This article is devoted to the study of the influence of the bladed stator on pump characteristics by means of hydrodynamic modeling, which has been gaining popularity recently.

Methods

To solve problems of continuum mechanics hydrodynamic modeling was carried in a software package STAR CCM + for two compact high-speed centrifugal pumps (Fig. 1): the first pump with a spiral stator, the second with a bladed one. Hydrodynamic modeling capabilities are discussed in [1-15].

IOP Conf. Series: Materials Science and Engineering **589** (2019) 012029 doi:10.1088/1757-899X/589/1/012029

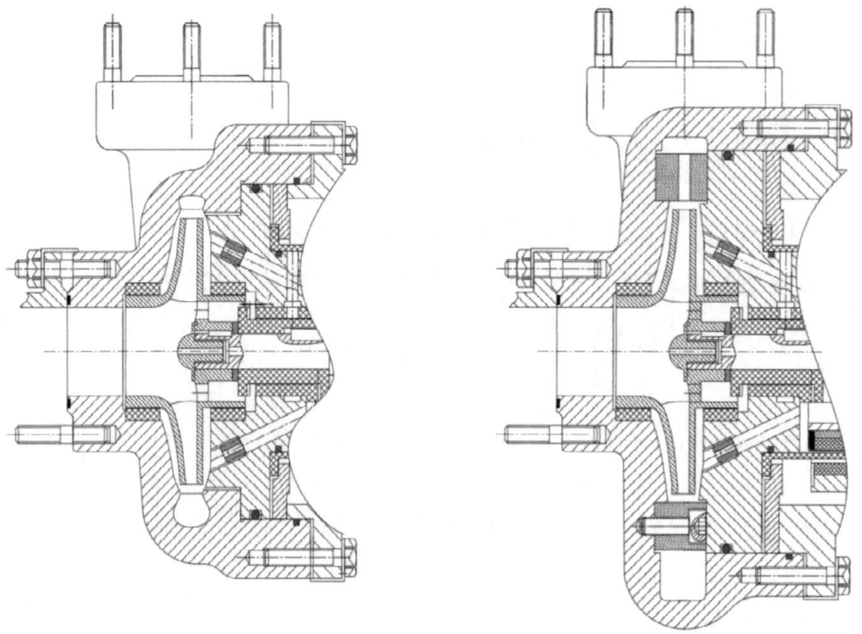

Fig.1. Configuration of the pumps under consideration (on the left – the pump with the spiral stator,on the right – the pump with the bladed stator)

Mathematical model:

- The continuity equation (mass conservation equation):

$$\frac{\partial \rho}{\partial t} + div(\rho \cdot \vec{u}) = 0, \text{where}$$

ρ – density;
t – time;
$\vec{u}$ – velocity vector.

- Reynolds equations (equations of conservation of momentum averaged over time):

$$\frac{\partial}{\partial t}(\rho \cdot u_j) + \frac{\partial}{\partial x_j}(\rho \cdot u_i \cdot u_j) = -\frac{\partial p}{\partial x_i} + \frac{\partial}{\partial x_i}\left[T_{ij} - \rho \cdot \overline{u'_i u'_j}\right], \text{where}$$

$T_{ij} = 2\mu s_{ij} - \frac{2}{3}\frac{\partial u_j}{\partial x_j}$ - viscous stress tensor;

μ – dynamic-viscosity coefficient, Pa·s;

$s_{ij} = \frac{1}{2}\left(\frac{\partial u_i}{\partial x_j} + \frac{\partial u_j}{\partial x_i}\right)$-strain velocity tensor;

$\rho\overline{\langle u'_i u'_j \rangle}$ – Reynolds stresses;

u' - pulsation velocity component;

p – averaged pressure.

- Equation of turbulent kinetic energy transfer:

$$\frac{\partial}{\partial t}(\rho k) + \frac{\partial}{\partial x_j}(\rho u_j k) = P_k - \rho\beta^* k\omega + \frac{\partial}{\partial x_j}\left[(\mu + \sigma_k \mu_t)\frac{\partial k}{\partial x_j}\right], \text{ where}$$

k – turbulent kinetic energy;

P_k – turbulent energy generation by velocity shift;

σ_k, β^* – closure factors;

μ_t – turbulent viscosity coefficient.

- Equation of the relative dissipation rate of this energy transfer:

$$\frac{\partial}{\partial t}(\rho\omega)+\frac{\partial}{\partial x_j}(\rho u_j\omega)=a\frac{\omega}{k}p_k-\rho\beta\omega^2+\frac{\partial}{\partial x_j}\left[(\mu+\sigma_\omega\mu_t)\frac{\partial\omega}{\partial x_j}\right]+2\cdot(1-F_i)\sigma_{\omega2}\cdot\frac{1}{\omega}\frac{\partial k}{\partial x_i}\frac{\partial\omega}{\partial x_i},$$

where

ω – turbulent kinetic energy dissipation rate;

a, β, σ_ω – closure factors;

F_i – weight function.

Mathematical model parameters:

1. Meshing models:
1.1. Polyheral;
1.2. Surface remesher;
1.3. Prism layer mesher;
1.4. Extrusion (constant velocity in the normal direction).
2. Mesh default controls:
2.1. Base size 1,5 mm;
2.2. Prism layer total thickness 0,75 mm;
2.3. Number of prism layers 5;
2.4. Prism layer stretching 1,3;
Extrusion:
2.5. Magnitude 90 mm;
2.6. Number of layers 100;
2.7. Stretching factor 5.
3. Physical models:
3.1. Three-dimensional;
3.2. Material: water;
3.3. Constant density;
3.4. Reynolds-averaged Navier-Stokes equation;
3.5. k-omega SST turbulence;
3.6. Implicit unsteady.
4. Boundary conditions:
4.1. Rotation rate 10800 rpm;
4.2. Inlet stagnation 0 Pa;
4.3. Outlet velocity 3.148 m/s.

Computational mesh for both pumps is shown on Fig.2 and Fig.3. The grid for the pump with the spiral stator consists of 159114 cells, for the pump with the bladed stator - 382623 cells.

IOP Conf. Series: Materials Science and Engineering **589** (2019) 012029 doi:10.1088/1757-899X/589/1/012029

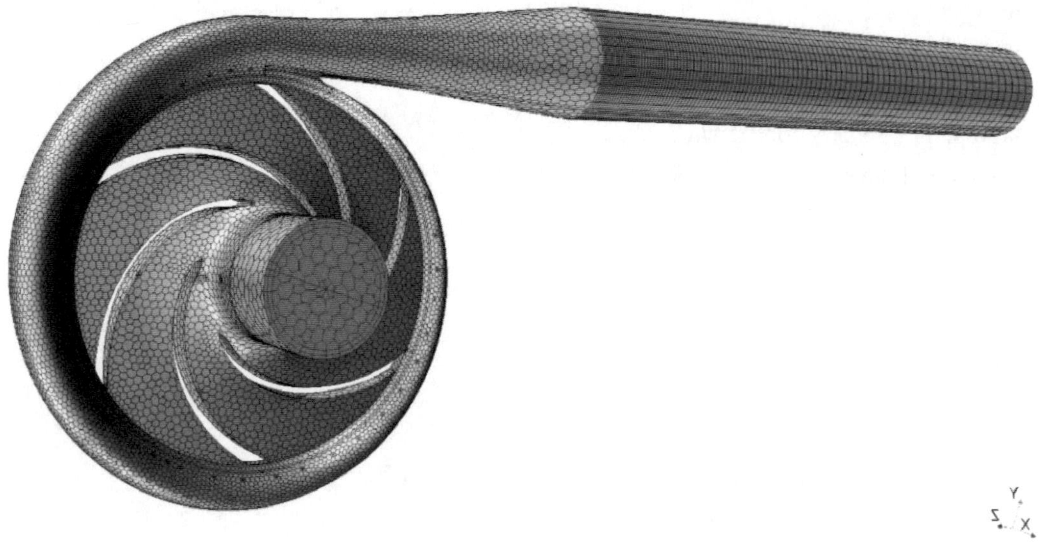

Fig.2. Mesh for the pump with the spiral stator

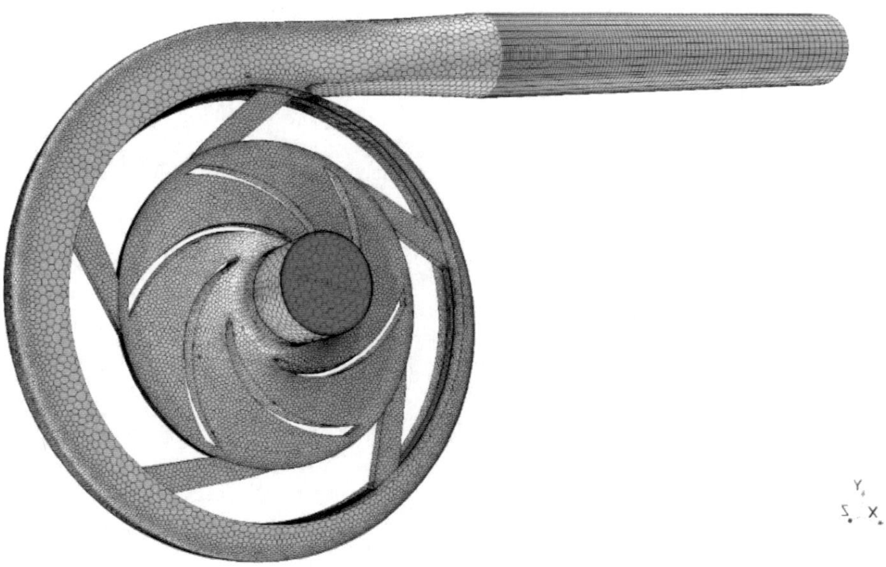

Fig.3. Mesh for the pump with the bladed stator

Results

Values of head, efficiency, radial force and power at different loading points were computed. The results are shown in table 1.

IOP Conf. Series: Materials Science and Engineering **589** (2019) 012029 doi:10.1088/1757-899X/589/1/012029

Table 1. Results of the simulation

Parameter		The pump with the spiral stator	The pump with the bladed stator
D_2, mm		45	48
Dimensions H·R·L, mm		94·77·82	105·83,5·82
Mass, kg		2	2,6
Q=0.25 Qnom	H, m	35,5	42,5
	Efficiency, %	51	59
	Radial force, N	12	5
	N, W	100	111
Q=0.5 Qnom	H, m	37	40
	Efficiency, %	74	58
	Radial force, N	9	1,6
	N, W	142	172
Q=0.75 Qnom	H, m	38	39.5
	Efficiency, %	81	67,3
	Radial force, N	5,5	1,2
	N, W	190	238
Q=Qnom	H, m	37	35
	Efficiency, %	84	62
	Radial force, N	0,9	1,1
	N, W	230	296
Q=1.25·Qnom	H, m	28,5	25,5
	Efficiency, %	74	50
	Radial force, N	5	2,2
	N, W	270	349
Q=1.5 Qnom	H, m	18	14
	Efficiency, %	50	24
	Radial force, N	10	4
	N, W	295	394

According to the results obtained, the graphs presented in Fig. 4 and Fig. 5 are build.

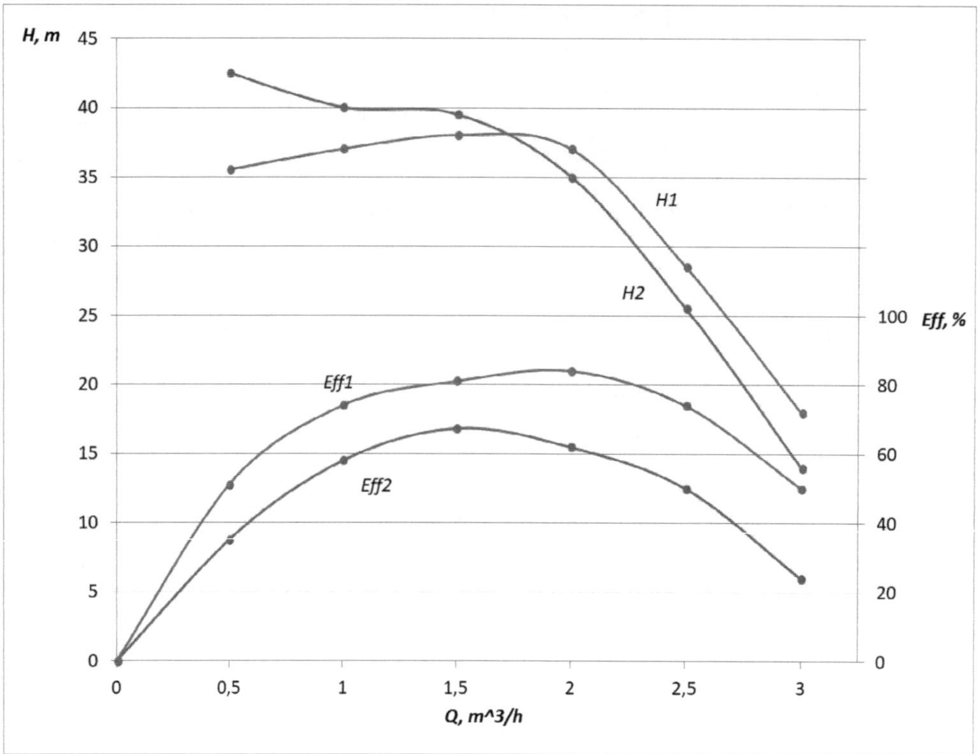

Fig.4. Dependence of head on the flow rate

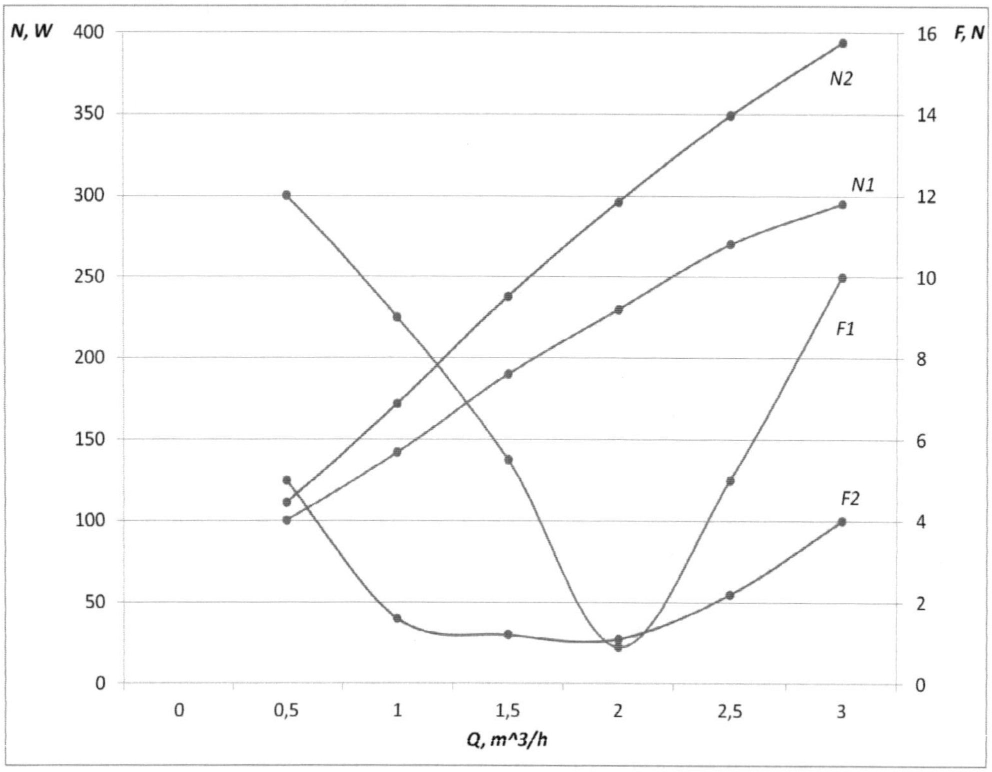

Fig.5. Dependence of efficiency and power on the flow rate

IOP Conf. Series: Materials Science and Engineering **589** (2019) 012029 doi:10.1088/1757-899X/589/1/012029

Discussions

The results of the study showed that introduction of a bladed stator can significantly reduce the radial force on the rotor when the pump is running in a non-optimal mode, but leads to a decrease in efficiency, head and increase in power and size. To minimize these effects, it is necessary to optimize the guide vane.

References

[1] A Protopopov and D Bondareva 2019 *IOP Conf. Ser.: Mater. Sci. Eng.* **492** 012002
[2] A Protopopov and V Vigovskij 2019 *IOP Conf. Ser.: Mater. Sci. Eng.* **492** 012003
[3] P Chaburko and Z Kossova 2019 *IOP Conf. Ser.: Mater. Sci. Eng.* **492** 012011
[4] V Lomakin et al 2019 *IOP Conf. Ser.: Mater. Sci. Eng.* **492** 012012
[5] A Gouskov et al 2019 *IOP Conf. Ser.: Mater. Sci. Eng.* **492** 012013
[6] N Egorkina and APetrov 2019 *IOP Conf. Ser.: Mater. Sci. Eng.* **492** 012015
[7] K Dobrokhodov and APetrov 2019 *IOP Conf. Ser.: Mater. Sci. Eng.* **492** 012016
[8] N Isaev 2019 *IOP Conf. Ser.: Mater. Sci. Eng.* **492** 012026
[9] S Korsakova and AProtopopov 2019 *IOP Conf. Ser.: Mater. Sci. Eng.* **492** 012032
[10] A Protopopov and C Jakovich 2019 *IOP Conf. Ser.: Mater. Sci. Eng.* **492** 012034
[11] A Shablovskiy and E Kutovoy 2019 *IOP Conf. Ser.: Mater. Sci. Eng.* **492** 012035
[12] A Petrov et al 2019 *IOP Conf. Ser.: Mater. Sci. Eng.* **492** 012036
[13] V Lomakin and O Bibik 2019 *IOP Conf. Ser.: Mater. Sci. Eng.* **492** 012037
[14] T Valiev and APetrov 2019 *IOP Conf. Ser.: Mater. Sci. Eng.* **492** 012038
[15] V Cheremushkin and V Lomakin 2019 *IOP Conf. Ser.: Mater. Sci. Eng.* **492** 012039

IOP Conf. Series: Materials Science and Engineering **589** (2019) 012030 doi:10.1088/1757-899X/589/1/012030

Investigation of hermetic low-speed pumps performance when operating on highly viscous liquids

A Petrov[1], A Morozov[1,2] and M Kuleshova[1]

[1]Bauman Moscow State Technical University, 5 Second Baumanskaya Street, Moscow, 105005, Russian Federation

[2]E-mail: alexeymorozov12@mail.ru

Annotation
During the operation of hermetic low-speed vane pumps in a wide range of viscosities of the working fluid, their characteristics change significantly. Experimental dependencies of the effect of viscosity on these characteristics for hermetic vane pumps do not exist. To determine these changes, 3D-models of the flow parts for various pumps are designed. By means of hydrodynamic modeling characteristics for each model are calculated when operating on highly viscous liquids. The results obtained are summarized in the form of relative head ratio and relative efficiency, which determine the performance characteristics of a hermetic pump for various operating modes. A comparison of the results with existing ones is carried.

Introduction

To date, there is no single method for designing centrifugal pumps operating in a wide range of the working fluid viscosities. First, the fundamental works of D.Ya. Sukhanov. [1] should be noted. In his work Sukhanov D.Ya. conducted a series of experiments on typical models of centrifugal pumps operating on viscous fluids. The results were presented in the form of dimensionless coefficients of performance recalculation, allowing based on the existing characteristics of the pump running on water to obtain a new characteristic when working on another viscosity of the working fluid. Also, Sukhanov D.Ya. formulated recommendations for the design of pumps operating in a wide range of viscosities in order to improve its energy efficiency. High efficiency is one of the key requirements imposed on manufactured machines [2]-[4]. Since Sukhanov studied the performance of typical centrifugal pumps these methods are not applicable for designing low-speed pumps. In such pumps hydraulic friction in relatively narrow channels of the stators significantly affects the performance of the device.

Recently, technical requirements to modern mobile installations have increasingly stiffened and can only be satisfied with introduction of up-to-date simulation methods [5]-[8]. Lately the development of computer technology in the field of CFD modeling has made it possible to accurately predict hydrodynamic processes occurring during the operation of centrifugal pumps [9]-[15]. With the help of hydrodynamic modeling, it became possible to design pumps without conducting multiple full-scale tests.

Method

In this work the main task was to study changes in the characteristics of low-speed hermetic pumps when operating on highly viscous liquids by means of hydrodynamic modeling with the aim of reducing the number of full-scale tests [16]. In order to fulfill it, a two-stage pump design with shrouded impellers and a discharge device in a form of a three-channel guide vane was developed. A 3D - model of the pump flow part is shown in Fig. 1.

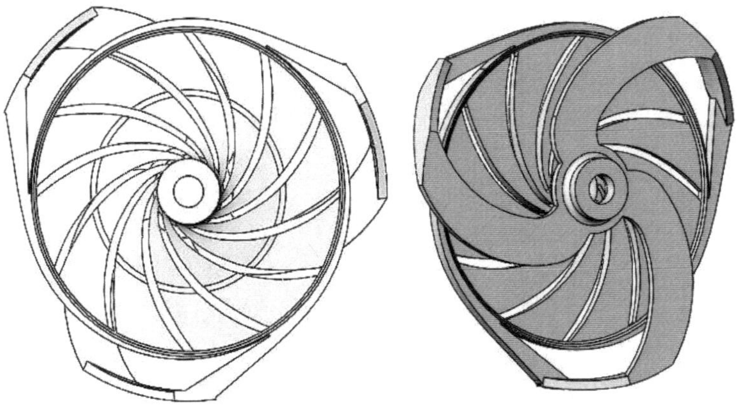

Fig. 1. 3D – model of the pump flow part.

Nominal specifications of the pump are the following:

$$H = 40\ \text{m};\ Q = 10\frac{1}{\min} = 1.67\cdot 10^{-4}\text{m}^3/\text{s};\ n = 2900\ \text{rpm} = 48.3\ \text{rps}$$

Working fluid: ethylene glycol antifreeze.
The required range of the kinematic viscosity of the working fluid:

$$\nu = [4;800]\cdot 10^{-6} m^2/s$$

As the object of the study 4 pump models with extended impellers and guide vanes were additionally designed [17], [18]. Parameters of the configurations are presented in Table 1 and Table 2

Table 1. Impellerdimensions: D2 – diameter of the impeller at the outlet; b2 –widthofthemeridionalsectionchannelontheD2 diameter; D1 –diameter of the impeller at the inlet; db – diameter of the barrel; z – number of impeller blades.

Model #	D2, mm	b2, mm	D1, mm	db, mm	z
1	122	2	25	12	12
2	122	3	30	12	12
3	122	4	35	12	12
4	122	5	40	12	12
5	122	6	45	12	12

Table 2. Guidevanedimensions: b3 –width of the spiral channel; h1 –height of the spiral channel;h2 –height of the diffuser; h3 – height of the shifting channel; h4 – height of the return channel.

Model #	b3, mm	h1, mm	h2, mm	h3, mm	h4, mm
1	4	4,2	8,4	3	15
2	5	4,2	8,4	4,6	16,3
3	6	4,2	8,4	5,7	18,3

IOP Conf. Series: Materials Science and Engineering **589** (2019) 012030 doi:10.1088/1757-899X/589/1/012030

4	7	4,2	8,4	6,5	19,9
5	8	4,2	8,4	7,4	21,5

Numerical modeling is based on the equation of continuity, the Navier-Stokes Reynolds averaged equations and two additional differential equations that are responsible for modeling turbulence structures. In this case k-ω SST turbulence model was used [19]. A structured prismatic computational grid with 5 layers near solid walls and an unstructured multi-faceted mesh in the flow core were used to solve the equations. The number of mesh cells for each model was more than 450 000, Fig.2.

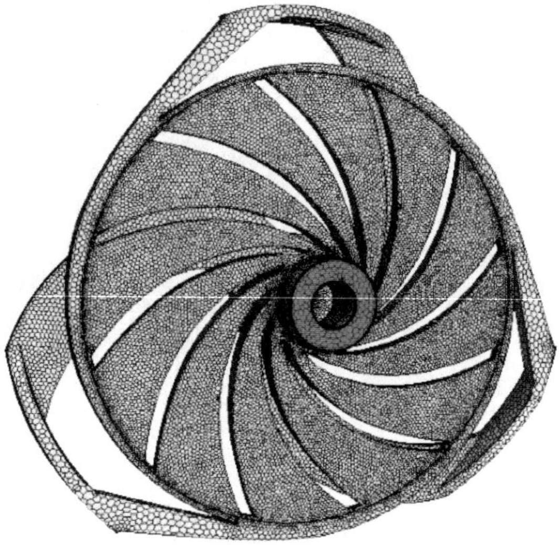

Fig. 2. Calculation grid for the pump under consideration.

Results

For all models normal characteristics were calculated. Obtained data was represented as the following dependencies h(q) andη(q), where $h = H/(nD)^2$ - dimensionless head, $q = Q/nD^3$-dimensionless flow rate. These dependences are presented in Fig. 3 and Fig. 4.

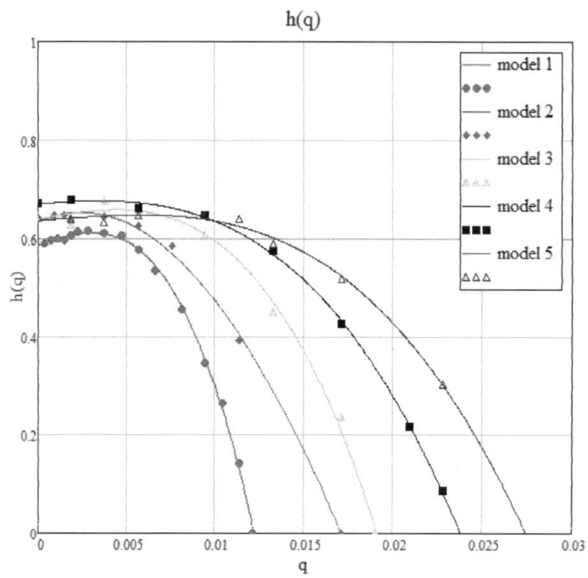

IOP Conf. Series: Materials Science and Engineering **589** (2019) 012030 doi:10.1088/1757-899X/589/1/012030

Fig. 3. Dependence of the dimensionless head on the dimensionless flow rate for the models under consideration with kinematic viscosity $6 \cdot 10^{-6} m^2 / s$

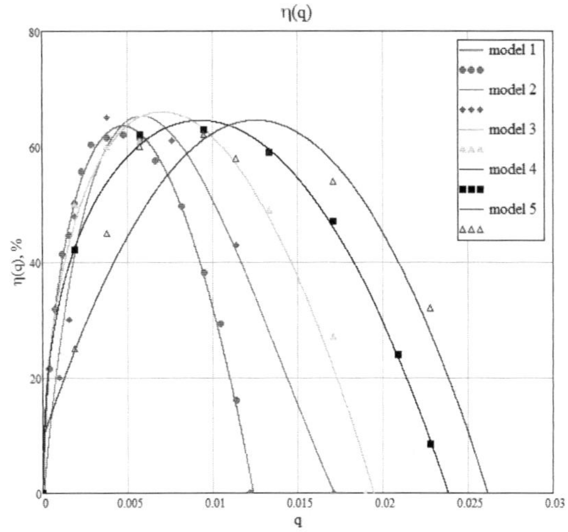

Fig. 4. Dependence of the efficiency on the dimensionless flow rate for the models under consideration with kinematic viscosity $6 \cdot 10^{-6} m^2 / s$

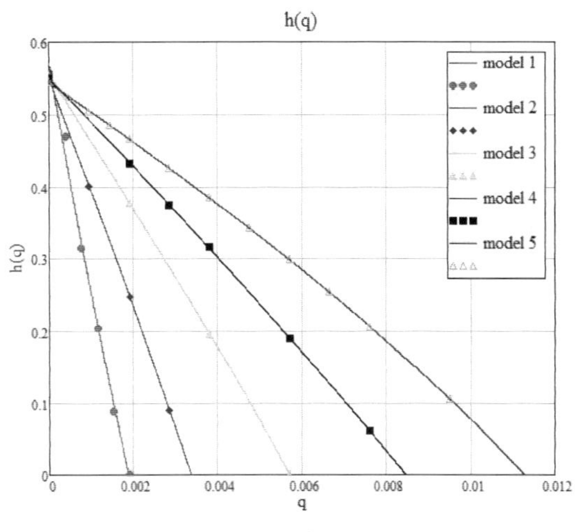

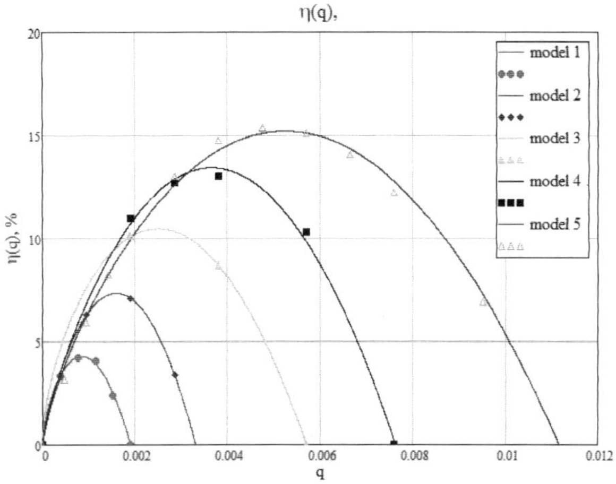

IOP Conf. Series: Materials Science and Engineering **589** (2019) 012030 doi:10.1088/1757-899X/589/1/012030

Fig. 5. Dependence of the dimensionless head and efficiency on the dimensionless flow rate for the models under consideration with kinematic viscosity $\nu = 400 \cdot 10^{-6} m^2 / s$

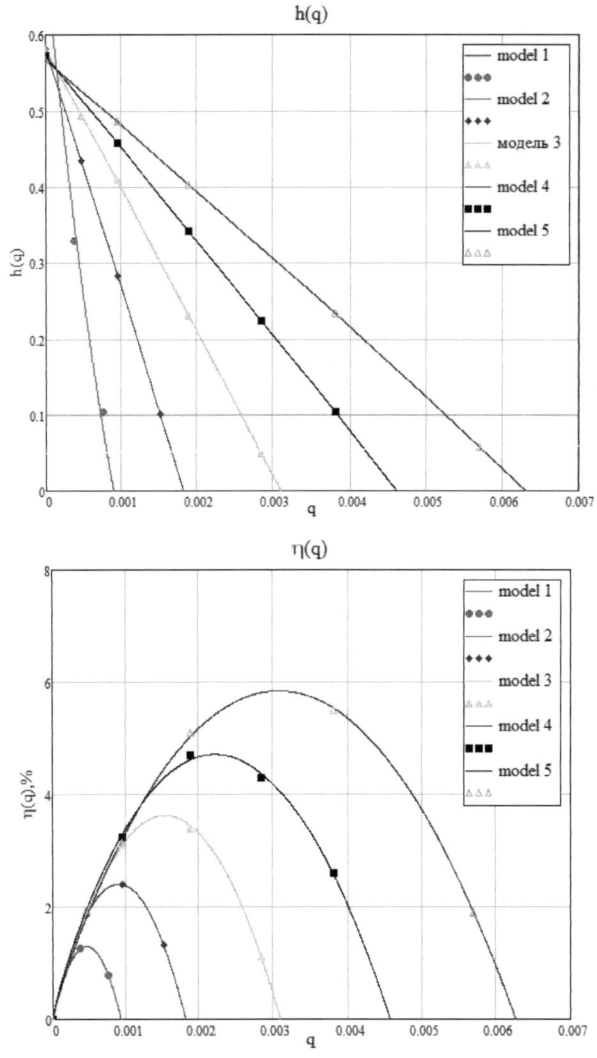

Fig. 6. Dependence of the dimensionless head and efficiency on the dimensionless flow rate for the models under consideration with kinematic viscosity $\nu = 800 \cdot 10^{-6} m^2 / s$

By viscosity of the working fluid increasing, pump characteristics begin to drop significantly, Fig. 5 and Fig. 6.

H'(Re) characteristic (Fig. 7)represents the dependence of the relative head ratio on the blade Reynolds number: $h' = h_\nu / h_\omega$, where h_ν — head calculated when the pump is operating on a viscous fluid, h_ω - head calculated when the pump is running on water.

η'(Re) characteristic (Fig. 8) represents the dependence of the relative efficiency on the blade Reynolds number: $\eta' = \eta_\nu / \eta_\omega$, where η_ν –efficiency calculated when the pump is operating on a viscous fluid, η_ω - efficiency calculated when the pump is operating on water.

It can be noted that the characteristic of Sukhanov D.Ya. differs from the results of the simulations, Fig. 7 and Fig. 8. Since the width of the impeller and the guide vane flow part of low-speed pump are relatively small, hydraulic losses are more significant there [20] - [22].

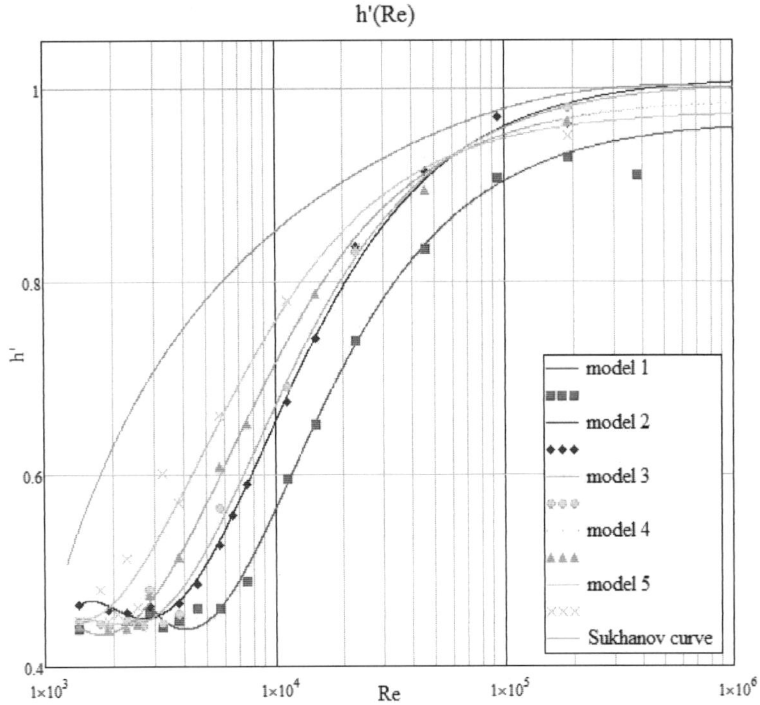

Fig. 7. The dependence of the relative head ratio on the blade Reynolds number.

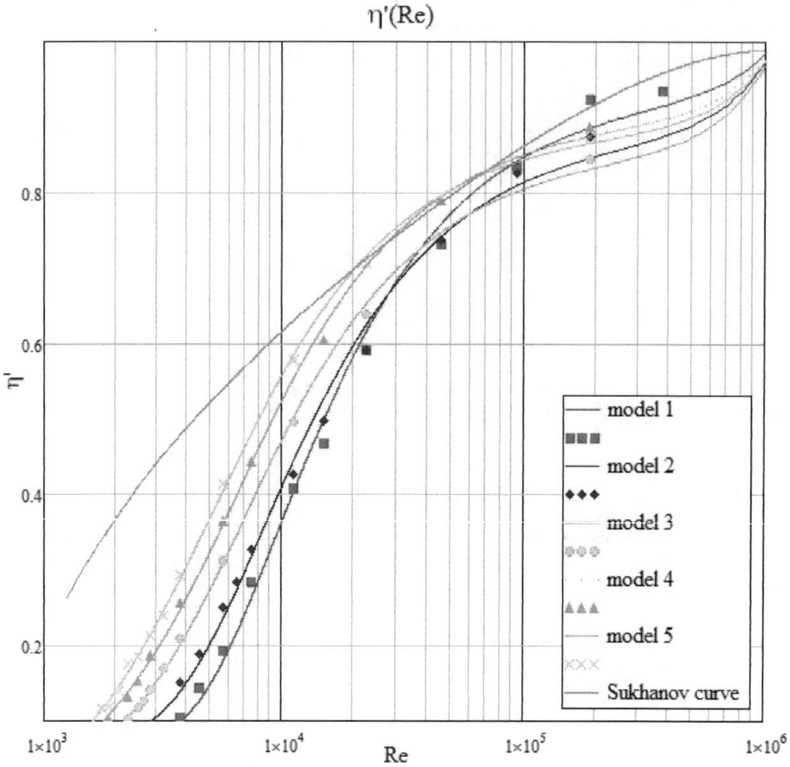

Fig. 8. The dependence of the relative efficiency on the blade Reynolds number.

Discussion

Simulation has shown that with an increase in the viscosity of the working fluid, the pressure characteristic of an ultra-low-speed vane pump drops significantly faster than that of pumps with greater specific speed. At the same time, the pump efficiency in the design mode decreased from 55 % to 21 %. Increasing the guide vanes channel swidth managed to extend the pump operation range by 116 % with the viscosity of the working fluid up to $800 \cdot 10^{-6} m^2 / s$.

References

[1] Sukhanov D.Ya. The operation of vane pumps on viscous liquids. Moscow, Mashgiz, 1952, pp. 1-34

[2] P Chaburko and Z Kossova 2019 *IOP Conf. Ser.: Mater. Sci. Eng.* **492** 012011

[3] A Petrov et al 2019 *IOP Conf. Ser.: Mater. Sci. Eng.* **492** 012040

[4] A Protopopov and C Jakovich 2019 *IOP Conf. Ser.: Mater. Sci. Eng.* **492** 012034

[5] Hao, H., Xinkai, L., Bo, G. Hydraulic optimization of multiphase pump based on CFD and genetic algorithm (2015) International Journal of Grid Distribution Computing, 8 (6), pp. 161-170

[6] Ji, C., Wang, Y., Yao, L. Numerical analysis and optimization of the volute in a centrifugal compressor (2007) Challenges on Power Engineering and Environment - Proceedings of the International Conference on Power Engineering 2007, ICOPE 2007

[7] Lomakin, V.O., Petrov, A.I., Stepanyuk, A.I. Optimization of geometrical parameters of tap of oil main pump of NM type (2012) Science and Education: Electronic Scientific Publication of BMSTU (accessed 10.11.2018) http://technomag.bmstu.ru/doc/347727.html

IOP Conf. Series: Materials Science and Engineering **589** (2019) 012030 doi:10.1088/1757-899X/589/1/012030

[8] Lomakin, V.O., Chaburko, P.S., Kuleshova, M.S. Multi-criteria Optimization of the Flow of a Centrifugal Pump on Energy and Vibroacoustic Characteristics (2017) Procedia Engineering, 176, pp. 476-482
[9] N Egorkina and A Petrov 2019 *IOP Conf. Ser.: Mater. Sci. Eng.* **492** 012015
[10] T Valiev and A Petrov 2019 *IOP Conf. Ser.: Mater. Sci. Eng.* **492** 012038
[11] A Protopopov and V Vigovskij 2019 *IOP Conf. Ser.: Mater. Sci. Eng.* **492** 012003
[12] A Protopopov and D Bondareva 2019 *IOP Conf. Ser.: Mater. Sci. Eng.* **492** 012002
[13] A Gouskov et al 2019 *IOP Conf. Ser.: Mater. Sci. Eng.* **492** 012013
[14] Lomakin, V.O., Kuleshovav, M.S., Bozh'eva, S.M. Numerical Modeling of Liquid Flow in a Pump Station (2016) Power Technology and Engineering, 49 (5), pp. 324-327
[15] Lomakin, V., Cheremushkin, V., Chaburko, P. Investigation of vortex and hysteresis effects in the inlet device of a centrifugal pump (2018) 2018 Global Fluid Power Society PhD Symposium, GFPS 2018, № 8472374
[16] K Dobrokhodov and A Petrov 2019 *IOP Conf. Ser.: Mater. Sci. Eng.* **492** 012016
[17] Sobol, I.M. (1973) Monte Carlo Numerical Methods, p. 311 (Moscow: NaukaPubl)
[18] A Petrov et al 2019 *IOP Conf. Ser.: Mater. Sci. Eng.* **492** 012036
[19] Wilcox D.C. Turbulence modeling for CFD. 3rd ed. DCW industries, 2006. pp. 1-515
[20] V Lomakin et al 2019 *IOP Conf. Ser.: Mater. Sci. Eng.* **492** 012012
[21] S Korsakova and A Protopopov 2019 *IOP Conf. Ser.: Mater. Sci. Eng.* **492** 012032
[22] Lomakin, V.O. (2015) Proceedings of 2015 International Conference on Fluid Power and Mechatronics

IOP Conf. Series: Materials Science and Engineering **589** (2019) 012031 doi:10.1088/1757-899X/589/1/012031

Study of the corrosion rate of the paths walls of aluminum heat exchangers in the ethylene glycol antifreezes environment under developed aeration of power fluid

A Petrov[1,2], G Saatashvili[1], A Averianov[1] and I Bilenko[1]

[1]Bauman Moscow State Technical University, 5 Second Baumanskaya Street, Moscow, 105005, Russian Federation

[2]E-mail: alexeypetrov@bmstu.ru

Annotation. An experimental and theoretical research work on the corrosion processes of heat exchanger channels made from different aluminum alloys in an ethylene glycol antifreeze environment conducted by the authors is described in the article. The hydrodynamic simulation of the flow in the paths was performed. Significant effects of the aeration of the power fluid, the possibility of electrochemical corrosion are proved. The comparative of anti-corrosion property of different aluminum alloys is shown when operating in a wide range of power fluid velocities and temperatures.

Introduction

Currently, heat exchangers made of aluminum alloys are widely used in different technical fields. Aluminum alloys, due to their low density, simple machining and high thermal conductivity, are used to manufacture shell and tube heat exchangers of all types that are used, among others, in engine cooling systems, communications-electronics equipment, hydraulic actuators, etc. Often, ethylene glycol antifreezes of different formulations, such as Tosol-40, Tosol-65 «Lena», etc., are used as power fluids for operation of such systems at low ambient temperatures.

The practice of using such hydraulic systems has shown that in some instances the paths of heat exchangers are intensively corroding, up to the formation of open pittings, and this process develops quite quickly, within a few hundred operation hours. Possible causes of this corrosion may be the presence of a large amount of dissolved air in the liquid (aeration), the presence of copper or bronze elements in the system (electrochemical corrosion), and a surface defects of the heat exchanger paths. This article includes the results of theoretical and experimental studies of corrosion processes in such heat exchangers made of various aluminum alloys.

Methods

Originally estimated, that in the heat exchanger paths developed cavitation erosion can occur, in order to determine the possible zones of which it was necessary to carry out a hydrodynamic simulation of flow. In the future, the flow model in the paths was dilated to atwo-phase flow model, taking into account the occurrence of gas cavitation [1-5].

The 3-D model [6-15] of the heat exchanger paths was created and the flow simulation in the paths, based on score equations and boundary condition described below was carried out.

In general Navier-Stokes vector equation for fluid is written as:

$$\frac{\partial \vec{v}}{\partial t}+(\vec{v}\cdot\nabla)\vec{v}=-\frac{1}{\rho}\nabla p+\nu\Delta\vec{v}+\vec{f}$$

Wheret - is time, ∇- is Nabla, Δ- vector Laplace operator, ρ- is density, p - is pressure, ν - is kinematic viscosity, $\vec{v}$ - velocity field, $\vec{f}$ - field of body forces.

The continuity equation is also applied:

$$\frac{\partial \rho}{\partial t}+\operatorname{div}\rho v=0$$

For the simulation of aeration fluid flow was created a two-phase flow model, consisting of water and air under specified values of density and viscosity. Proportional ratio: 95% of water and 5% of air. For the calculation, physical models of gravity force and a k-ω turbulence model were used. The velocity inlet was set in accordance with the required flow rate; the outlet pressure was set as zero.

For creation of mesh the following models were used: surface remesher, polyhedral mesher (base size 0,5 mm), prism layer mesher (number of layers – 5, stretching– 1,3, total thickness – 45% of base size) [1-3]. The simulation results were compared with experimental results.

For replication of conditions for the corrosion occurrence, the test bench was developed and created, simulating a real system.

The created test bench all owed to pass through the paths required fluid flow rate (ethylene glycol antifreeze Tosol-65) at the temperature up to 70°C and at the same time to provide aeration of power fluid (through the spout). In which case, accumulated operating time of heat exchangers made from different aluminum compositions (heat exchangers made from alloy D16T [aged-hardening duralumin] and AMg6 [93,5% Al, 5,5% Mg]was up to 140 hours.

Heat exchangers were connected as one by one, as well two and two in series through quick disconnect coupling (QDC), similar to using in real system. The diagram of the test bench is shown in figure 1. To simulate the presence of bronze elements in the real system, inserts and fittings made from bronze without galvanic isolation (flexible pipe in a metal braid were used) between them and the heat exchangers were used in the test bench. The pressure at the inlet of the heat exchanger (before the QDC) was maintained within 3…4 bar, which corresponds to the pressure in the real system.

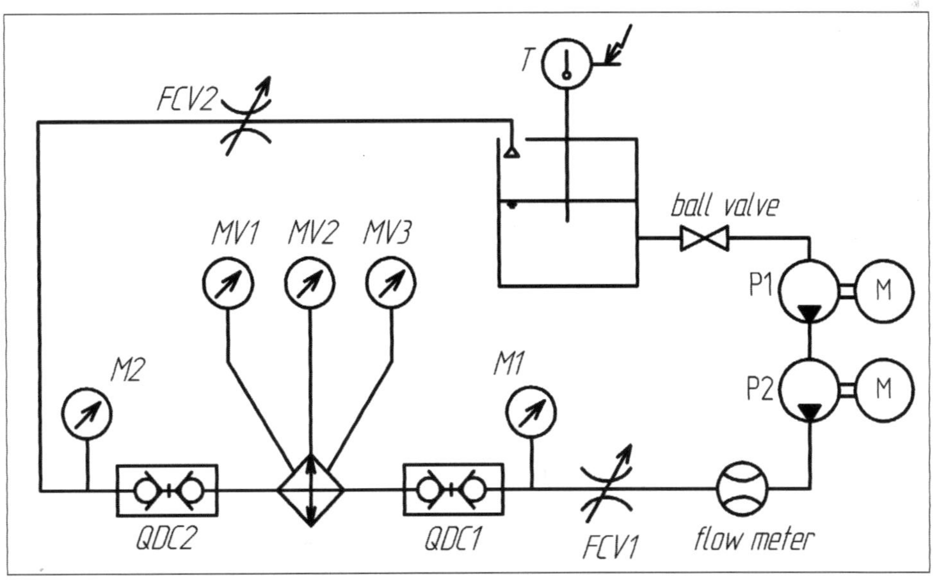

Figure 1. Diagram of the test bench

During the experiments, the temperature and flow rate of power fluid, pressure difference in the heat exchanger elements, operating time, noise and vibration levels on the heat exchanger surfaces

IOP Conf. Series: Materials Science and Engineering **589** (2019) 012031 doi:10.1088/1757-899X/589/1/012031

were monitored. After operating time, the condition of heat exchanger paths was estimated through factory-made endoscope, then by cutting a part of paths to access the inner surface.

Results

According to the hydrodynamic simulation results, the following conclusions can be drawn: the occurrence of steam cavitation in the heat exchanger paths is unlikely, because zones of pressure reduction equals to saturated vapor pressure were not detected (figures 2, 3). This pressure distribution pattern was also corroborated by measuring pressures in the holes (about 20 holes were made) along the heat exchanger paths.

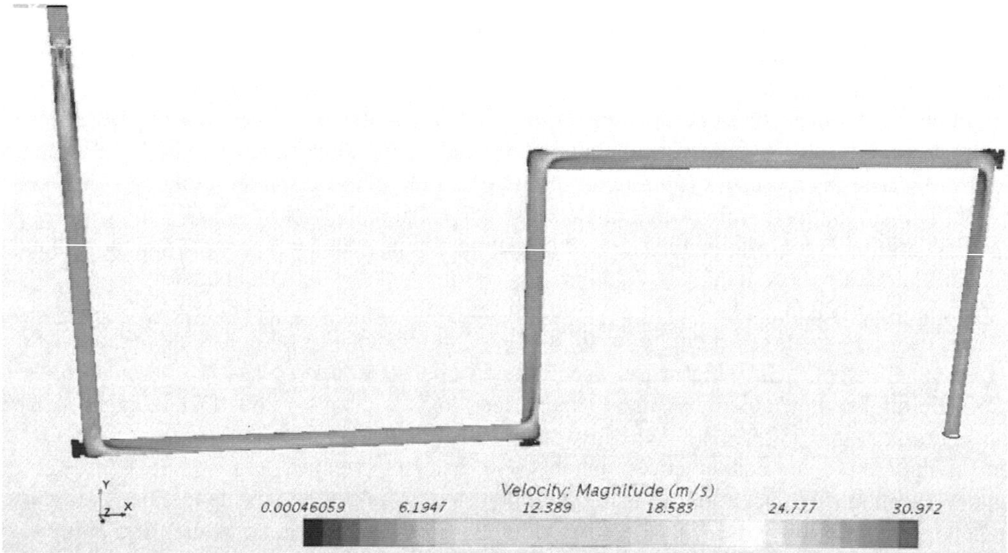

Figure 2. Velocity field distribution pattern in the heat exchanger paths at the flow rate 14 l/min

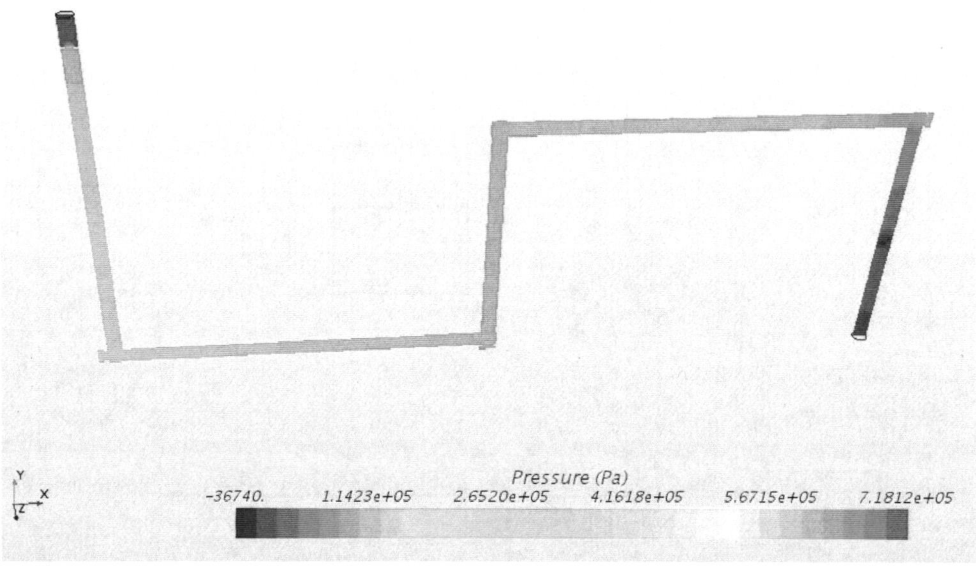

Figure 3. Pressure distribution pattern in the heat exchanger paths at the flow rate 14 l/min

There are no also anomaly high velocity zones, except of zone proximately after the QDC and particularly after flow deflection. However, the rapid decay of pressure after QDC, including the

IOP Conf. Series: Materials Science and Engineering **589** (2019) 012031 doi:10.1088/1757-899X/589/1/012031

laminar flow (figure 4), which causes a high probability of gas cavitation, i.e. release of dissolved gas, with a significant aeration of the fluid.

The presence of low flow velocity zones along path centerline of heat exchanger (figure 5), which combined with gas releases and low flow velocity in general can create a local high oxygen content zones in the upper medium paths zones at some distance of flow deflections. The hydraulic simulation of two-phase laminar flow (flow rate is 2 l/min) is shown in figure 6, a zones of high gas content in the upper path part are visible.

It may be concluded that the presence of sharp narrowing and increase of flow, and flow deflection, which combined with high aeration level of fluid, cause intensive gas releases and laminar flow with low velocity results to movement of gas bubbles to upper part of path, causing a large amount of oxygen in wall layers.

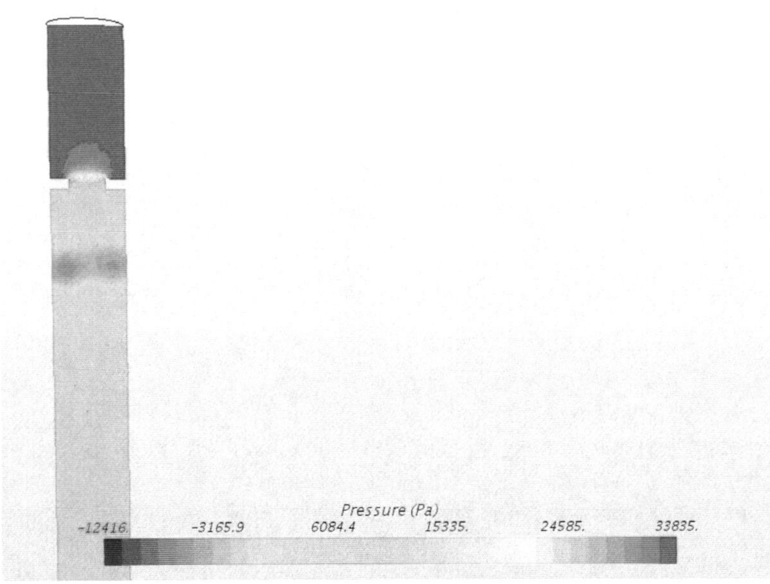

Figure 4. Pressure reduction zone after local flow narrowing in the heat exchanger inlet

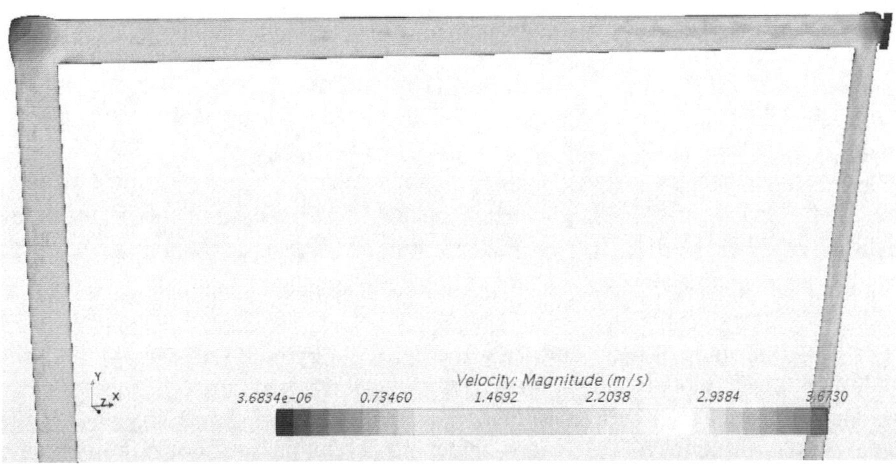

Figure 5. Local zones of velocity reduction

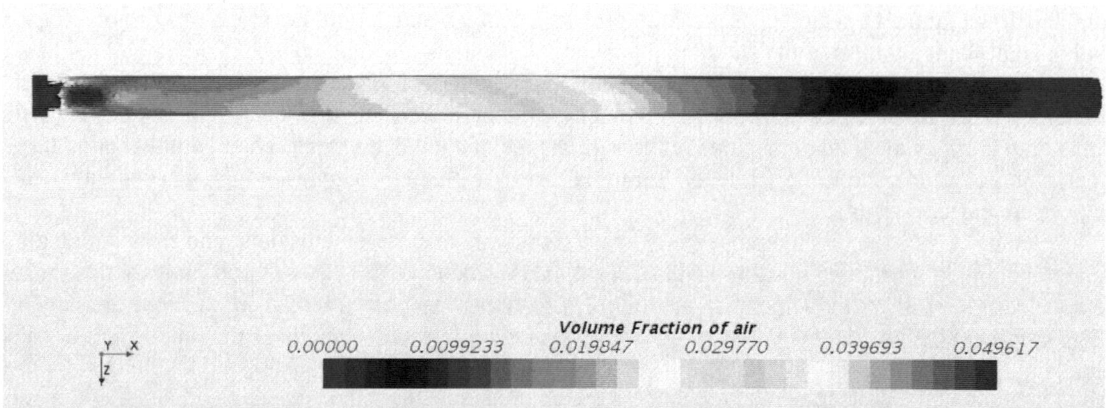

Figure 6. Two-phase laminar flow of fluid motion

The occurrence of intensive gas cavitation during the operating is confirmed by acoustic image of fluid flow in the heat exchanger [2]. The acoustic spectrum of the results of noise level measurement among the path (spectrum corresponds to zone after inlet part) is shown in figure 7. White shows the acoustic background in the room, orange - the noise in the path. Visible peak at frequency of about 1 kHz, arising due to the occurrence of steam cavitation. When the flow rate decreases, the peak amplitude decreases, but it does not disappear completely as long as there is a flow process.

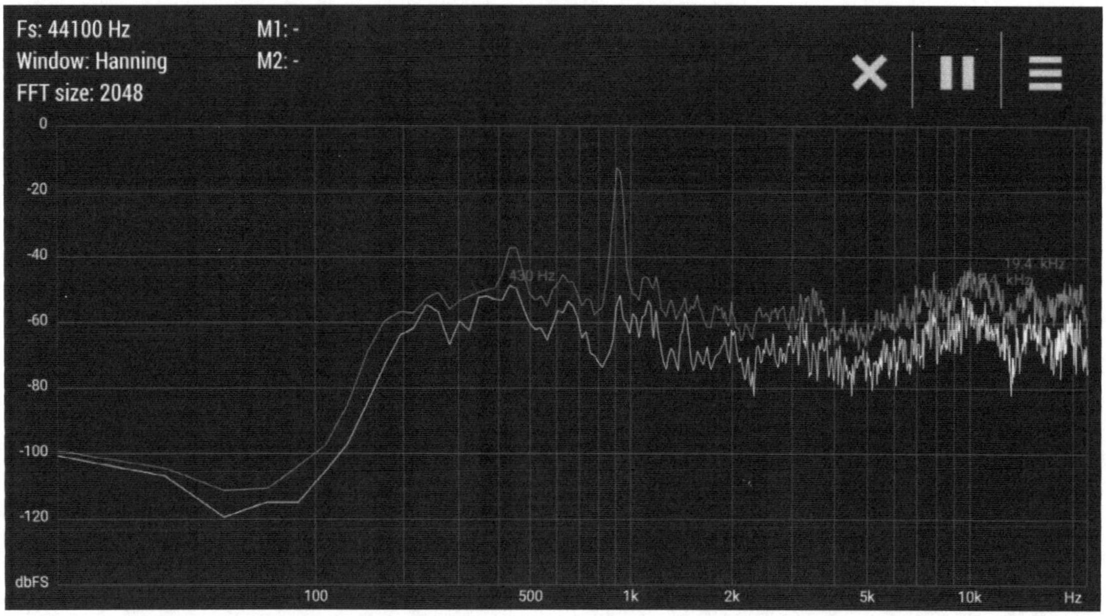

Figure 7. Noise levels measurement during the testing

After the testing, the study of paths surface was carried out by bore scope and milling for opening of some path parts. Among other things, was created the fluid precipitation in the reservoir, where numerous aluminum particles up to 0,1 mm in size were found out. Signs of corrosion in the opening path of heat exchanger made from D16T [aged-hardening duralumin] are shown in figures 8 and 9, the inner surface of the same path is shown in figure 10.

Figure 8. Corrosion of the path surface

Figure 9. Open pitting corrosion on inside of heat exchangerpath

IOP Conf. Series: Materials Science and Engineering **589** (2019) 012031 doi:10.1088/1757-899X/589/1/012031

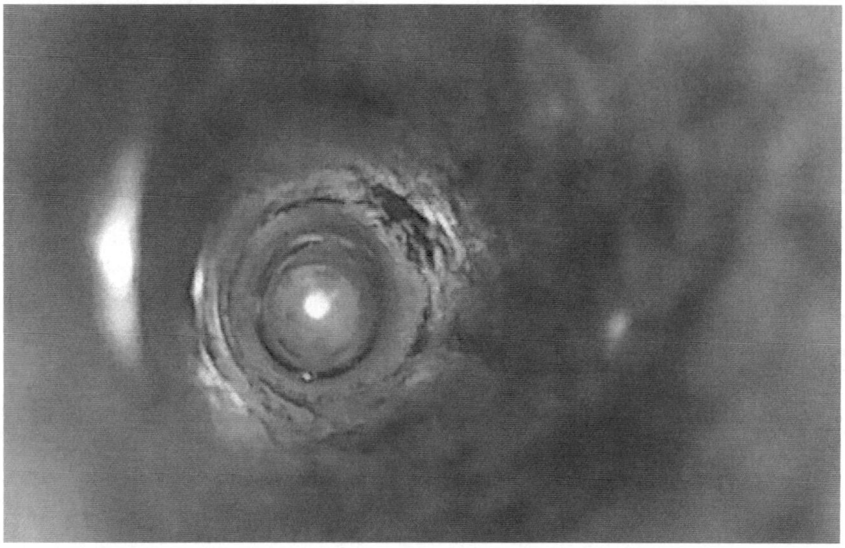

Figure 10. The path of heat exchanger made from D16T [aged-hardening duralumin], operating time is 130 hours, corrosion is visible

The effect of spray coating Sn-Bi on the corrosion developing is interesting. Zone of the path interface with and without spray coating that, perhaps, proves electrochemical type of corrosion [10,11].

Figure 11. Corrosion zones in the path with and without spray coating

The testing of heat exchanger made from AMg6 [93,5% Al, 5,5% Mg] has shown the absence of corrosion signs on the surface (figure 12):

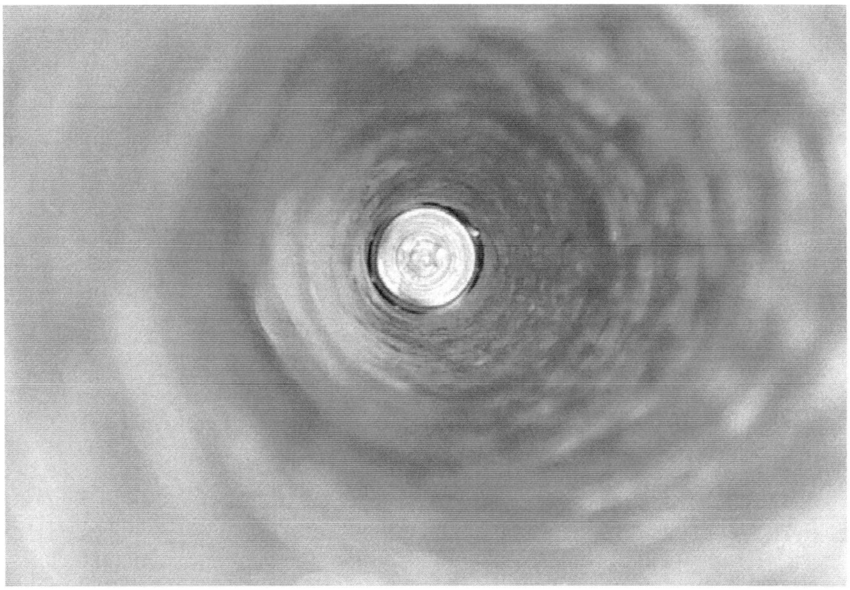

Figure 12. The path of heat exchanger made from AMg6 [93,5% Al, 5,5% Mg], operating time is 100 hours, there are no corrosion signs (there are signs of treatment)

Discussion

1. Hydrodynamic simulation and acoustic flow pattern showed that in the heat exchanger in the zone of rapid decay of pressure gas can be released. It appears to suggest that the released gas can (especially at low velocities of fluid in the paths) be accumulated in the upper parts of paths and thereby sharply accelerate the corrosion of the walls of the paths in the presence of excess oxygen;

2. The presence of copper color tone on the paths walls and a lot of corrosion rate mean the possible electrochemical corrosion in the system, especially in copper aluminum combination.

3. D16T [aged-hardening duralumin]alloy in the circumstances is extremely unstable to corrosion, in any case, without additional coating the inner surface of the paths walls;

4. D16T [aged-hardening duralumin] alloy within the time of testing, proved to be corrosion resistant;

5. The detection of a significant amount of small aluminum pieces shows the developed cavitation of the paths walls with carry-over of metal from the affected zones.

List of references

[1] Lomakin V O, Chaburko P S, Kuleshova M S 2017 Multi-criteria Optimization of the Flow of a Centrifugal Pump on Energy and Vibroacoustic Characteristics Procedia Engineering 176 pp 476-482

[2] Lomakin V O, Kuleshova M S, Kraeva E A 2015 Fluid Flow in the Throttle Channel in the Presence of Cavitation Procedia Engineering 106 pp 27-35

[3] Ionaitis, R.R., Chain, P.L. Principles for engineering active-passive hydrodynamic control-and-safety systems for nuclear reactors (2009) Atomic Energy, 106 (3), pp. 175-184.

[4] Krapivtsev, V.G. Model Studies of Coolant Flow Hydrodynamics in VVER-1000 In-Reactor Pressure Channel (2017) Atomic Energy, 122 (5), pp. 304-310.

[5] Marchevskii, I.K., Puzikova, V.V. Numerical simulation of the flow around two fixed circular airfoils positioned in tandem using the LS-STAG method (2016) Journal of Machinery Manufacture and Reliability, 45 (2), pp. 130-136.

[6] K.A. Morch. Dynamics of Cavitation bubbles and Cavitating liquids //Treatise on Materials Science and Technology, 1976. 16. P. 309-335.
[7] Kendrick H. Light. Development of cavitation erosion resistant advanced material system. Orono (Maine, USA) The University of Maine, 2005. 76 p.
[8] Patankar S. Numerical methods in heat transfer and fluid dynamics // Translated from English (1980), Energoatomizdat publishing, 1984. - 152 p.
[9] Petrov A.I., Skobelev M.M., Khanychev A.G. Research of comparative resistance to cavitation erosion of material samples and coatings of the flow section of hydraulic machines // Herald of the BMSTU. “Mechanicalengineering”. 2015. - №2
[10] Keshe G. Corrosion of metals. Physical and chemical principles and actual problems. Translated from German. / G. Keshe. Edited by IA.M. Kolotyrkina. Metallurgypublishing, 1984. – 400 p.
[11] Siniavskii V. S. Corrosion and protection of aluminum alloys. 2nd publ., translated and completed by V.S. Siniavskii, V. D. Valkov, V. D. Kalinin. Metallurgypublishing, 1986. – 368 p.
[12] N Egorkina and A Petrov 2019 *IOP Conf. Ser.: Mater. Sci. Eng.***492** 012015
[13] N Isaev 2019 *IOP Conf. Ser.: Mater. Sci. Eng.***492** 012026
[14] A Shablovskiy and E Kutovoy 2019 *IOP Conf. Ser.: Mater. Sci. Eng.***492** 012035
[15] V Cheremushkin and V Lomakin 2019 *IOP Conf. Ser.: Mater. Sci. Eng.***492** 012039

IOP Conf. Series: Materials Science and Engineering **589** (2019) 012032 doi:10.1088/1757-899X/589/1/012032

The effect of combined processing schedules on the value and nature of residual stresses in the surface layer of cylindrical friction pairs

S Fedorov[1], A Albagachiev[2,4], M Isaenkova[3], A Yakovleva[1,5], V Zaripov[1] and R Minushkin[3]

[1]Bauman Moscow State Technical University, 5 Second Baumanskaya Street, Moscow, 105005, Russian Federation
[2]Mechanical Engineering Research Institute of the Russian Academy of Sciences (IMASH RAN)
[3]National Research Nuclear University MEPhI
[4]MIREA - Russian Technological University

[5]Email: yakovleva525@mail.ru

Abstract
The purpose of the paper was to study the macroscopic stressesafter the combined processing. The combined processing essentially consists in the sequential implementation of electromechanical processing and diamond smoothing. We obtained the formula for calculating the residual stresses arising in the surface layer after the combined processing. For convenience of calculations, a computer program was developed to calculate the parameter for estimating the residual stresses in the language Visual Studio 2017 C #. To determine the residual macrostresses, the standard x-ray method sin ^{2} ψwas used. The study was performed on samples of steel 40X.

Introduction

The operation of machines and mechanisms requires the work of friction pairs, i.e. rods, pistons, plugs, seats, spools, pins, rolls for grinding non-metallic materials. These are parts of engines, pumps, spinning machines including guide rollers, thread guides, tensioning devices, as well as cultivators and other machines, which are used in construction, automotive equipment, agricultural machinery, etc. Most of them are cylindrical [1-3]. Due to the low operational reliability of cylindrical friction pairs, the consumption of spare parts is high [2-7].

The reliability of the machines is directly related to the quality of the surface layer, which is characterized by geometrical and physico-mechanical parameters. The quality of the surface layer is crucial for the operational properties, i.e. wear resistance, corrosion resistance, contact fatigue resistance, etc. [2, 3, 8-12].

In many cases the level of residual stresses is an important parameter that determines the quality of the surface layer of products [4, 6, 13].

In the surface layer of the metal there are always various defects that serve as crack nuclei, therefore, the reliability of parts is determined mainly by the value of crack propagation work [1, 4, 8, 9, 14]. The compressive residual stresses appearing on the surfaces of parts due to various technological methods prevent the nucleation and propagation of fatigue microcracks, increasing their work. Therefore, for parts operating under conditions of slipping friction, especially in cases of fatigue and abrasive wear during operation, it is important to make a surface layer with compressive residual stresses [4, 5, 9, 10-17].

Description of the Experiment

In view of the foregoing, the technology of combined processing (CP) has been developed. The combined processing essentially consists in the sequential implementation of electromechanical processing and diamond smoothing [5, 12]. Using this technology, it is possible to technologically control the properties of the working surfaces of cylindrical parts and make compressive residual stresses on the surface [1, 4, 6, 12].

To test this assumption, the value of tangential residual stresses was theoretically calculated. In the present CP embodiment, the value of residual stresses will be composed of the following components:

$$\sigma_{res} = \sigma_{temp} + \sigma_{ph} + \sigma_{mech}, \tag{1}$$

where σ_{temp} is the residual temperature stress, σ_{ph} is the structural phase residual stress, σ_{mech} is the mechanical residual stress.

Examining in detail each component of formula (1), it is possible to derive a general view of the dependence of the tangential residual stress value on the combined processingschedules:

$$\sigma_{res} = \propto E \frac{2\alpha_m (P_g V_p + IU)}{\lambda_2 l_{tool}^2 \sqrt{\pi}} \left(\sqrt{ate}^{\left(\frac{-z^2}{4at}\right)} \right) - \frac{P_g}{\pi} * \left[\frac{3}{2\left(l_{tool}^2 + z^2\right)} - \frac{1,2}{l_{tool}^2} \ln\left(\frac{z}{2l_{tool}}\right) - \frac{3}{10 l_{tool}^2} \right] + \\ + E\left(1 - \sqrt[3]{\frac{\rho_1}{\rho_2}}\right) \tag{2}$$

z is the current coordinate, m; α_t is the heat sharing ratio (Sharon's heat absorption coefficient); P_g is thereduced pressure force of the tool during the combined processing, H; $V_p$is the sample rotation speed, m / min; I is the current strength, A; U is the voltage, B; λ_2 is the thermal conductivity of the sample material, $Wm^{-1}K^{-1}$; $l_{tool}$is the length of the tool contact with the work surface (the width of the roll electrode), m; ais the thermal conductivity of the sample material, m^2/s; t is the contact time, c; ρ_1, $\rho_2$is the density of the sample material and tool, respectively, kg/m^3; E is the modulus of elasticity of the sample material, Pa.

For convenience of calculations, a computer program was developed to calculate the parameter for estimating residual stresses by formula (1) in the language Visual Studio 2017 C #. All data is entered into a table, and at the output we obtain the value of the parameter of residual stresses depending on the combined processing schedules.

Experimental Result

An x-ray evaluation method was used to verify the theoretical values obtained.

A BRUKER D8 Discover X-ray diffractometer (Fig. 1) using chromium K-radiation was used to study the phase composition of the sample material and to measure the residual macrostress. Data processing was carried out using the software DIFRAC.EVA and DIFRAC.LEPTOS.

For this purpose, steel 40X samples were made with a diameter of 40 mm, a length of 200 mm and treated with a combined processing in the following schedules.

Electromechanical processing: current I is 2000 A; U voltage is 4 V; contact roller pressure P1 is 300 N; part rotation speed is 2.0 m / min; feed S1 is 1.2 mm / rev. Diamond smoothing: the pressure of the diamond smoother P2 is 300 N, feed S2 is 0.02 mm / rev, the radius of the diamond smoother is 1 mm.

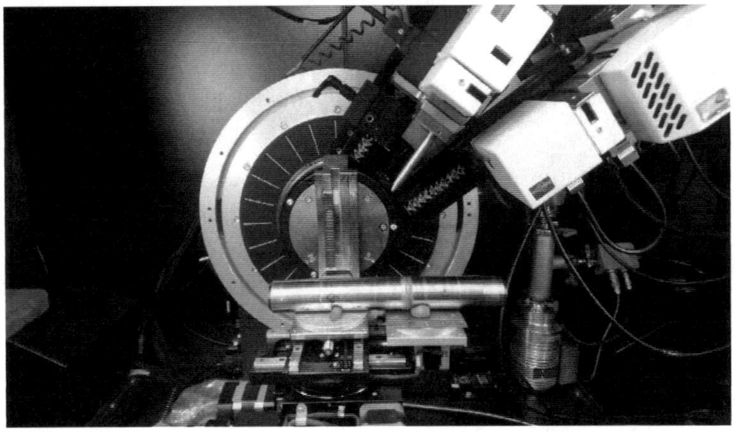

Fig. 1. X-ray diffractometer BRUKER D8 Discover

In accordance with GOST 16865-79 and with ASTM E2860-12, the most accurately residual macrostresses are measured in the precision X-ray spectrum area, i.e. the lines located at large angles of 2Θ, which also allow the use of large angles of rotation of the samples. To determine the residual macrostresses, the standard x-ray method $\sin^2\psi$ was used [18, 19]. The location of the investigated surface areas on the samples in which the measurement of stress was performed is schematically shown in Fig. 2. The size of the investigated area is determined by the size of the collimator and is ~ 5 mm^2.

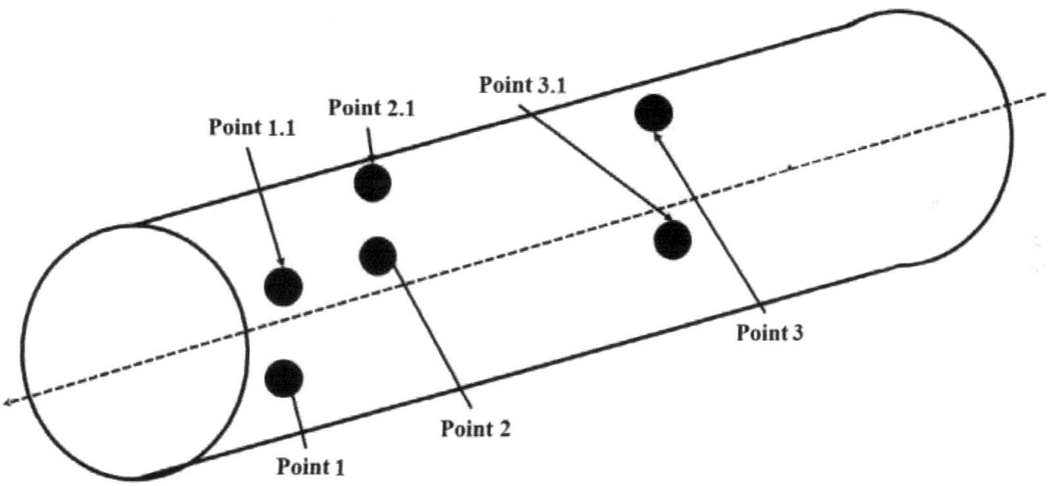

Fig. 2. The location of the investigated surface areas on the samples in which the measurement of stresses was done

In this case, to determine the value of the residual stresses, a line (211) (2Θ = 156.084 °) was recorded at different angles of rotation (Fig. 3, 4). To approximate and calculate the obtained lines, the Pearson VII function was used.

IOP Conf. Series: Materials Science and Engineering **589** (2019) 012032 doi:10.1088/1757-899X/589/1/012032

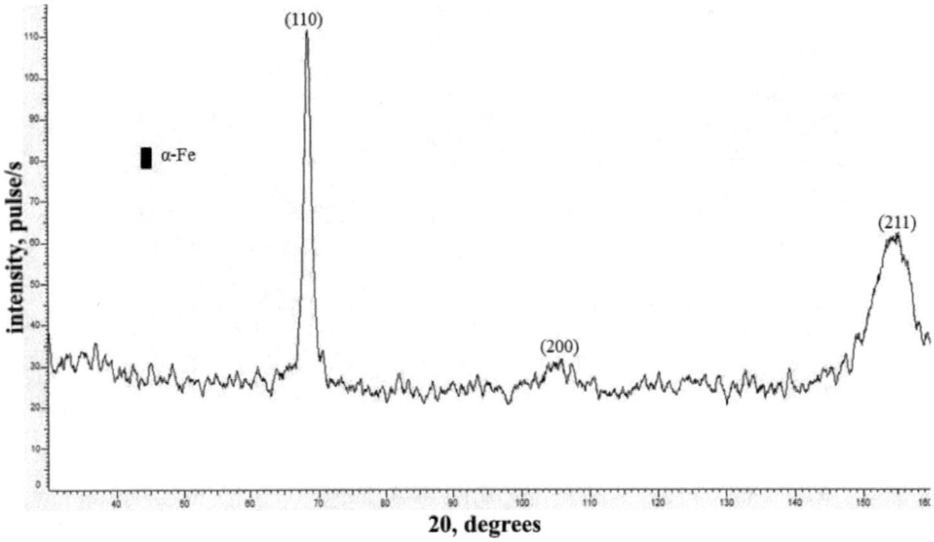

Fig.3. X-ray diffraction spectrum of the rod (the material of the anode of the x-ray tube is Cr)

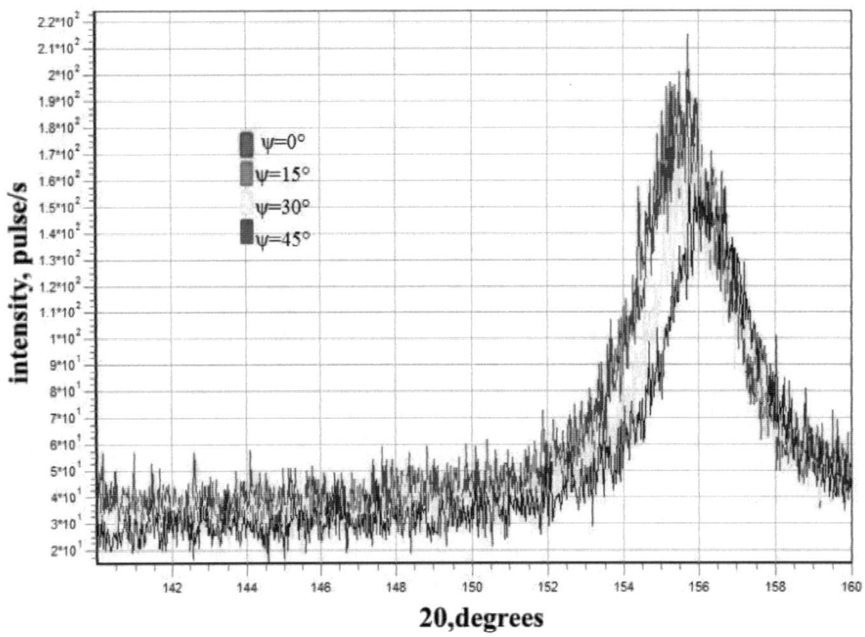

Fig. 4. X-ray line (211) for different angles of sample rotation

The interplanar distances, as well as the degree of deformations, are calculated from the angular positions of the X-ray lines, using the Wulff-Bragg formula. Further on, we plotted a graph of the dependence of the degree of deformations (ε) on sin2ψ (where ψ is the angle of sample rotation). In the general case, the dependence graph has a linear character ($\varepsilon=a*sin2\psi+b$), from the slope coefficient of which the value of the residual macrostress is calculated (Fig. 5). The error in the determination of macrostress is the deviation of the values obtained from the linear dependence [20, 21].

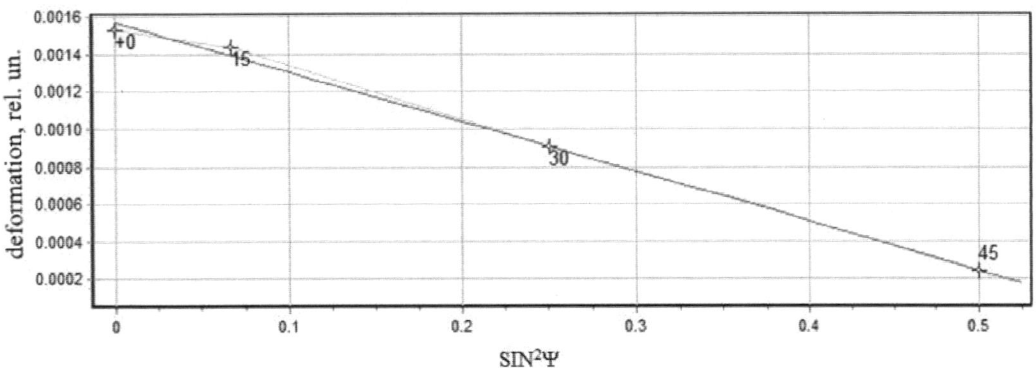

Fig. 5. A graph of the dependence of the value of deformations on sin2ψ

During the study, the following results were obtained:according to the obtained phase analysis data, the material of the studied samples is anα-phase (bcc-structure) of iron.

The stresses at the sample control points (Fig. 2) are presented in the table.

Stresses at sample control points

	Stress, MPa	
	Axial	**Tangential**
Point 1	-1901,2±87,7	-866,9±21,5
Point 1.1	-1774,2±91,5	-1409,1±44,1
Point 2	-1738,7±129,1	-1420,0±186,3
Point 2.1	-2019,8±172,5	-1127,5±114,9
Point 3	-760,5±14,6	-975,6±41,3
Point 3.1	-806,1±43,4	-455,7±17,8

This method allows us to determine the numerical value of axial and tangential residual stresses.

Conclusions

In this research, macrostresses have been studied, as theyappear to be the most significant from the point of view of practical application.

Comparing the results of theoretical and practical research, we arrive at a conclusion that the combined processing creates favorable compressive residual stresses on the surface of the samples under study. This is evidenced by the sign "-" before the numerical values of the axial and tangential stresses of the experimental studies and the tangential residual stresses in the theoretical study. The convergence of the results is 89%.

Furthermore, a positive feature of the combined processing is the fact that the value and nature of residual stresses obtained in previous operations do not affect the formation of the value and sign of residual stresses in the CP process, and the method presented may be a barrier to technological heredity.

Reference list

[1] L. Fedorova, S. Fedorov, A. Sadovnikov, Y. Ivanova, M .Voronina Abrasive wear of Hilong Bo TN hardfacings*VII International conference on mechanical engineering and applied composite materials*. Hong Kong, China, November 23-24, 2017.

[2] A.S. Vasilev Controlled Forming of Machine Components Operating Characteristics. *Procedia Engineering*, 2016, no. 150, pp. 975–979. DOI: 10.1016/j.proeng.2016.07.073

[3] A.I. Kondakov. Productivity as Measure of Multiproduct Metal Processing Production Efficiency. *Procedia Engineering*, 2016, Vol. 150, pp. 987–991. DOI: 10.1016/j.proeng.2016.07.151.

[4] L.V. Fedorova, S.K.Fedorov, Y.S.Ivanova, M.V.Voronina Increase of Wear Resistance of the Drill Pipe Thread Connection by Electromechanical Surface Hardening *International Journal of Applied Engineering Research* ISSN 0973-4562, Vol. 12, no 18 (2017) pp. 7485-7489© Research India Publications.

[5] A.P. Yakovleva Improving the Durability of Machine Parts Using a Combined Method. *Materials Science Forum,* 2019, Vol. 946, pp 37-41. DOI:10.4028/www.scientific.net/MSF.946.37.

[6] N.N. Zubkov, I.V. Bezin, M.Y. OshchepkovSurface reinforcement of carbon composites with microstructural metal materials*Polymer Science - Series D,* 2016, 9(1), c. 91-95.

[7] N.N. Zubkov, Yu.LBityutskaya Simulation of the Deformational Cutting and the Geometric Parameters of Pin Structures to Analyze the Thermohydraulic Characteristics of Heat-Removal Plates. *Russian Metallurgy (Metally)*. 2018, no. 13.- p.1202-1207. DOI:10.1134/S003602951813027X (Scopus, WoS).

[8] N.N. Zubkov, A.I. Ovchinnikov, S.G. Vasil'evTool–workpiece interaction in deformational cutting. *Russian Engineering Research*, 2016, no. 36(3), c. 209-212.

[9] Yu.G. ShneyderEkspluatatsionnyesvoystvadetaley s regulyarnymmikrorel'yefom [Performance characteristics of parts with regular microrelief]. Moscow, Mashinostroenie, 1982, 245 p. (in Russian).

[10] N.N. Zubkov, V. Poptsov, S. Vasiliev., A. Batako Steel Case Hardening Using Deformational Cutting. *Journal of Manufacturing Science and Engineering*, 2018, vol. 140, no. 6, art. 061013. DOI: 10.1115/1.4039382.

[11] N. Zubkov, V. Poptsov., S. Vasiliev Surface Hardening by Turning without Chip Formation. *Jordan Journal of Mechanical and Industrial Engineering*, 2017, vol. 11, no. 1, pp. 13–19.

[12] A.P. Yakovleva Strengthening of coarse pitch gears. *Aviatsionnayapromyshlennost'* [Aviation Industry], 2014, no. 2, pp. 31–33. (in Russian).

[13] A.G. SuslovKachestvopoverkhnostnogosloyadetaleymashin [Property of machine parts surface layer]. Moscow, Mashinostroenie, 2000, 320 p. (in Russian).

[14] P.H. Mayrhofer, C. Mitterer, J. Musil Structure-property relationships in single- and dual-phase nanocrystalline hard coatings. *Surface and Coatings Technology*, 2003, Vol. 174-175, pp. 725–731. DOI: 10.1016/S0257-8972(03)00576-0.

[15] A. Leyland, A. Matthews Design criteria for wear-resistant nanostructured and glassy-metal coatings. *Surface and Coatings Technology*, 2004, Vol. 177-178, pp. 317–324. DOI: 10.1016/j.surfcoat.2003.09.011.

[16] N.N. Zubkov, I.V. BezinSurface modification by deformational cutting for improving the bonding strength of polymers and polymer composite materials with metals and with each other*IOP Conference Series: Materials Science and Engineering* 2017, Vol. 213(1),012010.

[17] O. Fanidi, A.V. Shchedrin, A.A. Kostryukov, N.Yu. Chihacheva Improving Efficiency of the Surface Plastic Covering Deformation of Billets from Aluminum Alloys // Tribology in Industry , 2019, Vol. 41, No. 1. Pp, 50-55. DOI: 10.24874/ti.2019.41.01.06

[18] Perlovich Yu., Bunge H.J., Isaenkova M.G. The fullest description of the structure of textured metal materials with generalized pole figures: The example of rolled Zr alloys // *Materials Science Forum, vols. 378-381, 2001, pp.180-185.*

[19] Isaenkova M., Perlovich Yu., Fesenko V. Use of generalized pole figures in the X-raystudy of textured metal materials // *Z. Kristallographie*, 2007, suppl. 26, s. 327–332.

[20] Perlovich Yu., Isaenkova M. and Fesenko V. General Principles of Substructure Inhomogeneity, Arising in Metal Materials under Plastic Deformation // *Mat. Sci. Forum*, 2007,v. 550, p. 253–258.

[21] Perlovich Yu., Isaenkova M. and Fesenko V. Principles of microstress equilibrium in textured metal materials*//Advances in X-ray Analysis, 2009, 53, December, pp. 125-140.*

IOP Conf. Series: Materials Science and Engineering **589** (2019) 012033 doi:10.1088/1757-899X/589/1/012033

Investigation of the performance of the oil gas separator depending on the gas content of the pumped liquid. Development of methods for optimizing the geometry of flow-through parts of centrifugal gas separators

M Ershov[1] and V Lomakin[1,2]

[1]BMSTU, Moscow, Russian Federation

[2]Email: Lomakin@bmstu.ru

Annotation

This article formulates the problems of using oil separators in the current conditions of well operation, as well as research and pilot implementation of innovative oil production technologies. A variant of the improved flow part of the separator was proposed, a mathematical model was written for its further calculation, and the hydrodynamic calculation itself was carried out, the results of which will later be used to study methods for optimizing the geometry of flow parts of centrifugal gas separators.

Introduction

At the moment, there is a steady tendency to the deterioration of the operating conditions of wells in the fields of the Russian Federation - deposits with favorable geological field parameters are entering a late stage of development, the relative share of oil production from marginal wells with difficultly recoverable reserves is increasing.

Work is underway to intensify the flow from the bottomhole zone and increase oil recovery, with the result that most of the wells were transferred to operating modes with bottomhole pressures of 3–5 MPa. But the growth rate of the average equipment failure time began to decline, and it became clear that most of the ESP wells with the wear resistance group used could no longer provide a significant increase in the time between failures, since they do not correspond to the changed operating conditions [1]. Serial submersible pumping equipment for well operation is often unable to operate in such difficult conditions.

The share of marginal wells in the Russian Federation is about 30-40%, and most of them have complicating factors: high content of free gas, mechanical impurities, corrosion, salt deposition. An acute problem is the introduction of an effective technology for their operation with an average operating time of at least 600 days [2].

In accordance with this, the task of developing, researching and pilot-industrial introduction of innovative technologies for oil production in complicated conditions with the use of submersible pumping systems becomes urgent. New complete equipment with special design of steps, pump and other elements of the installation is necessary.

In oil-producing wells with a high gas content, an increase in the efficiency of centrifugal pumps (ESPs) is achieved by completing them with centrifugal gas separators, in which free gas is drawn from the pumped fluid to the annular space before the pump reaches the pump inlet.

It was established experimentally [3, 4] that the efficiency of a centrifugal gas separator significantly depends on the dispersion of the gas-liquid structure of the pumped fluid. In a real oil well, on reception of a centrifugal gas separator, a bubble flow regime is usually observed, with the diameter of the gas phase bubbles ranging from 80 to 300 μm [5], [6] - [11].

In order to increase the efficiency of the oil production process at low costs, a number of new conceptual equipment was developed at Rimera Group of Companies, including gas separators with interchangeable augers and multiphase modules with a new work concept. The gas separator is considered to be an unreliable element of a pumping unit due to hydroabrasive cutting of the body due to reverse currents that occur if the pump feed is more than half the estimated gas separator feed and the optimum auger feed installed at the inlet [12]. Reverse currents are a trap for mechanical impurities - their concentration is rapidly increasing, and a rotating abrasive ring can cut the casing and other elements of the gas separator. [13] - [16] Traditionally, they try to solve these problems by reducing the angle of the blades at the screw inlet, but this unnecessarily increases the diffusivity flow-through part and reduces the efficiency of the screw. Together with the specialists of RSU NG, the Research and Development Center of the Rimera Group of Companies developed vortex and centrifugal gas separators with replaceable elements of the flow part. It was proposed to reduce the screw diameter in them, keeping the optimal flow path, the inlet angles and the range of components. In addition, it was decided to separate the inlet part of the gas separator with an axial stage and the separation chamber with a special conical sleeve. In the separation chamber, a separation process takes place, and a rotating ring with a partially separated liquid is formed at the periphery. Previously, this ring pressed on the blades of the stage, supplying the gas-liquid mixture, which led to reverse currents and dispersion. Now the ring presses against the fixed axial support in the form of a conical sleeve and a high pressure gradient occurs, due to which efficient gas separation is carried out.

The obvious advantages of the gas separators developed by the Rimera Group for low-cost installations should be attributed to lower power - this saves energy. In addition, the gas separators are ahead of the well-known counterparts for separation properties and wear resistance.

Currently, there are no methods for calculating the separation characteristics of centrifugal gas separators, taking into account the influence of the degree of dispersion of the gas-liquid mixture pumped by the pump. [17] - [20]

The aim of the presented work is a numerical study of the effect of the dispersed gas-liquid mixture structure on the efficiency of the gas separation process in a centrifugal gas separator, calculation of the gas separator performance with different gas contents and the development of methods for improving the geometry of the flow-through part of the equipment.

Mathematical model

In this article, we use the model of a multiphase incompressible fluid flow (ρ = const). Numerical simulation is based on solving discrete analogs of the basic hydrodynamic equations.

The calculation is carried out on the basis of a mathematical model of a divided multiphase flow. That is, for each phase, the equations of mass and momentum transfer are solved separately, but the pressure field is the same for all phases.

The mathematical model consists of a set of differential and algebraic equations:

The volume of the i-th phase in each calculated cell is calculated as:

$$V_i = \int_V \alpha_i dV$$

where α_i – is the concentration of the i-th phase in the cell.

The sum of the concentrations of all phases in the cell is one.

$$\sum_{i6}^{n} \alpha_i = 1$$

The equation of conservation of mass (continuity equation):

$$\frac{\partial}{\partial t}\int_V \alpha_i \rho_i dV + \int_A \alpha_i \rho_i \vec{V_i} d\vec{a}$$

where ρ_i – i-th phase density

$\vec{V_i}$ – i-th phase velocity (in the case of turbulent flow modeling by a RANS-type model averaged over time)

The equation for the change in the amount of motion:

$$\frac{\partial}{\partial t}\int_V \alpha_i \rho_i \vec{V_i} dV + \int_A \alpha_i \rho_i \left(\vec{V_i}\vec{V_i}\right) d\vec{a} = -\int_V \alpha_i \nabla p dV +$$

$$+\int_V \alpha_i \rho_i \vec{g} dV + \int_A \left[\alpha_i \left(T_i + T_i^t\right)\right] d\vec{a} + \int_V \overrightarrow{M_i} dV$$

where $\left(\vec{V_i}\vec{V_i}\right)$ – tensor product of the velocity vectors of the i-th phase.

p– pressure

$\vec{g}$ – mass intensity vector (in this case, the gravity force is 9.81 m / s2 and the inertial force due to the rotation of the computational domain)

T_i - molecular viscosity stress tensor

T_i^t – turbulent stress tensor

$\vec{M}$ –vector of total intensity of interfacial interaction forces per unit volume, for a vector $\overrightarrow{M_i}$ fair equality:

$$\sum_i \overrightarrow{M_i} = 0$$

Vector $\overrightarrow{M_i}$ characterizes all the forces that separate phases interact with each other.

$$M_i = \sum_{j \neq i}\left(F_{ij}^D + F_{ij}^{VM} + F_{ij}^L + F_{ij}^{TD} + F_{ij}^{WL}\right)$$

$$\overrightarrow{M_i} = \sum_{i \neq j}\left(\overrightarrow{F_{ij}^D} + \overrightarrow{F_{ij}^{VM}} + \overrightarrow{F_{ij}^L} + \overrightarrow{F_{ij}^{TD}} + \overrightarrow{F_{ij}^{WL}}\right)$$

Where

$\overrightarrow{F_{ij}^D}$ – resistance force,

$\overrightarrow{F_{ij}^{VM}}$ – power of the virtual mass

$\overrightarrow{F_{ij}^L}$ – lifting force

$\overrightarrow{F_{ij}^{TD}}$ – turbulent dispersion force

$\overrightarrow{F_{ij}^{WL}}$ – turbulent dispersion force

Power of resistance

In the case of modeling the flow of continuous and dispersed media, the resistance force acting on the dispersed medium I from the side of the continuous medium j is equal to:

$$\overrightarrow{F_{ij}^{D}} = A_D \overrightarrow{V_r}$$

Where

A_D – linearized drag coefficient

$\overrightarrow{V_r} = \overrightarrow{V_j} - \overrightarrow{V_i}$ – relative velocity of one medium relative to another

$$A_D = C_D \frac{1}{2} \rho_C \left|\overrightarrow{V_r}\right| \left(\frac{a_{cd}}{4}\right)$$

Where

C_D – drag coefficient

ρ_C – continuous phase density

a_{cd} – interaction area of the phases (in this case, the member $\frac{a_{cd}}{4}$ is the projected area of the spherical particle on the plane).

The drag coefficient is based on the relation:

$$C_D = f_D C_{D\infty}$$

Where

$C_{D\infty}$ - coefficient of resistance of a single spherical particle moving in an infinite flow

f_D – coefficient taking into account the concentration of particles

$C_{D\infty}$ is out of relationship:

$C_{D\infty} = \frac{24}{Re_d}\left(1 + 0.15 Re_d^{0.687}\right)$ при $Re_d < 1000$

And

$C_{D\infty} = 0{,}44$ при $Re_d > 1000$

$$Re_d = \frac{\rho_C \left|V_r\right| l}{\mu_C}$$

l– characteristic interaction length or bubble size

μ_C – dynamic viscosity coefficient of continuous medium

Coefficient f_D is the ratio:

$$f_D = \alpha_C{}^{n_D}$$

α_C – continuous phase concentration

$n_D = -8.3$ for spherical particle

Virtual mass

The inertia of the surrounding fluid affects the acceleration of the particle immersed in the fluid. This effect is modeled by adding mass to the dispersed particle.

The force from the virtual mass acting on phase I, moving rapidly with respect to phase j is found as:

$$\overrightarrow{F_{ij}^{VM}} = C_{VM}\rho_C\alpha_C\left(\overrightarrow{a_j} - \overrightarrow{a_i}\right)$$

$\overrightarrow{a_{j,i}}$ – acceleration of jth and ith phase

$C_{VM} = 0,5$ – virtual mass coefficient for a spherical particle

Lift force

In the case where the flow surrounding the dispersed particle is non-uniform or the particle is swirling, it experiences a force perpendicular to the relative velocity.

$$\overrightarrow{F_{ij}^{L}} = C_{Leff}\rho_C\alpha_d\left(\overrightarrow{V_r} \times \left(\nabla \times \overrightarrow{V_c}\right)\right)$$

$\overrightarrow{V_r}$ – relative phase velocity

$\overrightarrow{V_c}$ – continuous phase speed

α_d – dispersed phase concentration

$C_L = 0,25$ – lift force coefficient

Turbulent dispersion force

Additional change in phase concentrations caused by flow turbulence is modeled as the force of turbulent dispersion

$$\overrightarrow{F_{ij}^{TD}} = A_D\overrightarrow{V_{TD}}$$

A_D – drag force coefficient

$\overrightarrow{V_{TD}}$ – relative slip speed

$$\overrightarrow{V_{TD}} = D_{TD}\left(\frac{\nabla\alpha_d}{\alpha_d} - \frac{\nabla\alpha_c}{\alpha_c}\right)$$

$D_{TD} = C_0\dfrac{\nu_c^t}{\sigma_\alpha}I$ – turbulent diffusion tensor

$$C_0 = 1$$

ν_c^t – turbulent viscosity coefficient kinematic

σ_α – turbulentPrandtl number

I– unit tensor

$$\sigma_\alpha = \sigma_0\sqrt{1 + C_\beta\xi^2}\,\frac{1+\eta}{b+\eta}$$

$\sigma_0 = 1$ – unmodified turbulent Prandtl number

$C_\beta = 1,8$ – correction factor

ξ – particle sliding velocity related to the speed of turbulent fluctuations

η – particle interaction time related to relaxation time

b– the ratio of the accelerations of the continuous / dispersed phase

$$b = \frac{1 + C_{VM}}{\frac{\rho_d}{\rho_c} + C_{VM}}$$

$$\eta = \frac{\tau_I}{\tau_R}$$

τ_I – characteristic time of interaction of a particle and a turbulent vortex

τ_R – particlerelaxationtime

$$\tau_I = \frac{\tau_T}{\sigma_0} \frac{1}{\sqrt{1 + C_\beta \xi^2}}$$

$$\tau_T = \frac{2}{3} C_\mu \frac{k_c}{\varepsilon_c}$$

ε_c — *continuous phase turbulence energy dissipation rate*

$$\tau_R = \tau_D \left(1 + \frac{\rho_c}{\rho_d} C_{VM} \right)$$

$\tau_D = \frac{\rho_d d^2}{18\mu_c}$ –*characteristic time scale for a dispersed particle*

$$\xi = \frac{\left| \vec{V_r} \right|}{\sqrt{\frac{2}{3} k_c}}$$

k_c – kinetic turbulence energy of the continuous phase

Wall strength

Being located near the solid wall, the gas bubble undergoes an asymmetrical action from the liquid. The force per unit volume that a gas bubble experiences is equal to:

$$\overrightarrow{F_{ij}^{WL}} = -C_{WL} y_w \alpha_D \rho_C \frac{\left| \overrightarrow{V_{r,\tau}} \right|^2}{d_p} \vec{n}$$

Where

y_w – distance from the wall

α_D – dispersed phase concentration

ρ_C – dispersed phase density

d_p – bubble diameter

C_{WL} –coefficient is a function of the distance from the wall and decreasing with increasing distance

$\vec{n}$ – unit normal to the wall at the point closest to the bubble

$\overrightarrow{V_{r,\tau}}$ – tangent to the wall component of the relative velocity

Coefficient C_{WL} is like:

$$C_{WL} = max\left(C_{W1} + \left(\frac{C_{W2}}{y_w} \right) d_p, 0 \right)$$

IOP Conf. Series: Materials Science and Engineering **589** (2019) 012033 doi:10.1088/1757-899X/589/1/012033

Coefficients are$C_{W1} = -0{,}01$, C_{W2}=0,05. Thus the force caused by the influence of the wall disappears at a distance from the wall equal to the five diameters of the bubble.

Mathmodelling Results

To carry out hydrodynamic calculations, the geometry of the flow section of the oil gas separator was constructed.

At the entrance there is a grid, then a screw, which gives additional turbulence to the flow and creates pressure, then the stator is located, straightening the direction of fluid flow. Next is the drum, which discards the fluid to the periphery, and the gas continues to move closer to the axis of the separator due to the distribution of centrifugal force. On a larger diameter, there is one more straightening flow stator, gas is diverted through a channel into the annular space.

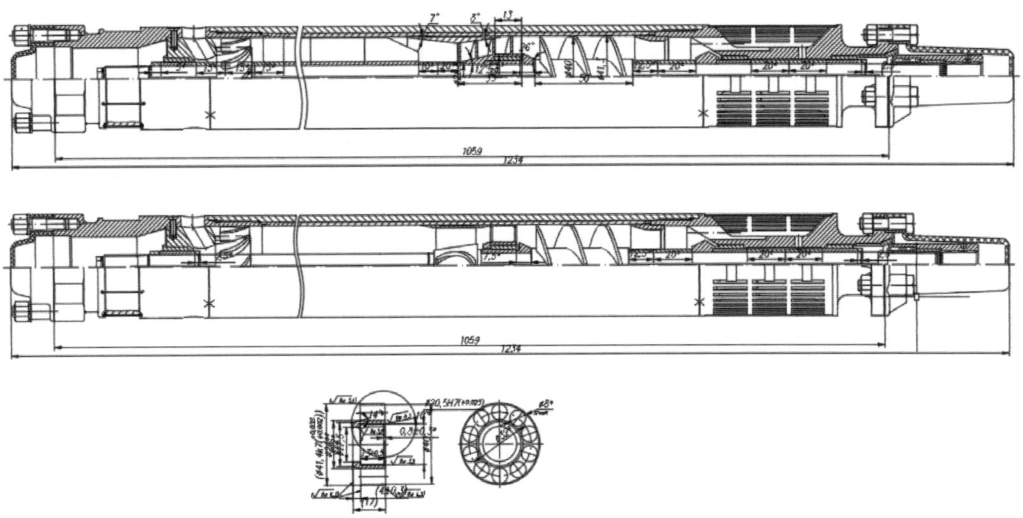

Pic.1. Drawing of oil separator

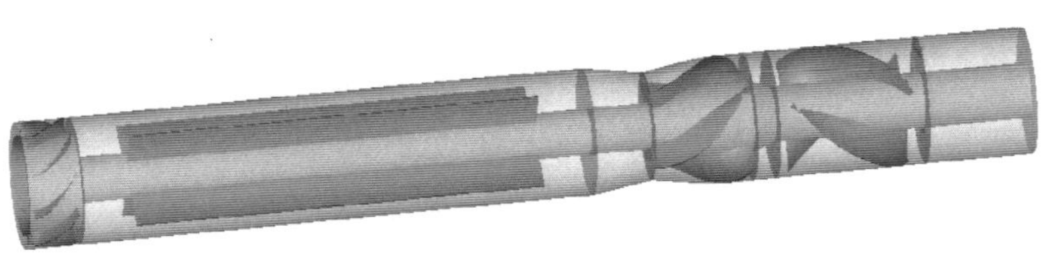

Pic.2. 3D model of oil gas separator

For the calculation, the StarCCM software package was used. As the surface mesh, the previously constructed geometry of the flow part was imported.

The following parameters are used as a physical model:

- Multiphase interaction, scale of interaction length 0. 1mm.
- H2O and Air gas (air) are taken as phases.
- Initialliquid / gasdistribution 1/0
- The inlet is set to stagnation at the inlet with a total pressure of 0 Pa and a liquid / gas ratio of 4/1
- At the exit boundary of the fluid, a negative speed is set at 4.5 m / s.
- Atmospheric pressure is set at the gas outlet boundary.

The following parameters were taken as the studied parameters: head, separation coefficient (water fraction at the separator outlet) and power depending on the flow rate.

IOP Conf. Series: Materials Science and Engineering **589** (2019) 012033 doi:10.1088/1757-899X/589/1/012033

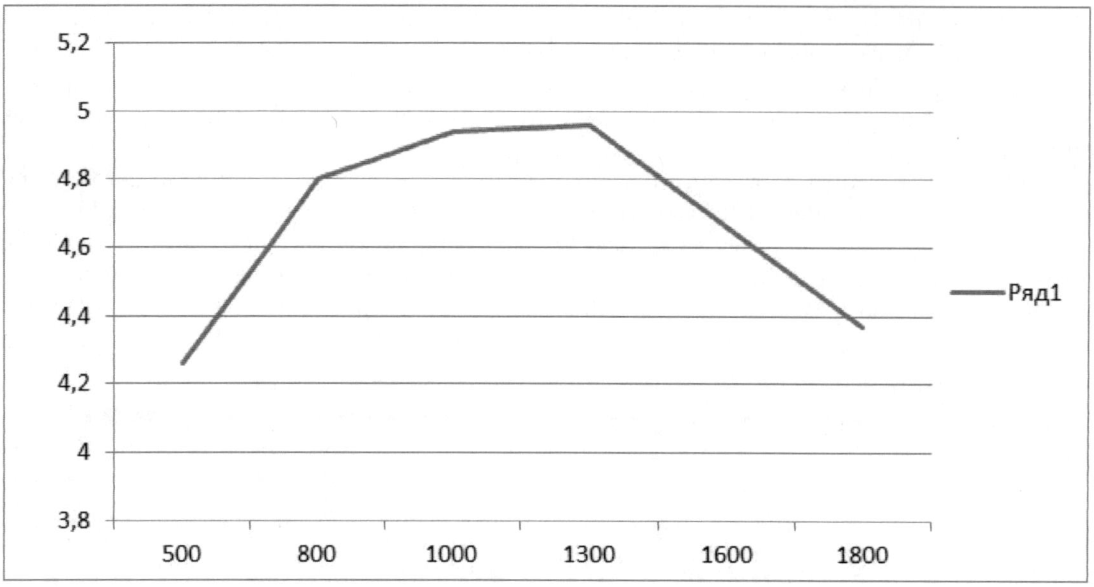

Figure 1. Head (m) versus flow rate (m3 / s)

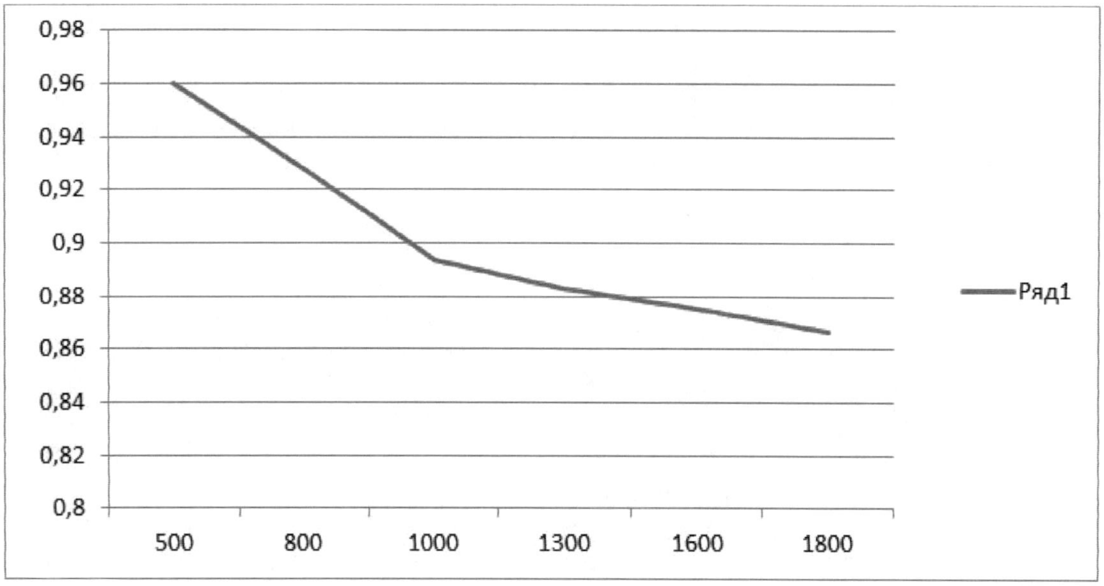

Figure 2. The value of the separation coefficient depending on the flow rate (m3 / s)

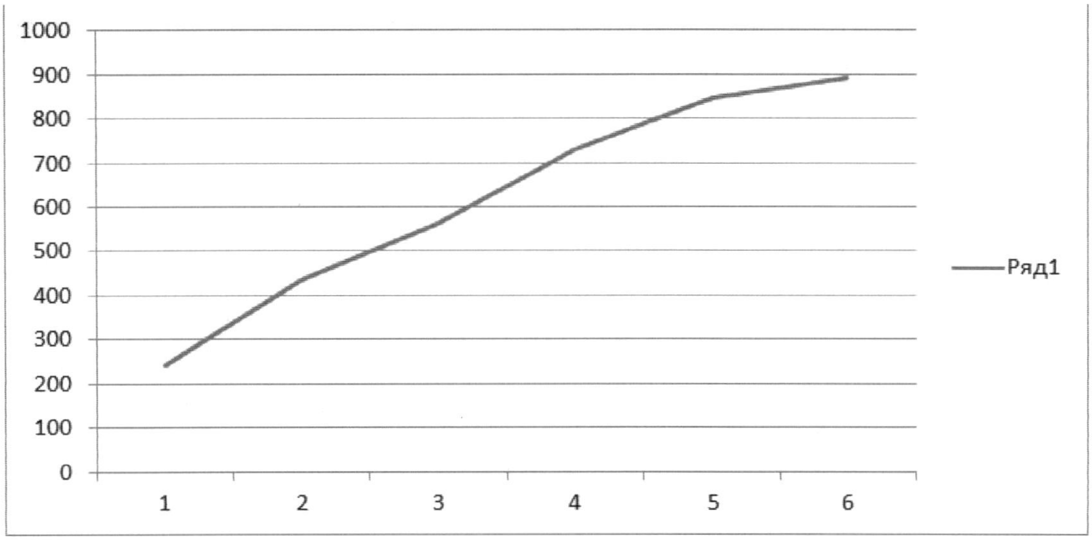

Figure 3. Power value (N * m) depending on the flow rate (m3 / s)

Findings

Based on the results, we can draw the following conclusions:

1. Studies of the separator at various values of head from 500 m3 / day to 1800 m3 / day showed that the head graph has a certain peak at a value of 1300 m3 / day, with an increase in flow, the head value begins to fall sharply, with a decrease, the drop occurs smoothly.
2. The value of the separation effect is inversely proportional to the feed, while the graph noticeably changes its rate of decrease approximately at the optimal feed point.
3. The more feed, the more power at the separator shaft. The power at the separator shaft gradually increases in proportion to the increase in the feed value without any jumps.

The obtained characteristics, as well as the constructed geometry of the flow section of the oil gas separator, will be used in future to study methods for optimizing the geometry of the flow sections of the centrifugal gas separators.

List of references:

[1] Yakimov S. B. On the prospects of using radially stabilized compression electric centrifugal pumps to improve the efficiency of reservoir operation of the AB group of Samotlor field // Territory "NEFTEGAZ". 2016. No. 6. P. 78-86.

[2] Kosilov D. A. Improving the efficiency of wells ' mechanical Fund management in the current macroeconomic conditions/ Mechanized production // Materials of the specialized section of the Rosneft conference. – M: 2015. – Pp. 8-11.

[3] Iglewski L. V. improving the efficiency of operation of submersible pump-ejector systems for oil: Diss. kand. Techn. sciences'. M., 2002. 216 p.

[4] Dengaev A.V. Improving the efficiency of well operation by submersible centrifugal pumps when pumping gas-liquid mixtures: dis. kand. Techn. sciences'. M., 2005. 212 s

[5] Vasiliev Yu. N., Maksutov R. A., Bashkirov A. I. Experimental study of the structure of oil and gas flow in the fountain well // Oil industry. 1961. No. 4. P. 41-44.

[6] Lomakin V O, Chaburko P S, Kuleshova M S 2017 Multi-criteria Optimization of the Flow of a Centrifugal Pump on Energy and Vibroacoustic Characteristics *Procedia Engineering* **176** pp 476-482

[7] Lomakin V O, Kuleshova M S, Bozh'eva S M 2016 Numerical Modeling of Liquid Flow in a Pump Station *Power Technology and Engineering* **5** pp 324-327
[8] Lomakin V O 2015 Proceedings of 2015 International Conference on Fluid Power and Mechatronics
[9] Lomakin V O, Kuleshova M S, Kraeva E A 2015 Fluid Flow in the Throttle Channel in the Presence of Cavitation*Procedia Engineering* **106** pp 27-35
[10] Shargatov, V.A., Gorkunov, S.V., Il'ichev, A.T. Dynamics of front-like water evaporation phase transition interfaces (2019) Communications in Nonlinear Science and Numerical Simulation, 67, pp. 223-236.
[11] Arefyev, K.Y., Prokhorov, A.N., Saveliev, A.S. Study of the breakup of liquid droplets in the vortex wake behind pylon at high airspeeds (2018) Thermophysics and Aeromechanics, 25 (1), pp. 55-66. DOI: 10.1134/S0869864318010055
[12] Borovsky B. I., Petrov V. I. high-Speed shovel pumps. M., "Mechanical Engineering", 1975.
[13] Gouskov, A.M., Lomakin, V.O., Banin, E.P., Kuleshova, M.S. Minimization of Hemolysis and Improvement of the Hydrodynamic Efficiency of a Circulatory Support Pump by Optimizing the Pump Flowpath (2017) Biomedical Engineering, 51 (4), pp. 229-233. DOI: 10.1007/s10527-017-9720-9
[14] Arefyev, K.Y., Voronetsky, A.V., Suchkov, S.A., Ilchenko, M.A. Computational and experimental study of the two-phase mixing in gas-dynamic ignition system (2017) Thermophysics and Aeromechanics, 24 (2), pp. 225-237. DOI: 10.1134/S086986431702007X
[15] Lomakin, V.O., Kuleshovav, M.S., Bozh'eva, S.M. Numerical Modeling of Liquid Flow in a Pump Station (2016) Power Technology and Engineering, 49 (5), pp. 324-327. DOI: 10.1007/s10749-016-0623-9
[16] Arefyev, K.Y., Voronetsky, A.V. Modelling of the process of fragmentation and vaporization of non-reacting liquid droplets in high-enthalpy gas flows (2015) Thermophysics and Aeromechanics, 22 (5), pp. 585-596. DOI: 10.1134/S0869864315050078
[17] Belov, P.A., Kobets, L.P., Borodulin, A.S. Impregnation kinetics of fibers with liquids: Simulation within the generalization of Navier-Stokes equations (2014) Inorganic Materials: Applied Research, 5 (4), pp. 403-406. DOI: 10.1134/S2075113314040182
[18] Loginov V. F. Compression pumps for oil production in complicated conditions.// Oilandgasjournal. 2016, № 6. C. 54-56.
[19] D. C. Wilcox (1994). "Turbulence modeling for CFD" // DCW Industries, Inc., 460 pages.
[20] Aref'Ev, K.Yu., Voronetskii, A.V., Il'Chenko, M.A. Dynamic characteristics of a resonant gas-dynamic system for ignition of a fuel mixture (2013) Combustion, Explosion and Shock Waves, 49 (6), pp. 657-661. DOI: 10.1134/S0010508213060038

Left ventricular assist devices of a new generation

V Grishakov[1], S Shein[1,2] and V Lomakin[1]

[1]Bauman Moscow State Technical University, 5 Second Baumanskaya Street, Moscow, 105005, Russian Federation

[2]E-mail:sheinsvyatoslav@gmail.com

Abstract
The most competitive left ventricular assist devices (LVAD) in the world were identified in this article. 3D models of two samples were constructed for pumping blood based on existing pumps. The models were imported into the software package, where the flow of blood through these models was modeled. For this, a computational grid was constructed. According to the results of calculations, the main characteristics of the samples were obtained, as well as pictures of the distribution of velocities and pressures in the flow part. The calculation used a number of basic hydraulic equations that form a mathematical model.

Introduction
LVADsare becoming increasingly popular in medicine. They resolve the problem of heart failure and other heart diseases. Such devices can be divided into 3 generations. The most modern pumps of the 3rd generation have a continuous flow, are most compact and cause minimal damage to blood bodies.[1]-[4]

As part of this study, two centrifugal pumpsto support the left ventricle of the heart were calculated.

The size of the first pump is 69mm in diameter, 35mm in height, weight – 475g. A flow rate of 7L/min was obtained at135mmHg pressure with 8 W of power consumption. At a rotational speed of 2000$5500rpm, 10L/min can be produced. [4]

The second pump is made of titanium with an anti-thrombogen coating and weighs 370 g. It has dimensions of 55x64 mm. Impeller with a diameter of 36 mm open type, designed to prevent blood stagnation in the working chambers. A flow rate is 12 L/min against 100 mmHg pressure at 2600 rpm.The hemolysis rate is 0.005±0.002 g/100 L. [5]

The main objective of the study is to improve the characteristics of devices in comparison with prototypes, as well as reduce their effects on the human body.

The calculations of two prototypes were carried out in order to further improve them on parameters such as head and efficiency. In modern blood pumps, the efficiency is very low. Therefore, one of the objectives of the study is to improve the pumps in this parameter as much as possible.

Mathematical model
The prototypes for the calculated pumps are the Evaheart and HeartMate III LVADs (Fig. 1, 2).

IOP Conf. Series: Materials Science and Engineering **589** (2019) 012034 doi:10.1088/1757-899X/589/1/012034

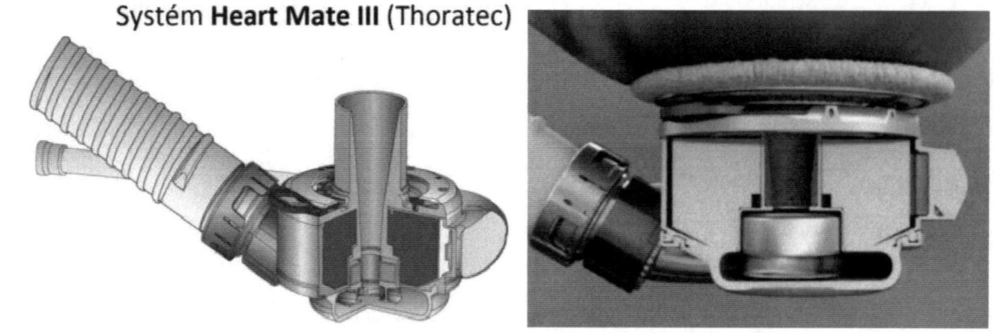

Fig. 1. HeartMateIII LVAD

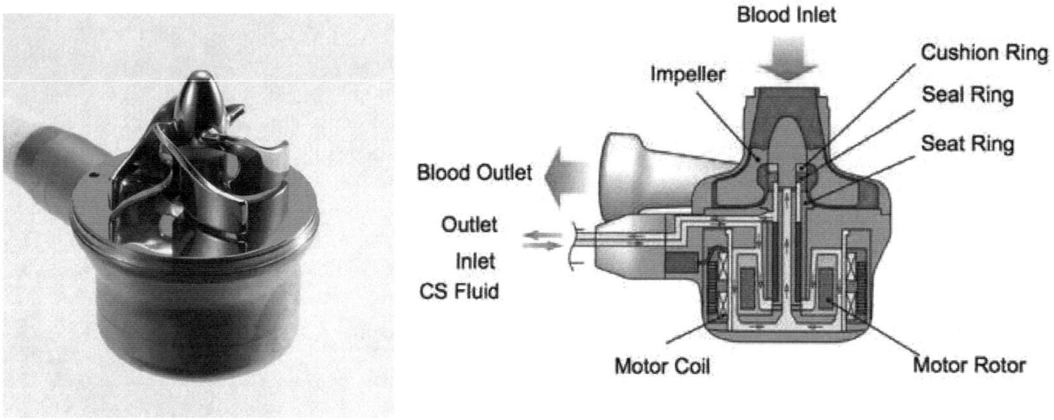

Fig. 2. Evaheart LVAD.

On the basis of the above-mentioned prototypes, initial models of two LVADs were created (Fig.3, 4).

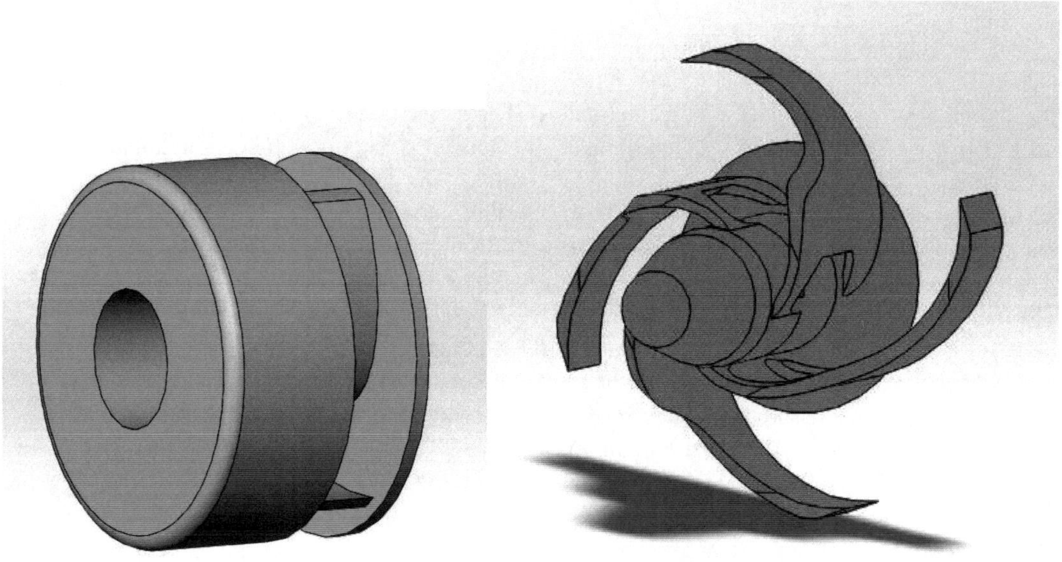

Fig. 3. ImpellersbuiltinSolidWorks

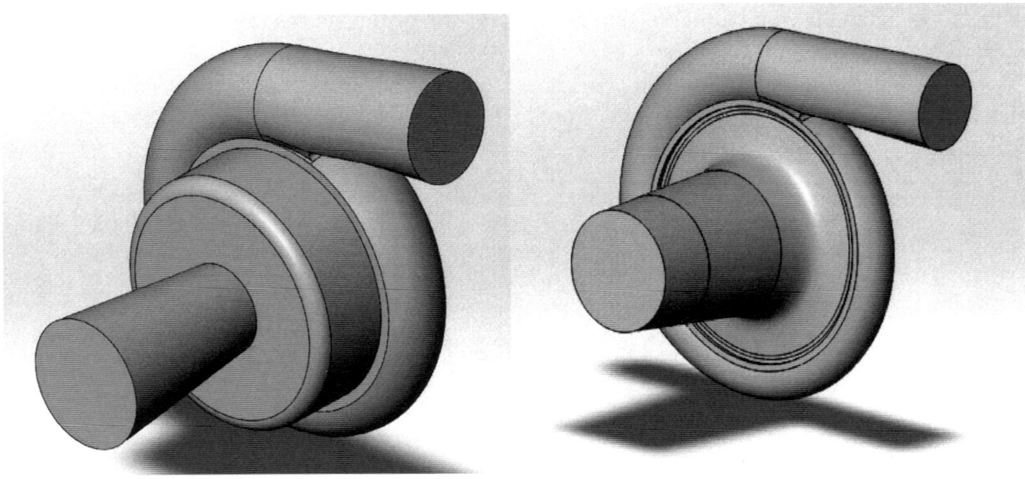

Fig.4. Flow parts built in SolidWorks

In the case of calculating the flow of incompressible fluid inside the pump, the following equations are applied:

The continuity equation for incompressible fluid.

$$\frac{\partial V_i}{\partial x_i} = 0$$

The equation of conservation of momentum averaged over time (Navier – Stokes equation averaged over Reynolds).

$$\rho\left(\frac{\partial V_i}{\partial t} + V_j\frac{\partial V_i}{\partial x_j}\right) = -\frac{\partial p}{\partial x_i} + \frac{\partial}{\partial x_i}\left(\mu\left(\frac{\partial V_i}{\partial x_j} + \frac{\partial V_j}{\partial x_i}\right) - \rho V_i^{'}V_j^{'}\right) + F_{mi}$$

Terms of the form $\rho\langle V_i^{'}V_j^{'}\rangle$ are called Reynolds stresses. They appear in the equations as a result of the formal averaging procedure and make the system of equations open. As a result, it becomes necessary to introduce additional assumptions and relationships in the original system of equations.

One of the main assumptions is the adoption of the Boussinesq hypothesis:

$$-\rho V_i^{'}V_j^{'} = \mu_t\left(\frac{\partial V_i}{\partial x_j} + \frac{\partial V_j}{\partial x_i}\right) - \frac{2}{3}\rho k\delta_{ij}$$

The combination of new relationships and assumptions forms a model of turbulence.

The model is complemented by two additional equations for the transfer of the kinetic energy of turbulence and the relative dissipation rate of this energy.

$$\frac{\partial k}{\partial t} + V_j\frac{\partial k}{\partial x_j} = P_k - \beta^* k\omega + \frac{\partial}{\partial x_j}\left(\left(\nu + \sigma_k\nu_T\right)\frac{\partial k}{\partial x_j}\right)$$

$$\frac{\partial \omega}{\partial t} + V_j\frac{\partial \omega}{\partial x_j} = \alpha S^2 - \beta\omega^2 + \frac{\partial}{\partial x_j}\left(\left(\nu + \sigma_\omega\nu_T\right)\frac{\partial \omega}{\partial x_j}\right) + 2\left(1 - F_i\right)\sigma_{\omega 2}\frac{1}{\omega}\frac{\partial k}{\partial x_i}\frac{\partial \omega}{\partial x_i}$$

The simulation was carried out in the software package STAR-CCM + 13.06.012-R8.

IOP Conf. Series: Materials Science and Engineering **589** (2019) 012034 doi:10.1088/1757-899X/589/1/012034

In order to simulate the flow of blood in the devices, a computational grid and the most illustrative sections of the computational domain were created (Fig.5). The grid captures the entire computational area of the prototypes, which makes it possible to calculate not only the hydraulic component of the efficiency, but also the volume component. On the second model, 2 extruders were created at the pump inlet and at the outlet. This is done to smooth the inlet vortices and the unevenness of the velocity distribution field at the outlet.[6]-[9]

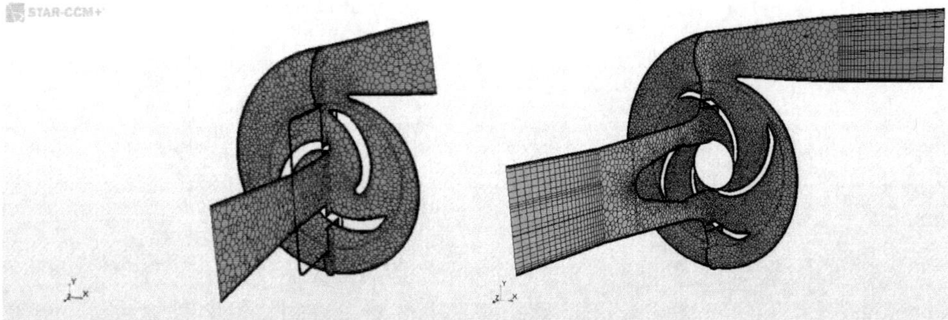

Fig. 5. Cross sections of the grid.

The input speed is set based on the flow rate values, the cross-sectional area and the rotor speed. It is 1.18 m/s and 5000 rpm for the first prototype and 0.52 m/s and 2600 rpmfor the second.

Simulation results

After the flow stabilized in the models, the following graphs and flow patterns in the flow parts were obtained[10] (Fig.6,7,8,9):

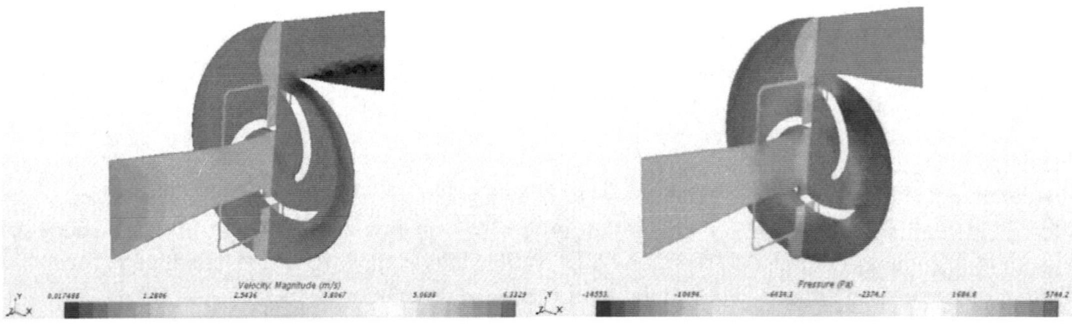

Fig.6. Velocity and pressure distribution scenes in the flow part of the first model.

In the pictures of the velocity and pressure distributions there are areas of fairly strong irregularity near the "cut-water" of the outlet (Fig.6). However, the stresses arising in these areas are permissible, and, presumably, will not have a large effect on hemolysis.

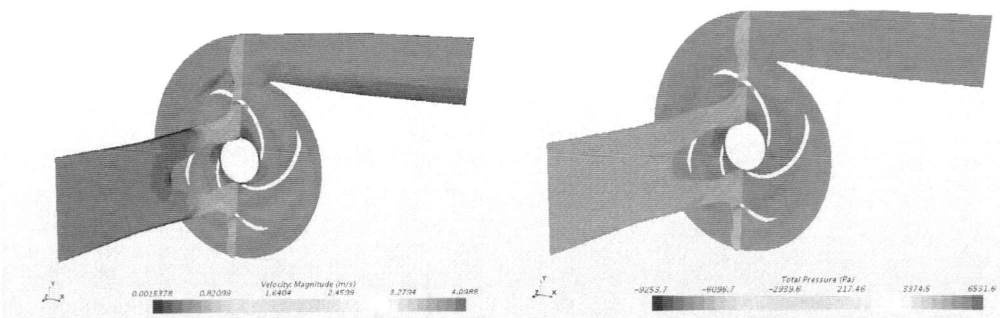

Fig.7. Velocity and pressure distribution scenes in the flow part of the second model

Despite the quite complicated shape of the impeller and the flow part, the fluid moves evenly and practically without turbulence, which indicates a successful shape of the flow part. The pressure in the flow part also varies uniformly, without obvious deviations (Fig. 7). All this suggests low damage to blood cells in the process of passing through the impeller and the possibility of testing this model.[11]-[15]

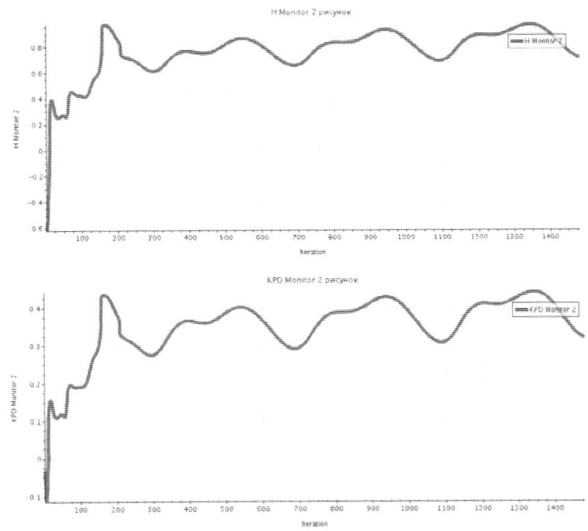

Fig.8. Head and efficiency graphs of the first model

During the calculation, the pump impeller turned about 1.3 times. The graphs show regularly repeated pulsations during the passage of each blade of a certain position (Fig.8). The results of calculations of the first prototype indicate some irregularity of flow in the pump [16]. The obtained average pressure value (0.75 m) can provide sufficient blood circulation during quiet life, the efficiency value of 35% is a good indicator for blood pumps.[17]-[18]

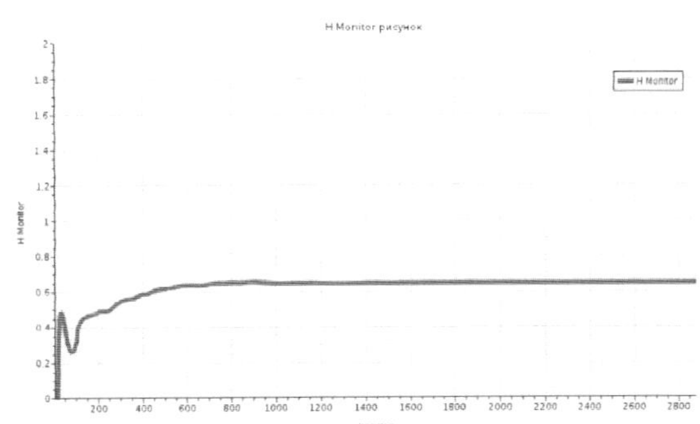

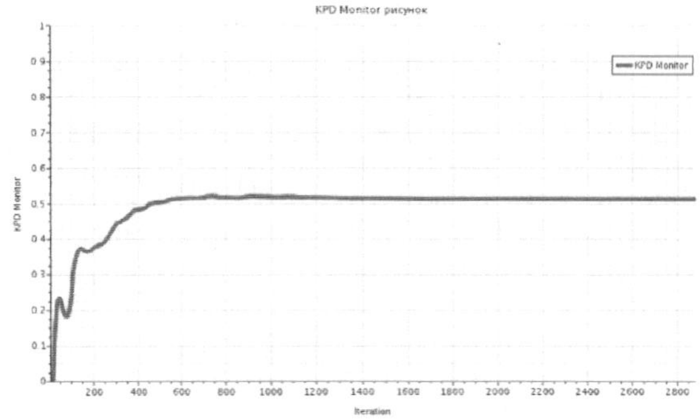

Fig.9. Head and efficiency graphs of the second model

The results of calculations of the second prototype showed that a well-received head of 0.65 m can ensure normal blood circulation in conditions of complete calmness of a person. Regarding the energy consumption of this model, a hydraulic efficiency of 51.5% was obtained, which is a good indicator, given the size of the device and the complicated shape of the flow part. The smoothness of the graphs shows that the prototype ensures continuous operation in this mode and maintains the necessary head for a long time (Fig.9).[19]-[20]

Conclusion

Based on the obtained simulation results, it can be concluded that the developed prototypes are quite promising, which allows us to continue working on their optimization. Already at this design stage, we can say that the first model outperforms the second in the head, but is inferior to it in efficiency. These models allow further calculation of hemolysis, which will give a more complete picture of the flow of blood in the flow parts.

Reference

[1] FrancisD. Pagani. Continuous-FlowRotaryLeftVentricular Assist Devices with "3rd Generation" Design. *SeminThoracCardiovascSurg 2008, 20: 255-263.*

[2] Steven M. Parnis, Mehmet HakanAkay, O.H.Frazier. Newer-Generation Rotary Blood Pumps. *Springer Science+Business Media New York 2013.*

[3] SvetankAgarway, Kane M. High. Newer-genertion ventricular assist devices. *Best Practice & Clinical Anesthesiology 26 (2012) 117-130.*

[4] Hideo Hoshi, TadahikoShinshi, SetsuoTakatani. Third-generation Blood Pumps With Mechanical Noncontact Magnetic Bearings. *Artificial Organs 30(5):324–338, Blackwell Publishing, Inc.*

[5] Kenji Yamazaki, Shinichiro Kihara, Takehide Akimoto, Osamu Tagusari, Akihiko Kawai, MitsuoUmezu, Jun Tomioka, Robert L. Kormos, Bartley P. Griffin, Hiromi Kurosawa. EVAHEART: An Implantable Centrifugal Blood Pump for Long-Term Circulatory Support. *Jpn J ThoracCardiovaseSurg 2002, 50: 461-465.*

[6] Lomakin, V.O., Kuleshova, M.S., Bozh'eva, S.M. Numerical Modeling of Liquid Flow in a Pump Station (2016)*Power Technology and Engineering, 49 (5), pp. 324-327.*

[7] JeerasitBumrungpetch, Andy Chit Tan, Shu-Hong Liu, Xian-Wu Luo, Qing-Yu Wu, Jian-Ping Yuan, and Ming-Kui Zhang. Flow Evaluation and Hemolysis Analysis of BVAD Centrifugal Blood Pump by Computational Fluids Dynamics. *International Journal of Fluid Machinery and Systems Vol. 7, No. 1, January-March 2014.*

[8] M. ErtanTaskin, Tao Zhang, Katharine H. Fraser, Bartley P. Griffith, Zhongjun J. Wu. Design Optimization of a Wearable Artificial Pump-Lung Device With Computational Modeling. *Journal of Medical Devices SEPTEMBER 2012, Vol. 6 / 031009-1.*

[9] Valeria Marinova, ImanKerroumi, Andreas Lintermann, Jens Henrik Gobbert, Charles Moulinec, Sebastian Rible, Yvan Fournier, Mehdi Behbahani, Numerical Analysis of the FDA Centrifugal Blood Pump. *NIC Series 2016, Vol. 48, ISBN 978-3-95806-109-5, pp. 355.*

[10] Loktionov, E.Y., Mikhaylova, M.G., Sitnikov, D.S. Experimental setup for studying dynamics of the calcium interaction in cells (2016)*Instruments and Experimental Techniques, 59 (4), pp. 601-606.*

[11] Christoph Gross, Christiane Marko, Johann Mikl, Johann Altenberger, Thomas Schlöglhofer , Heinrich Schima, Daniel Zimpfer, Francesco Moscato. LVAD Pump Flow Does Not Adequately Increase with Exercise. *Artificial Organs 2018, 0(0):1–7.*

[12] Gouskov, A.M., Lomakin, V.O., Banin, E.P., Kuleshova, M.S. Minimization of Hemolysis and Improvement of the Hydrodynamic Efficiency of a Circulatory Support Pump by Optimizing the Pump Flowpath (2017) *Biomedical Engineering, 51 (4), pp. 229-233.*

[13] BenteThamsen, Bastian Blümel, Jens Schaller, Christian O. Paschereit, Klaus Affeld, Leonid Goubergrits, Ulrich Kertzscher. Numerical Analysis of Blood Damage Potential of the HeartMate II and HeartWare HVAD Rotary Blood Pumps. *Artificial Organs 2015, 39(8):651–659.*

[14] Vladimirov, I.Y., Korchagin, N.N., Savin, A.S. Effects of the Influence of a Wave on an Underwater Obstacle by a Stratified Flow of Finite Depth (2018) *Doklady Earth Sciences, 478 (1), pp. 132-136.*

[15] Gouskov, A.M., Lomakin, V.O., Banin, E.P., Kuleshova, M.S. Assessment of Hemolysis in a Ventricular Assist Axial Flow Blood Pump (2016) *Biomedical Engineering, 50 (4), pp. 233-236.*

[16] M. Behbahani, M. Behr, M. Hormes, U. Steinseifer, D. Arora, O. Coronado and M. Pasquali. A review of computational fluid dynamics analysis of blood pumps. *Euro. Jnl of Applied Mathematics (2009), vol. 20, pp. 363–397.*

[17] Gordon Paul, Amin Rezaienia, EldadAvital, TheodosiosKorakianitis. Machinability and optimisation of shrouded centrifugal impellers for implantable blood pumps. *Journal of Medical Devices, March 17, 2017.*

[18] GuskovA.M.,SorokinF.D., BaninE.P. Simulation of the input design of an implantable axial blood pump. *Medical Equipment 2016.- № 1.- pp. 11 – 14.*

[19] Yurechko, V.N., Omel'yanchuk, A.M., Yurechko, R.V., Kuznetsova, E.I., Fadeev, A.A., Radkevich, F.A. Study of the Functional Characteristics of Artificial Cardiac Valves (2003)*Biomedical Engineering, 37 (2), pp. 101-110.*

[20] BaninE.P.,GuskovA.M., SorokinF.D. Analysis of modern approaches to the design of artificial ventricles of the heart of the rotor type. *Science and education: a scientific publication of BMSTU 2015.- № 02.*

IOP Conf. Series: Materials Science and Engineering **589** (2019) 012035 doi:10.1088/1757-899X/589/1/012035

Research of centrifugal gas-liquid separator

G Budaev[1], D Danilov[1], A Kuznechov[1], V Lomakin[1,2] and V Cheremushkin[1]

[1]Bauman Moscow State Technical University, 5 Second Baumanskaya Street, Moscow, 105005, Russian Federation

[2]E-mail: Lomakin@bmstu.ru

Abstract. The article is devoted to research of centrifugal gas-liquid separator.The fields of application of gas separators are described. Parameters on which depends the separation process are revealed. The mathematical model used in computer simulation of the two-phase flow is presented . The figures of the scalar physical quantities distribution in the flow part of the separator are shown. Graphs of the gas separator's characteristics are given.

1.Introduction

A separator is a device that separates a product into fractions with different characteristics. These devices are widely used in various spheres of human activity, such as agriculture, mining, medicine, food industry, etc.Figure 1.

In this article the gas separator intended for reduction of volume content of gas to admissible values in a mix arriving in the centrifugal pump as at the raised gas content its work becomes unstable and is characterised by fluctuations of working parameters: feed, created pressure and consumed power is considered [1,2]. Fluctuations in pump parameters lead to increased vibration of the unit and, consequently, to low operating hours.

Fig. 1. Example of a gas separation plant

A further increase in gas content leads to a feed failure. There are three types of gas separators: gravity, eddy and centrifugal. The last one is the most effective and was discussed further.Figure 2.

The principle of operation of such separator is based on the action of centrifugal force separating gas from water. The arriving mixture enters the intake ports and moves to the screw type inducer, from where it moves to the centrifuge with increased pressure, where it is divided into phases (under the influence of inertia forces the heavier fraction is pressed against the walls, and the lighter one remains near the shaft). After that the liquid is fed to the pump and the gas goes into the annulus.

Numerous studies have shown that the efficiency of the separator depends on parameters such as flow rate and gas content of the mixture [3,4,5,6]. With the help of modern methods of hydromodelling, the influence of these factors on the pressure, power and separation coefficient of the unit has been considered in this article.[7,8,9,10]. The mathematical model of running processes was also described. In addition, the negative influence of backflow was revealed. [11,12,13].

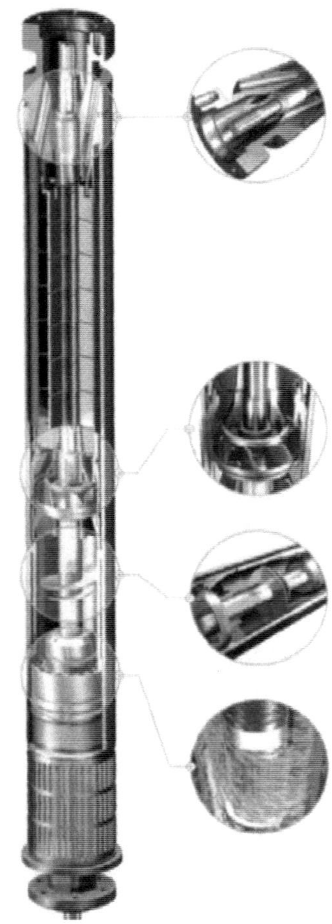

Fig. 2. Centrifugal gas separator design scheme

The main purpose of this work is to study the influence of input parameters on the efficiency of the gas separator.

2. Mathematical modeland methods.

This paper uses the model of multiphase flow of incompressible liquid (ρ=const). Numerical modeling is based on solving discrete analogues of basic equations of hydrodynamics.

Calculation is carried out on the basis of mathematical model of divided multiphase flow.[14,15,16] That is, the equations of mass transfer and the amount of motion are solved separately for each phase, but the pressure field is the same for all phases.

The mathematical model consists of a set of differential and algebraic equations:

1. The phase volume in each calculation cell is calculated as follows:

$$V_i = \int_V \alpha_i dV$$

where α_i – the concentration of the i-phase in the cell.

The sum of concentrations of all phases in the cell is equal to one

$$\sum_{i6}^{n} \alpha_i = 1$$

2. Mass conservation equation (continuity equation):

$$\frac{\partial}{\partial t}\int_V \alpha_i \rho_i dV + \int_A \alpha_i \rho_i \vec{V_i} d\vec{a} = 0$$

where ρ_i – i-phase density

$\vec{V_i}$ – i-phase velocity (in case of modeling of turbulent flow by RANS-type model - time averaged)

3. Equation of the change in the amount of motion:

$$\frac{\partial}{\partial t}\int_V \alpha_i \rho_i \vec{V_i} dV + \int_A \alpha_i \rho_i \left(\vec{V_i}\vec{V_i}\right) d\vec{a} = -\int_V \alpha_i \nabla p dV +$$

$$+\int_V \alpha_i \rho_i \vec{g} dV + \int_A \left[\alpha_i \left(T_i + T_i^t\right)\right] d\vec{a} + \int_V \overrightarrow{M_i} dV$$

where $\left(\vec{V_i}\vec{V_i}\right)$ – tensor product of velocity vectors of the i-nd phase.

p – pressure

$\vec{g}$ – mass force intensity vector (in this case, gravity force 9.81 m/s2 and inertia force from the rotation of the calculation area)

T_i - molecular viscosity stress tensor

T_i^t – turbulent stress tensor

$\vec{M}$ – vector of total intensity of interphase interaction forces per unit volume, for the vector $\overrightarrow{M_i}$:

$$\sum_i \overrightarrow{M_i} = 0$$

Vector $\overrightarrow{M_i}$ characterizes all the forces by which the individual phases interact with each other.

$$M_i = \sum_{j \neq i}\left(F_{ij}^D + F_{ij}^{VM} + F_{ij}^L + F_{ij}^{TD} + F_{ij}^{WL}\right)$$

$$\overrightarrow{M_i} = \sum_{i \neq j}\left(\overrightarrow{F_{ij}^D} + \overrightarrow{F_{ij}^{VM}} + \overrightarrow{F_{ij}^L} + \overrightarrow{F_{ij}^{TD}} + \overrightarrow{F_{ij}^{WL}}\right)$$

Where

$\overrightarrow{F_{ij}^{D}}$ – resistance force

$\overrightarrow{F_{ij}^{VM}}$ – virtual mass strength

$\overrightarrow{F_{ij}^{L}}$ – lifting force

$\overrightarrow{F_{ij}^{TD}}$ – force caused by turbulent dispersion

$\overrightarrow{F_{ij}^{WL}}$ – the force caused by the wall effects

The semi-empirical model k-ω SST turbulence model well-proved in the calculation of dynamic machines was used for this task[17,18,19,20]. The flow part of the gas separator was simulated on a grid consisting of 580 thousand cells. The cells in the flow core have a multi-faceted shape, while the solid walls of the pipe have a prismatic shape. The design grid for fluid body is shown in Figure 3.

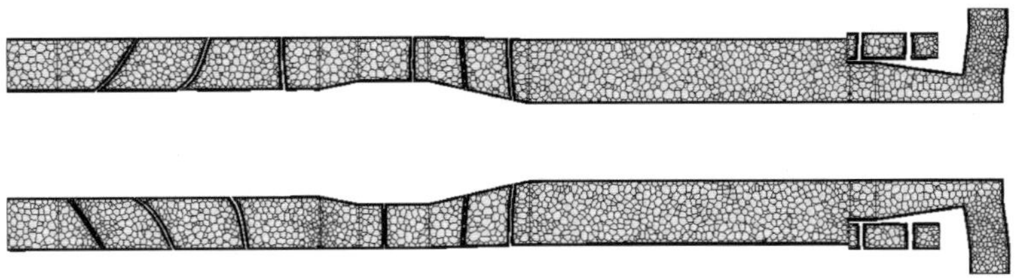

Fig.3. Mesh

3. Results and analysis

The calculated speed as an initial condition was set at the outlet boundary of the separated water in order to comply with the condition of equal pressure at the inlet and outlet of the gas separator into the pipe with unseparated fluid. Therefore, in order to compare the data at different phase ratios, graphs of the gas separation factor (Fig. 4), power (Fig. 5), head (Fig. 6) on the flow rate at the boundary where the speed was specified were made.

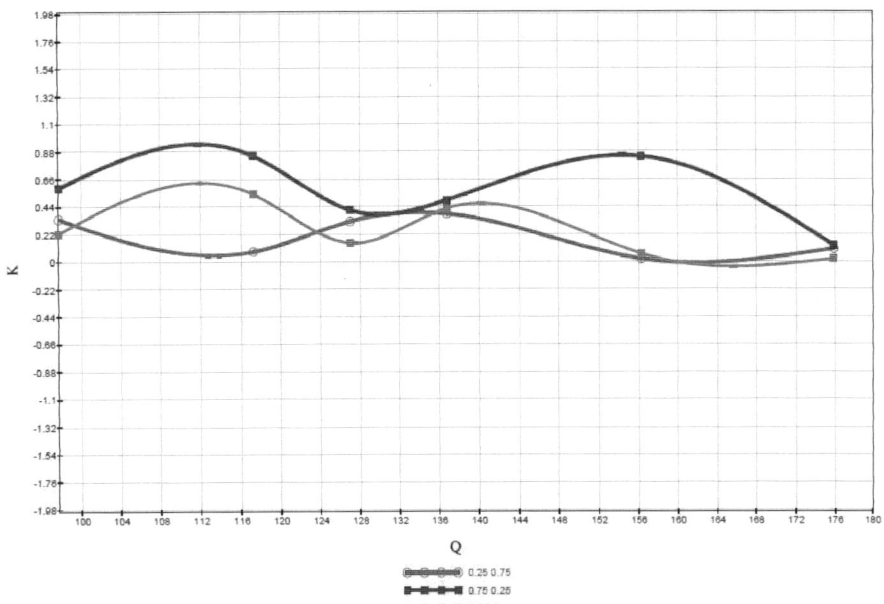

Fig.4. Gas separation factor

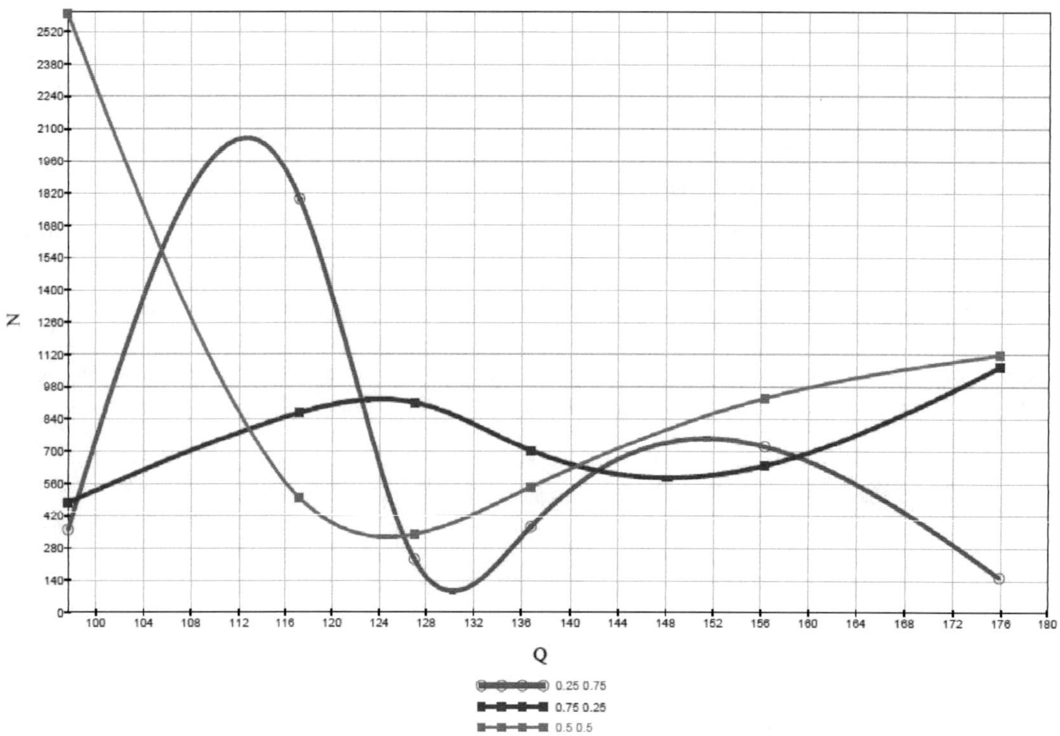

Fig.5. Power

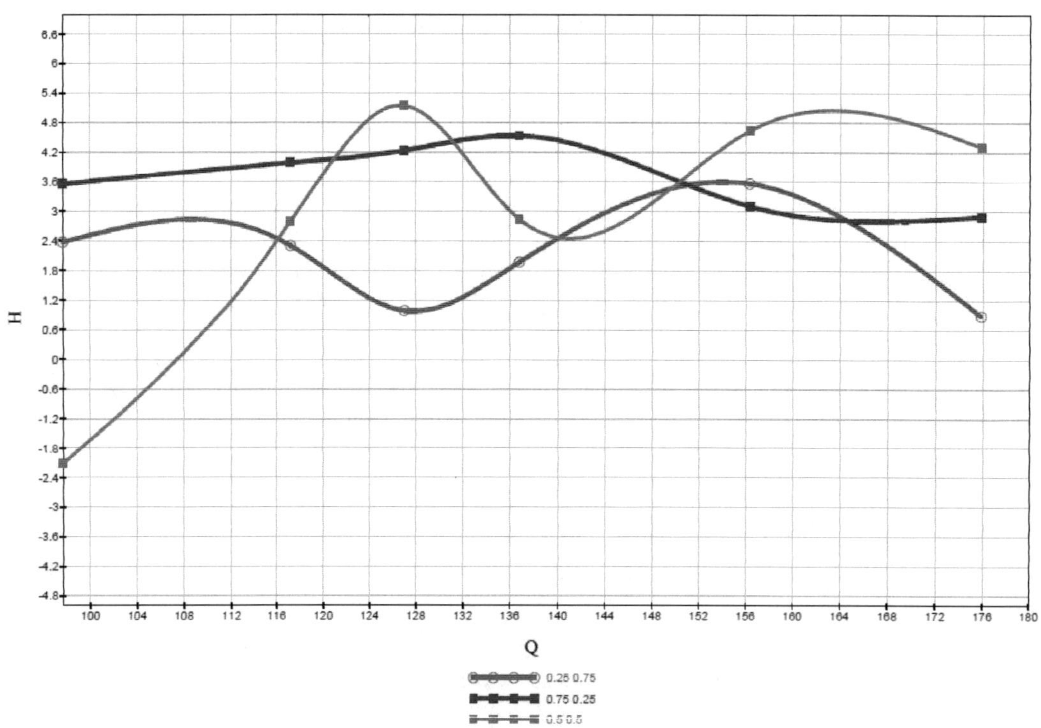

Fig.6. Head

IOP Conf. Series: Materials Science and Engineering **589** (2019) 012035 doi:10.1088/1757-899X/589/1/012035

Figure 7 shows the scene of volume fraction distribution. One can observe that the heavier fraction is pressed against the walls under the influence of inertial forces, while the lighter fraction remains near the shaft.Significant phase separation occurs in the centrifuge.

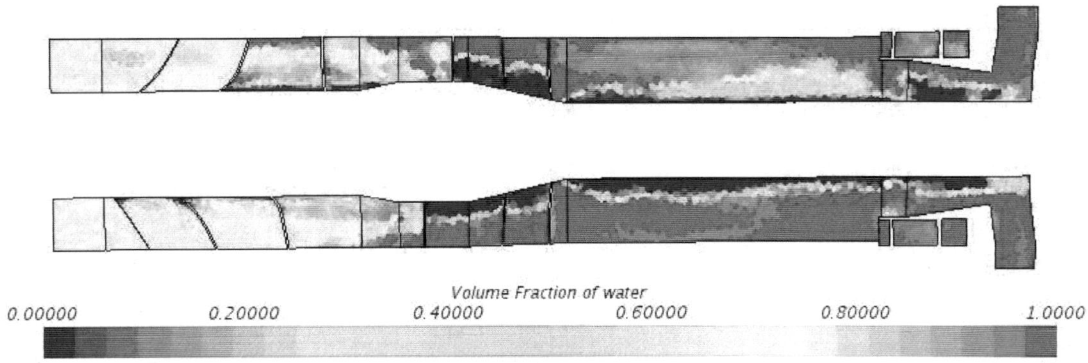

Fig.7. Volume Fracrion of water

For the most visual representation of a picture of a flow in a gas separator flowing part in figure 8,9 scalar scenes of pressure and speed distribution at Q=136,7m^3/day and a phase ratio on an input 75 % water, 25 % gas are shown.

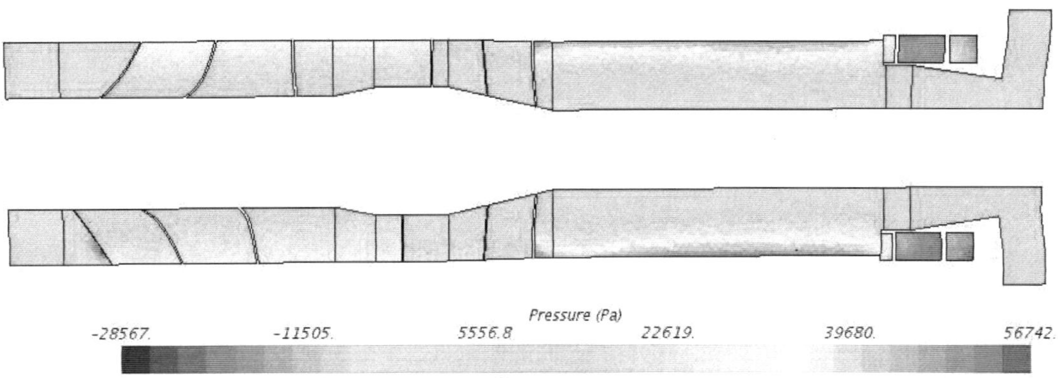

Fig.8. Pressure

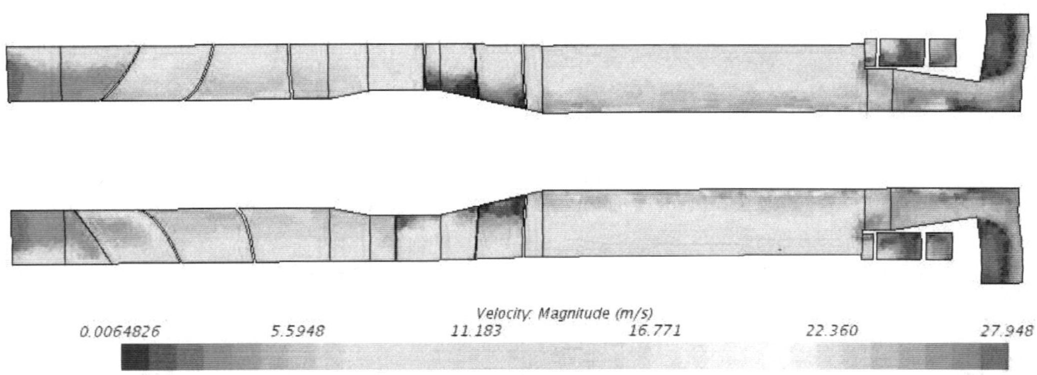

Fig.9. Velocity

IOP Conf. Series: Materials Science and Engineering **589** (2019) 012035 doi:10.1088/1757-899X/589/1/012035

4. Conclusions.

According to the results of the performed calculations, ambiguous dependences of the gas separation coefficient on the flow rate at different gas contents were obtained. Similar behavior of the coefficient value is apparently related to the mechanism of interaction between the screw inducer and the centrifuge when working on a mixture of air and water. In this paper, the study of the entire gas separator without studying the nature of the flow in its individual elements was carried out. Also it can be noticed that separation coefficient is high at low gas content. In addition, there is a flow rate range in which the optimum operating mode is observed(the lowest power with a high separation factor, in this case at a flow rate of 130….150m^3/day). To obtain a more detailed picture and explanation of the dependencies type further research is requiredincluding theflow simulation of individual components of the unit.

References.

[1] SWAROOP KOLLA, S., MOHAN, R.S. and SHOHAM, O., 2018. Structural Integrity Analysis of Gas-Liquid Cylindrical Cyclone (GLCC) Separator Inlet. Journal of Solar Energy Engineering, Transactions of the ASME, 140(5),.

[2] HE, X., LIU, Q., ZHANG, H., KUI, M.-., TAN, X.-. and LI, X.-., 2019. Numerical investigation of the performance of moisture separators based on two-way coupling model by Lagrangian-Eulerian methodology. Annals of Nuclear Energy, 124, pp. 407-417.

[3] XU, W., LI, Q., WANG, J. and JIN, Y., 2016. Performance evaluation of a new cyclone separator - Part II simulation results. Separation and Purification Technology, 160, pp. 112-116.

[4] XIONG, Z., JI, Z. and WU, X., 2014. Development of a cyclone separator with high efficiency and low pressure drop in axial inlet cyclones. Powder Technology, 253, pp. 644-649.

[5] XU, M., YANG, L., SUN, X., WANG, J. and GONG, L., 2019. Numerical analysis of flow resistance reduction methods in cyclone separator. Journal of the Taiwan Institute of Chemical Engineers, 96, pp. 419-430.

[6] SELVAGANESH, P., RESHI, O. and SCHLAUTMAN, J., 2018. Numerical prediction of engine air oil separator efficiency, FISITA World Automotive Congress 2018 2018.

[7] A Protopopov and V Vigovskij 2019 *IOP Conf. Ser.: Mater. Sci. Eng.***492** 012003

[8] N Egorkina and A Petrov 2019 *IOP Conf. Ser.: Mater. Sci. Eng.***492** 012015

[9] T Valiev and A Petrov 2019 *IOP Conf. Ser.: Mater. Sci. Eng.***492** 012038

[10] PAREKH, J. and RZEHAK, R., 2018. Euler–Euler multiphase CFD-simulation with full Reynolds stress model and anisotropic bubble-induced turbulence. International Journal of Multiphase Flow, 99, pp. 231-245.

[11] ZHARKOVSKY,A.,2003. Mathematical modeling of working processes in centrifugal pumps of low and medium speed for the solution of problems of the automated designing . Dissertation and the abstract on HAC RF 05.04.13.

[12] LOYTSYANSKIY,L.,2003. Mechanics of Fluid and Gas Textbook for HEIs. - 7th edition, Prosp. - Moscow: Drofa.

[13] GUO, H., ZHOU, C., JIANG, Z. and ZHOU, X., 2018. Study on separation characteristics of two-phase flow in double helical separator, VibroengineeringProcedia 2018, pp. 274-278.

[14] MADHANIA, S., MUHARAM, Y., WINARDI, S., PURWANTO, W.W. Mechanism of molasses–water mixing behavior in bioethanol fermenter. Experiments and CFD modeling (2019) Energy Reports, 5, pp. 454-461.

[15] LIU, M., TAN, L., CAO, S. Dynamic mode decomposition of gas-liquid flow in a rotodynamic multiphase pump (2019) Renewable Energy, pp. 1159-1175.

[16] HOSANGADI, A., LIU, Z., WEATHERS, T., AHUJA, V., BUSBY, J. Modeling multiphase effects in CO 2 compressors at subcritical inlet conditions (2019) Journal of Engineering for Gas Turbines and Power, 141 (8)

[17] A Trulev et al 2019 *IOP Conf. Ser.: Mater. Sci. Eng.***492** 012041

[18] A Sinitsyna et al 2019 *IOP Conf. Ser.: Mater. Sci. Eng.* **492** 012007
[19] Loginov, V.F. Compression pumps for oil production in complicated conditions (Kompressionnyyenasosydlyaneftedobychi v oslozhnennykhusloviyakh) (2016) Oil and Gas Journal, pp. 54-56.
[20] YAKHOT, V. and ORSZAG, S.A., 1986. Renormalization-group analysis of turbulence. Physical Review Letters, 57(14), pp. 1722-1724.

IOP Conf. Series: Materials Science and Engineering **589** (2019) 012036 doi:10.1088/1757-899X/589/1/012036

Synthesis of the relay-linear controlwith the hysteresis region for the second order object example

B Kulakov[1,2] **and D Kulakov**[1,2,3]

[1]Peoples Friendship University of Russia (RUDN University), 6 Miklukho-Maklaya St, Moscow, 117198, Russian Federation;
[2]Bauman Moscow State Technical University,5 Second Baumanskaya str., Moscow 105005, Russian Federation

[3]E-mail: kulakov@bmstu.ru

Abstract. The purpose of this work is a relay-linear control synthesis with using of a hysteresis region, that contains a switching surface and is build with piecewise linear approximation of the switching surface. An application of the hysteresis region together with linear control region allows to avoid sliding mode. A second order object is discussed for better visual presentation of many geometrical concepts that are used in relay-linear control theory.

Introduction

The relay control principles are widely used within constraints of control value [1-5]. Relay-linear control laws description and synthesis are described in [6-8]. The purpose of this work is a relay-linear control synthesis and its application to a second order object, that becomes classical [9-12] in describing optimal control laws for the Pontryagin maximum principle [13,14] and allows visual presentation of many geometrical concepts that are used in relay-linear control theory. For example, state space for a system (1) is a 2-d order phase plane, and switching surface for some time optimal control law (3) is transformed into some curve on that plane.

$$\dot{x} = Ax + bu, \text{with } A = \begin{bmatrix} 0 & 1 \\ 0 & 0 \end{bmatrix}, b = \begin{bmatrix} 0 \\ 1 \end{bmatrix}, |u| \le 1 \, (1)$$

The aim is a time-optimal control that leads x to its origin. The ideal time-optimal control law for (1) is [13]:

$$u = -sign\left(x_2 - \sqrt{2 \cdot |x_1|} \cdot sign(x_1)\right). (2)$$

The optimal control law (2) is defined by an ideal analytical switching curve [13]:

$$x_2 = -\sqrt{2 \cdot |x_1|} \cdot sign(x_1), (3)$$

For practical realization of a suboptimal control we will represent the switching curve (3) by its piecewise linear approximation (Fig. 1). For that we will define some set of the interpolated points $\boldsymbol{\omega}$ of the switching curve inside some considered x domain area.

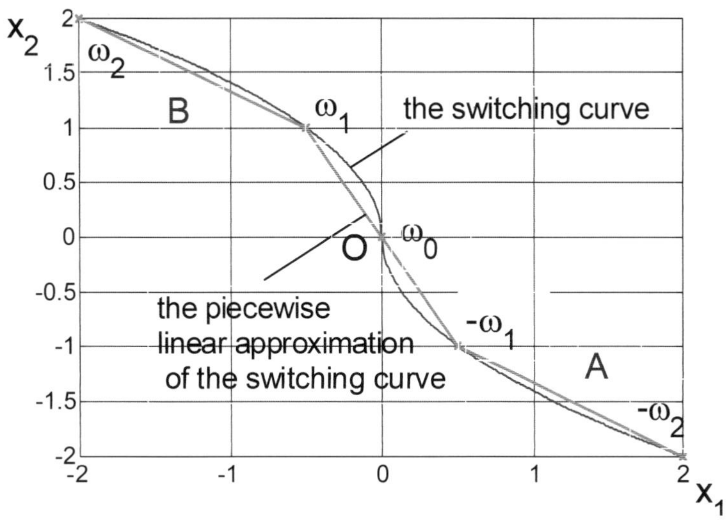

Fig. 1. The switching curve in the phase plane for the system

The interpolated points ω_i can be found by solving the system (1) in inverse time [15] from origin $x_0 = \begin{bmatrix} 0 \\ 0 \end{bmatrix}$ and $u = 1$ or $u = -1$; or by solving an equation (5):

$$x = -\int_0^{-r} \Phi(r,\varphi) b \cdot sign\left(g^T \cdot \Phi(r,\varphi) b\right) d\varphi, \quad (5)$$

where $\Phi(r,\varphi) = e^{A(r-\varphi)}$ – is a fundamental matrix of the system (1), r – corresponds to the expected transient time and the requirement conditions of g [6]:

$g^T b = 0$, $\|g\| = 1$. Then $g \in \{g_1, g_2\}$, where $g_1 = \begin{bmatrix} 1 \\ 0 \end{bmatrix}$, $g_2 = \begin{bmatrix} -1 \\ 0 \end{bmatrix}$.

The solutions (5) for our suggested case ($x_i \in [-2 \quad 2]$, $i = 1,2$) is shown in Fig.1: a switching curve AO - for a case $g = g_1$; a switching curve BO – for a case $g = g_2$.

Set of interpolated points $\boldsymbol{\omega} = \{-\omega_2, -\omega_1, \omega_0, \omega_1, \omega_2\}$ are chosen, where

$$\omega_0 = \begin{bmatrix} 0 \\ 0 \end{bmatrix}, \ \omega_1 = \begin{bmatrix} -0.5 \\ 1 \end{bmatrix}, \ \omega_2 = \begin{bmatrix} -2 \\ 2 \end{bmatrix}.$$

The relay-linear control synthesis basics

In a common case of $n-$ dimensional space the switching surface linear approximation can be produced by the $(n-1)$ dimensional hyperplanes, every of which can be defined by an equation [6]:

$$\mathrm{p}^T \cdot x - d = 0.$$

For $d \neq 0$ $\hat{\mathrm{p}} \overset{\Delta}{=} \mathrm{p} \cdot \frac{1}{d}$, $\hat{\mathrm{p}}^T \cdot x = 1$;

and for $d = 0$ $\hat{\mathrm{p}} \overset{\Delta}{=} \frac{\mathrm{p}}{|\mathrm{p}|}$, $\hat{\mathrm{p}}^T \cdot x = 0$.

In a case of a unit length vector p, d can be regarded as a distance along the vector p from the origin to the hyperplane that perpendicular to the vector p. According to the linear approximation of the switching curve a suboptimal [6] relay control law is:

$$u_{CR} = -sign\left({}^{l}\hat{p}^{T}x + sign\left({}^{l-1}\hat{p}^{T}x + ...sign\left({}^{1}\hat{p}^{T}x\right)\right)\right), \quad (6)$$

where l — is a quantity of approximating hyperplanes.

For our case quantity of approximating hyperplanes is two: $l = 2$. Then

$$u_{CR} = -sign\left({}^{2}\hat{p}^{T}x + sign\left({}^{1}\hat{p}^{T}x\right)\right), \text{ with } {}^{1}p = \begin{bmatrix}0.894 & 0.447\end{bmatrix}^{T}, {}^{2}p = \begin{bmatrix}1 & 1.5\end{bmatrix}^{T}.$$

Switching to the linear control law near the origin is intended to exclude the self-oscillation mode. In works [6,7] a law of switching between relay and linear control (Fig.2) is mathematically described as below (for a second order object in our case).

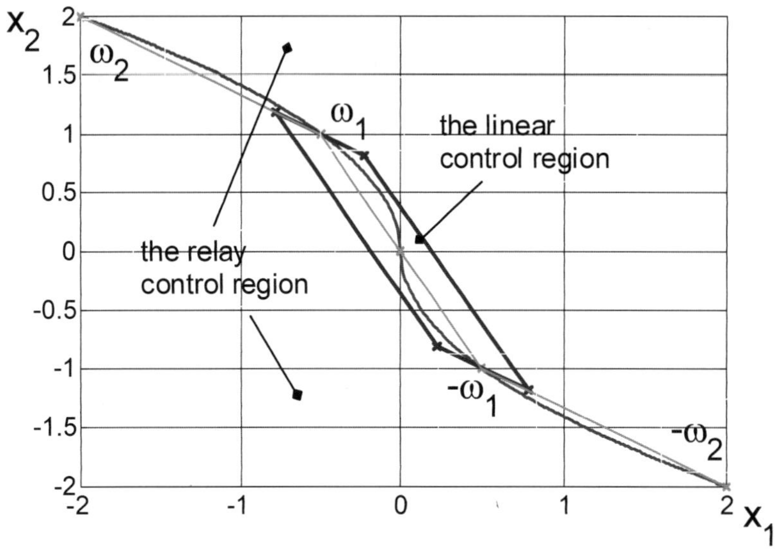

Fig. 2. "Classical" relay-linear control regions

Relay control law:

$F^{R}(x) = -sign\left({}^{2}\hat{p}^{T}x + sign\left({}^{1}\hat{p}^{T}x\right)\right)$ applicable at $\overline{y_1} \cup \overline{y_2} = 1$,

where

$$y_i = \begin{cases} 1: & \left|{}^{i}\hat{p}^{T}x\right| < 1 \\ 0: & \left|{}^{i}\hat{p}^{T}x\right| \geq 1 \end{cases}, \; i \in \overline{[1,2]},$$

Linear control law:

$F^{L}(x) = -k \cdot {}^{1}\hat{p}^{T}x$ applicable at $y_1 \cap y_2 = 1$,

where k is a coefficient of linear control law in the linear control law region.

Then final relay-linear control law can then be written as:

$$\begin{aligned} u &= F^{R}(x)\cdot\left(\overline{y_1} \cup \overline{y_2}\right) + F^{L}(x)\cdot\left(y_1 \cap y_2\right) = \\ &= -sign\left({}^{2}\hat{p}^{T}x + sign\left({}^{1}\hat{p}^{T}x\right)\right)\cdot\left(\overline{y_1} \cup \overline{y_2}\right) - k \cdot {}^{1}\hat{p}^{T}x \cdot \left(y_1 \cap y_2\right). \end{aligned} \quad (7)$$

A usage of the piecewise linear approximation of the switching curve can lead to a sliding mode on some sections of object phase trajectory (Fig.3).

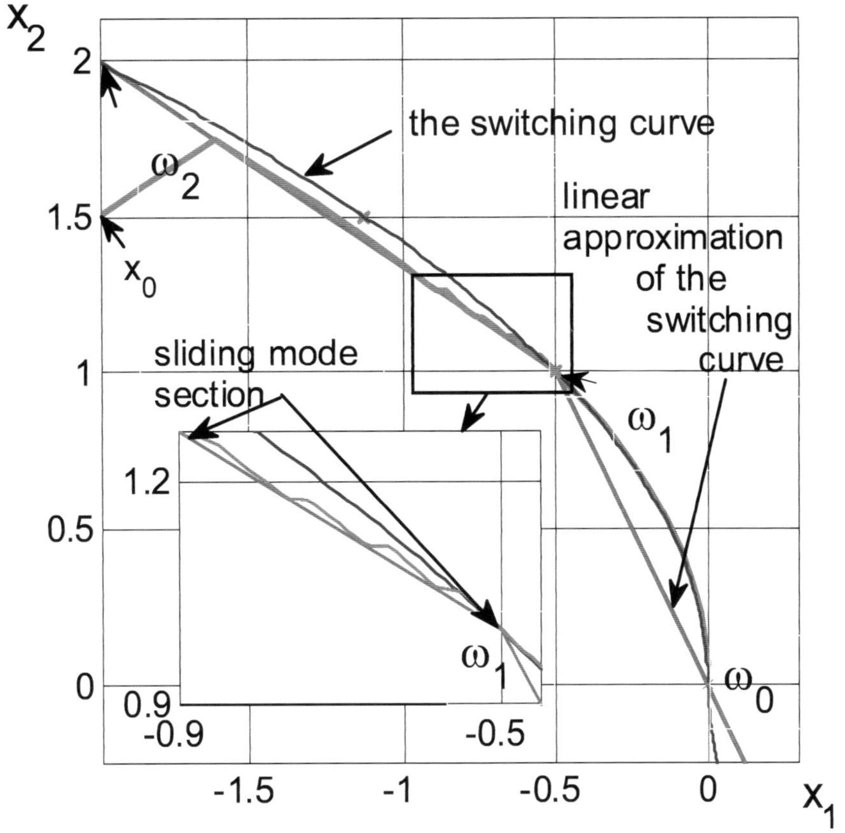

Fig. 3.A sliding mode section on the piecewise linear approximation of the switching curve for a given initial conditions.

The relay-linear control synthesis with the hysteresis region

If we want to avoid the sliding mode at all during movement we can introduce a hysteresis region that contains the switching curve. The hysteresis region is bounded by piecewise linear approximation of the switching curve and by the lines of support which are parallel to appropriate pieces of the switching curve linear approximation(Fig.3).

As a boundary of the linear law applicability a constant energy line may be chosen. Then an equation, that defines a region of the linear law applicability in our case (region, bounded by the ellipse):

$$\frac{k_1 x_1^2}{2} + \frac{x_2^2}{2} < c.$$

To define k_1 and c, an additional requirement to minimize the number of switches can be entered.

The belonging of a point in the phase plane to the regions: linear, hysteresis or relay, will be determined by a sequential check for the truth of the corresponding entered Boolean variables CL, CN, CR defined by expressions (for the second order case):

$$CL = \left(\frac{k_1 x_1^2}{2} + \frac{x_2^2}{2} < c \right); (8)$$

IOP Conf. Series: Materials Science and Engineering **589** (2019) 012036 doi:10.1088/1757-899X/589/1/012036

$$CN = \left(F_{R1} \cdot F_{R2} < 0\right), (9)$$

$$F_{R1} = -sign\left({}^{2}\hat{P}^{T}x + sign\left({}^{1}\hat{P}^{T}x + sign\left({}^{0}\hat{P}^{T}x\right)\right)\right),$$

$$F_{R2} = -sign\left({}^{2}\hat{p}^{T}x + sign\left({}^{1}\hat{p}^{T}x\right)\right),$$

where ${}^{i}P^{T}$, $i = 1,2$ is accordingly parallel to ${}^{i}\hat{p}^{T}$ and tangent to the switching line;

$$CR = \overline{CN} \cap \overline{CL}.$$

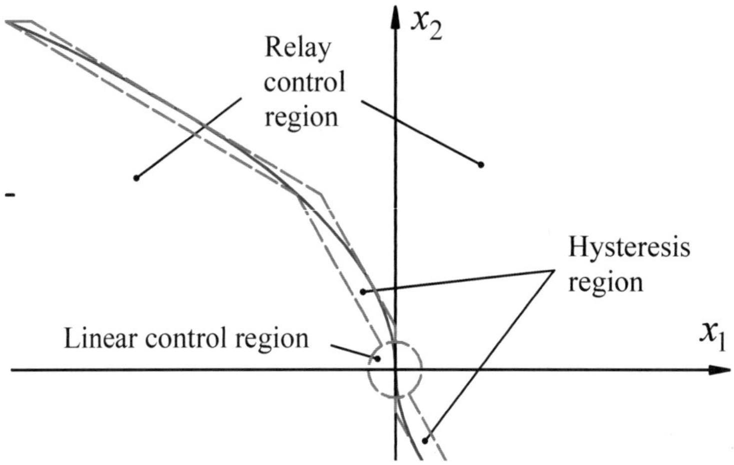

Fig. 4. The linear, relay and hysteresis control regions.

And control law with the hysteresis region defines as follows:
In the linear law region:

$$u = -Kx; (10)$$

In the relay law region:

$$u = -sign\left({}^{2}\hat{P}^{T}x + sign\left({}^{1}\hat{P}^{T}x + sign\left(x_1\right)\right)\right). (11)$$

In the hysteresis region:
The control is remaining the same as a value just before entering the hysteresis region and defined by the relay law. After x got out of the hysteresis region, the control will be defined by the appropriate relay or linear control laws.

Simulation results and discussion
The relay-linear control simulation results for the studied system mathematical model (1) was produces in Matlab with both control laws (7) and (8-11). The same coefficients were implemented for the corresponding regions of the linear control laws. The phase trajectories for the system with two control laws: with the hysteresis region and without it are considered (Fig.5).
In Figure 5:
1. – the linear control region for the control law (7) (without hysteresis);
2. – the hysteresis region border defined by (9);
3. – the linear region defined by (8);
4. – the phase trajectory of the system (1) with implemented (7) control law;
5. – the phase trajectory of the system (1) with implemented (8)-(11) control law.

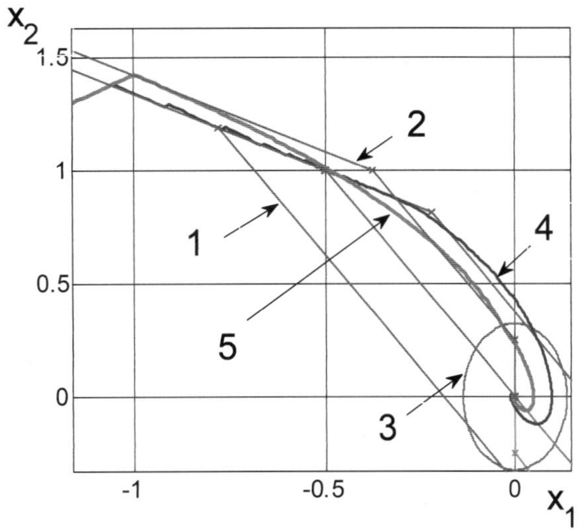

Fig. 5. The phase trajectories for the system with two control laws: with the hysteresis region and without it.

The phase trajectory 4 in Fig.5 contains sliding mode section. That is resulted in specific changes of control value during simulation (Fig. 6).

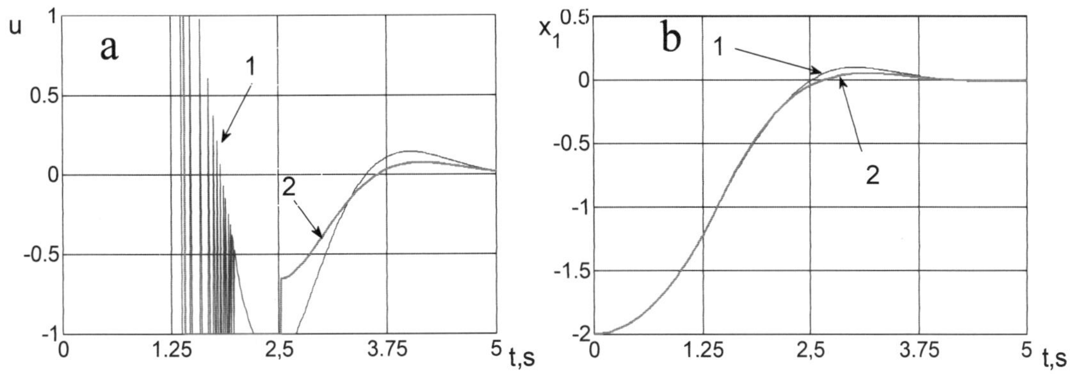

Fig. 6. The control u time dependence (a) and the transient processes (b) for the system with two control laws: without the hysteresis region (1) and with it (2).

The simulation results show the effectiveness of the hysteresis zone introduction in the linear-linear control law from the point of view of the absence of a large number of switches caused by the sliding mode section. In addition, the introduction of hysteresis in the formation of the control law does not lead to a deterioration of the transition process in this case.

References

[1] E. P. Popov, The Dynamics of Automatic Control Systems, Pergamon Press, 1961, 776 p., ISBN: 9781483184623.

[2] Reshmin, S.A. The Threshold Absolute Value of a Relay Control Bringing a Satellite to a Gravita-tionally Stable Position in Optimal Time (2018) Doklady Physics, 63 (6), pp. 257-261. https://www.scopus.com/inward/record.uri?eid=2-s2.0-85050142824&doi=10.1134%2fS1028335818060101&partnerID=40&md5=DOI: 10.1134/S1028335818060101

[3] Reshmin, S.A. Threshold Absolute Value of a Relay Control when Time-Optimally Bringing a Satel-lite to a Gravitationally Stable Position (2018) Journal of Computer and Systems Sciences International, 57 (5), pp. 713-722. https://www.scopus.com/inward/record.uri?eid=2-s2.0-85055038451&doi=10.1134%2fS106423071805012X&partnerID=40&md5=DOI: 10.1134/S106423071805012X

[4] Chkheidze, G.A. Synthesizing algorithms for the control of the motion of sliding-mode dynamic systems (1996) Journal of Computer and Systems Sciences International, 34 (3), pp. 1-8.

[5] Reshmin, S.A. Estimate of the control threshold value in the problem on a time-optimal satellite attitude transition maneuver (2017) Mechanics of Solids, 52 (1), pp. 9-17. Цитировано 2 раз. 1) https://www.scopus.com/inward/record.uri?eid=2-s2.0-85018451476&doi=10.3103%2fS0025654417010022&partnerID=40&md5=DOI: 10.3103/S0025654417010022

[6] Moroz, A.I., Pupkov, K.A. ON ONE METHOD OF SYNTHESIS OF RELAY-LINEAR CONTROL LAWS. (1978) Engineering Cybernetics (English translation of TekhnicheskayaKibernetika), 16 (5), pp. 138-144. https://www.scopus.com/inward/record.uri?eid=2-s2.0-0018017880&partnerID=40&md5=a7e0b7c45cbf1feba1529c32e66e045e

[7] Pupkov, K.A., Moroz, A.I. On the Theory of Relay-Linear Control Laws. [Article@K TEORII RELEINO-LINEINYKH ZAKONOV UPRAVLENIYA.] (1976) Probl Control Inf Theory, 5 (5-6), pp. 401-419 (engtransl, 1-18.

[8] Pupkov K A, KuiCh N 2006 Relay-linear control law in a system of a homing guidance. Problem imashinostroeniyaiavtomatizazii— Engineering and Automation Problems (4) pp 55–58 (in Russian)

[9] Ivanov, V.A., Kozhevnikov, S.A. A problem of synthesis of control optimal in fuel consumption for lin-ear second-order objects with control derivatives (1996) Journal of Computer and Systems Sciences International, 35 (4), pp. 574-580. 1) https://www.scopus.com/inward/record.uri?eid=2-s2.0-33746200105&partnerID=40&md5=84f90a239cf6a6978cb321a0d4548bf5

[10] Flugge-Lotz, I. Discontinuous automatic control (2015) Discontinuous Automatic Control, pp. 1-168 p. https://www.scopus.com/inward/record.uri?eid=2-s2.0-84954206760&partnerID=40&md5=98503b526955b4b6f685a993ea2a816

[11] Almuzara, JosL.G.,Flügge-Lotz, I. Minimum time control of a nonlinear system (1968) Journal of Differential Equations, 4 (1), pp. 12-39. DOI: 10.1016/0022-0396(68)90046-6

[12] Reshmin, S.A. Bifurcation in a time-optimal problem for a second-order non-linear system (2009) Journal of Applied Mathematics and Mechanics, 73 (4), pp. 403-410. Цитировано 8 раз. 1) https://www.scopus.com/inward/record.uri?eid=2-s2.0-71549116113&doi=10.1016%2fj.jappmathmech.2009.08.011&partnerID=DOI: 10.1016/j.jappmathmech.2009.08.011

[13] V. G. Boltyanskii Mathematical methods of optimal control (Balskrishnan-Neustadt series). Holt, Rinehart and Winston Publ.; 1st edition (1971). 272 p. ISBN-10: 0030722853

[14] PONTRYAGIN, L. S., BOLTANSKII, V. G., GAMKRELIDZE, R. V., AND MISHCHENKO, E. F., "The Mathematical Theory of Optimal Processes." Wiley (Interscience), New York, 1962.

[15] Pupkov K A, Egupov N D, et al. 2004 Metodyklassicheskoyisovremennoyteoriiavtomaticheskogoupravleniya v 5 tomakh. Tom 4. TeoriyaOptimisazii system avtomaticheskogoupravleniya [Methods of Classic and Modern Control Theory in 5 volumes Vol. 4 Automatic Control Systems Optimization Theory] (Moscow: Bauman MSTU Publ.) 744 p (in Russian)

Study of the corrosion rate of the paths walls of aluminum heat exchangers in the ethylene glycol antifreezes environment under developed aeration of power fluid

A Petrov[1,2], G Saatashvili[1], A Averianov[1] and I Bilenko[1]

[1]Bauman Moscow State Technical University, 5 Second Baumanskaya Street, Moscow, 105005, Russian Federation

[2]E-mail: alexeypetrov@bmstu.ru

Annotation. An experimental and theoretical research work on the corrosion processes of heat exchanger channels made from different aluminum alloys in an ethylene glycol antifreeze environment conducted by the authors is described in the article. The hydrodynamic simulation of the flow in the paths was performed. Significant effects of the aeration of the power fluid, the possibility of electrochemical corrosion are proved. The comparative of anti-corrosion property of different aluminum alloys is shown when operating in a wide range of power fluid velocities and temperatures.

Introduction

Currently, heat exchangers made of aluminum alloys are widely used in different technical fields. Aluminum alloys, due to their low density, simple machining and high thermal conductivity, are used to manufacture shell and tube heat exchangers of all types that are used, among others, in engine cooling systems, communications-electronics equipment, hydraulic actuators, etc. Often, ethylene glycol antifreezes of different formulations, such as Tosol-40, Tosol-65 «Lena», etc., are used as power fluids for operation of such systems at low ambient temperatures.

The practice of using such hydraulic systems has shown that in some instances the paths of heat exchangers are intensively corroding, up to the formation of open pittings, and this process develops quite quickly, within a few hundred operation hours [5,6]. Possible causes of this corrosion may be the presence of a large amount of dissolved air in the liquid (aeration), the presence of copper or bronze elements in the system (electrochemical corrosion), and a surface defects of the heat exchanger paths [7, 9-11]. This article includes the results of theoretical and experimental studies of corrosion processes in such heat exchangers made of various aluminum alloys.

Methods

Originally estimated, that in the heat exchanger paths developed cavitation erosion can occur, in order to determine the possible zones of which it was necessary to carry out a hydrodynamic simulation of flow. In the future, the flow model in the paths was dilated to a two-phase flow model, taking into account the occurrence of gas cavitation [1-5], [12-15].

The 3-D model of the heat exchanger paths was created and the flow simulation in the paths, based on score equations and boundary condition described below was carried out.

In general Navier-Stokes vector equation for fluid is written as:

$$\frac{\partial \vec{v}}{\partial t}+(\vec{v}\cdot\nabla)\vec{v}=-\frac{1}{\rho}\nabla p+\nu\Delta\vec{v}+\vec{f}$$

Where t - is time, ∇- is Nabla, Δ- vector Laplace operator, ρ- is density, *p*- is pressure, ν - is kinematic viscosity, $\vec{v}$ - velocity field, $\vec{f}$ - field of body forces.

The continuity equation is also applied:

$$\frac{\partial \rho}{\partial t}+\operatorname{div}\rho v=0$$

For the simulation of aeration fluid flow was created a two-phase flow model, consisting of water and air under specified values of density and viscosity. Proportional ratio: 95% of water and 5% of air. For the calculation, physical models of gravity force and a k-ω turbulence model were used. The velocity inlet was set in accordance with the required flow rate; the outlet pressure was set as zero.

For creation of mesh the following models were used: surface remesher, polyhedral mesher (base size 0,5 mm), prism layer mesher (number of layers – 5, stretching– 1,3, total thickness – 45% of base size) [1-3]. The simulation results were compared with experimental results.

For replication of conditions for the corrosion occurrence, the test bench was developed and created, simulating a real system.

The created test bench allowed to pass through the paths required fluid flow rate (ethylene glycol antifreeze Tosol-65) at the temperature up to 70°C and at the same time to provide aeration of power fluid (through the spout). In which case, accumulated operating time of heat exchangers made from different aluminum compositions (heat exchangers made from alloy D16T [aged-hardening duralumin] and AMg6 [93,5% Al, 5,5% Mg]was up to 140 hours.

Heat exchangers were connected as one by one, as well two and two in series through quick disconnect coupling (QDC), similar to using in real system. The diagram of the test bench is shown in figure 1. To simulate the presence of bronze elements in the real system, inserts and fittings made from bronze without galvanic isolation (flexible pipe in a metal braid were used) between them and the heat exchangers were used in the test bench. The pressure at the inlet of the heat exchanger (before the QDC) was maintained within 3…4 bar, which corresponds to the pressure in the real system.

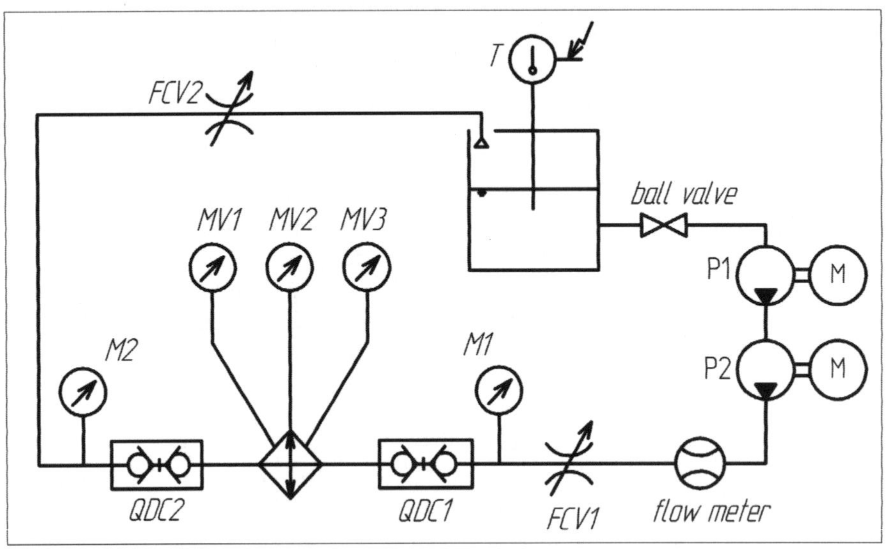

Figure 1. Diagram of the test bench

IOP Conf. Series: Materials Science and Engineering **589** (2019) 012037 doi:10.1088/1757-899X/589/1/012037

During the experiments, the temperature and flow rate of power fluid, pressure difference in the heat exchanger elements, operating time, noise and vibration levels on the heat exchanger surfaces were monitored. After operating time, the condition of heat exchanger paths was estimated through factory-made endoscope, then by cutting a part of paths to access the inner surface.

Results

According to the hydrodynamic simulation results, the following conclusions can be drawn: the occurrence of steam cavitation in the heat exchanger paths is unlikely, because zones of pressure reduction equals to saturated vapor pressure were not detected (figures 2, 3). This pressure distribution pattern was also corroborated by measuring pressures in the holes (about 20 holes were made) along the heat exchanger paths.

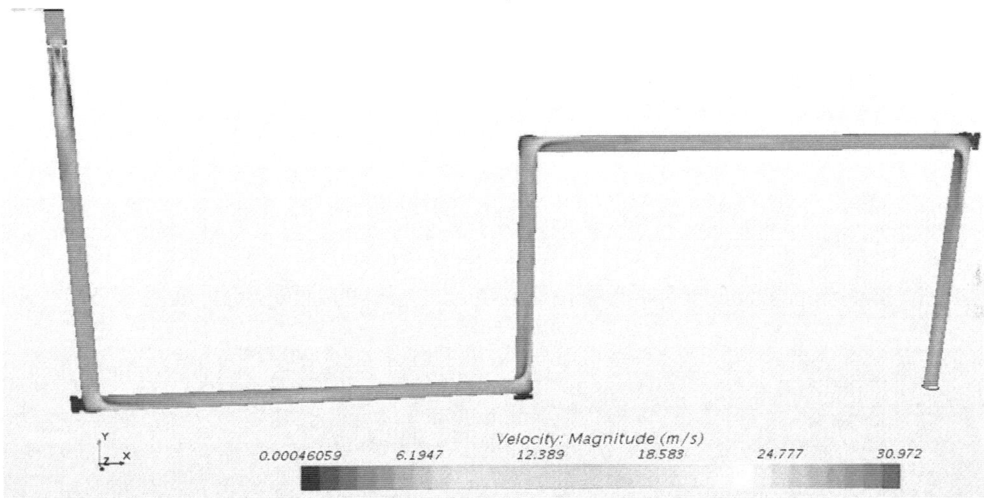

Figure 2. Velocity field distribution pattern in the heat exchanger paths at the flow rate 14 l/min

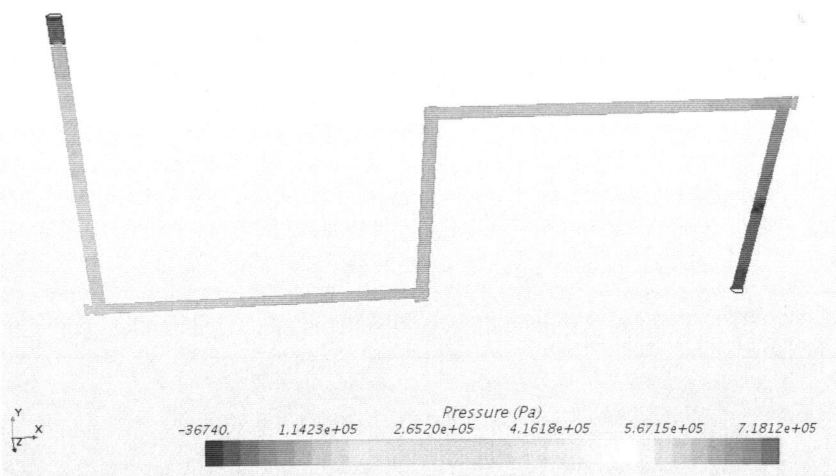

Figure 3. Pressure distribution pattern in the heat exchanger paths at the flow rate 14 l/min

There are no also anomaly high velocity zones, except of zone proximately after the QDC and particularly after flow deflection. However, the rapid decay of pressure after QDC, including the laminar flow (figure 4), which causes a high probability of gas cavitation, i.e. release of dissolved gas, with a significant aeration of the fluid.

The presence of low flow velocity zones along path centerline of heat exchanger (figure 5), which combined with gas releases and low flow velocity in general can create a local high oxygen content zones in the upper medium paths zones at some distance of flow deflections. The hydraulic simulation of two-phase laminar flow (flow rate is 2 l/min) is shown in figure 6, a zones of high gas content in the upper path part are visible.

It may be concluded that the presence of sharp narrowing and increase of flow, and flow deflection, which combined with high aeration level of fluid, cause intensive gas releases and laminar flow with low velocity results to movement of gas bubbles to upper part of path, causing a large amount of oxygen in wall layers.

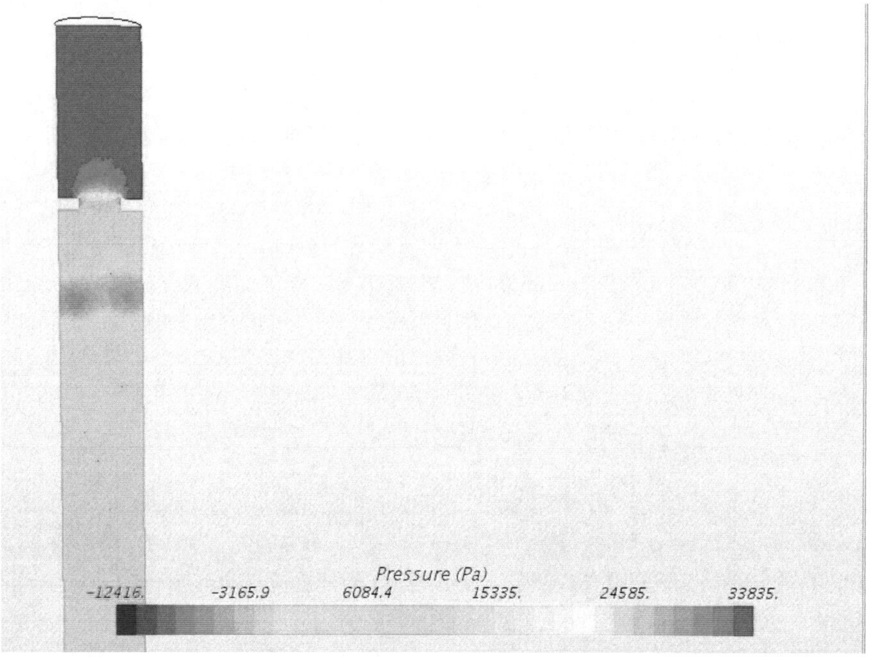

Figure 4. Pressure reduction zone after local flow narrowing in the heat exchanger inlet

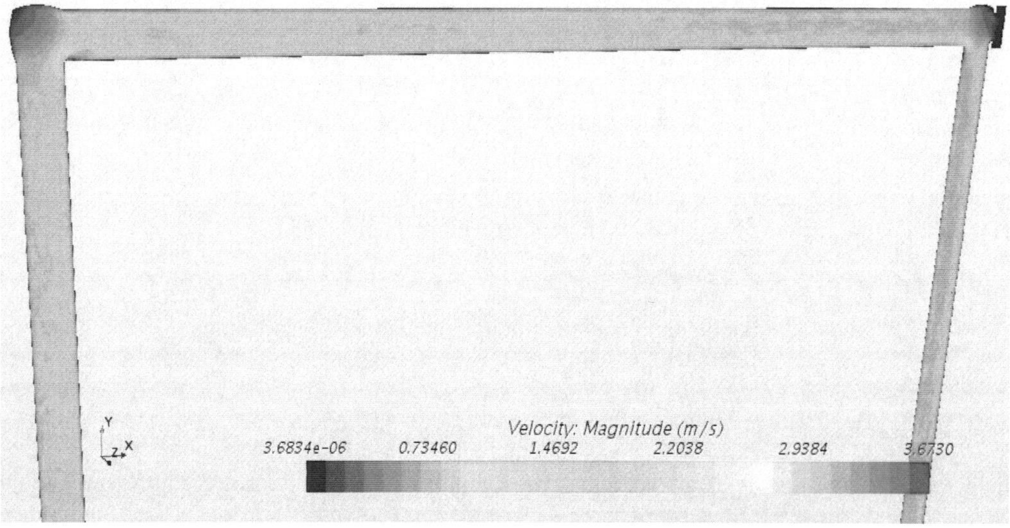

Figure 5. Local zones of velocity reduction

IOP Conf. Series: Materials Science and Engineering **589** (2019) 012037 doi:10.1088/1757-899X/589/1/012037

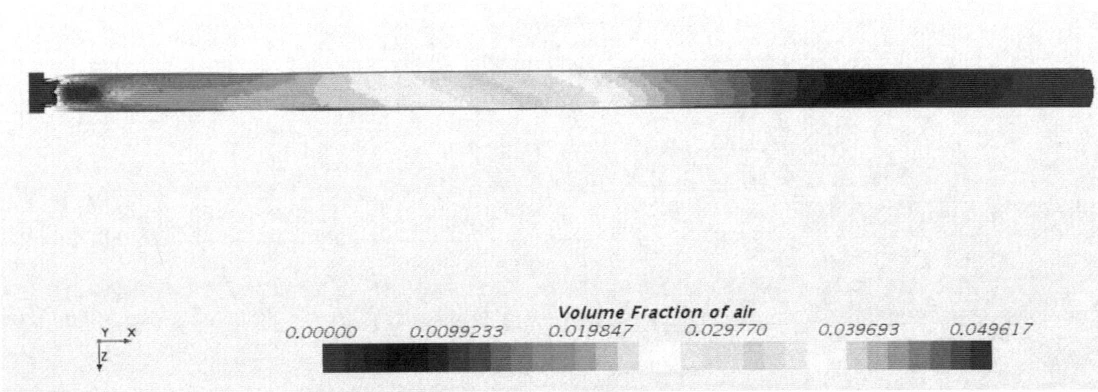

Figure 6. Two-phase laminar flow of fluid motion

The occurrence of intensive gas cavitation during the operating is confirmed by acoustic image of fluid flow in the heat exchanger [2]. The acoustic spectrum of the results of noise level measurement among the path (spectrum corresponds to zone after inlet part) is shown in figure 7. White shows the acoustic background in the room, orange - the noise in the path. Visible peak at frequency of about 1 kHz, arising due to the occurrence of steam cavitation. When the flow rate decreases, the peak amplitude decreases, but it does not disappear completely as long as there is a flow process.

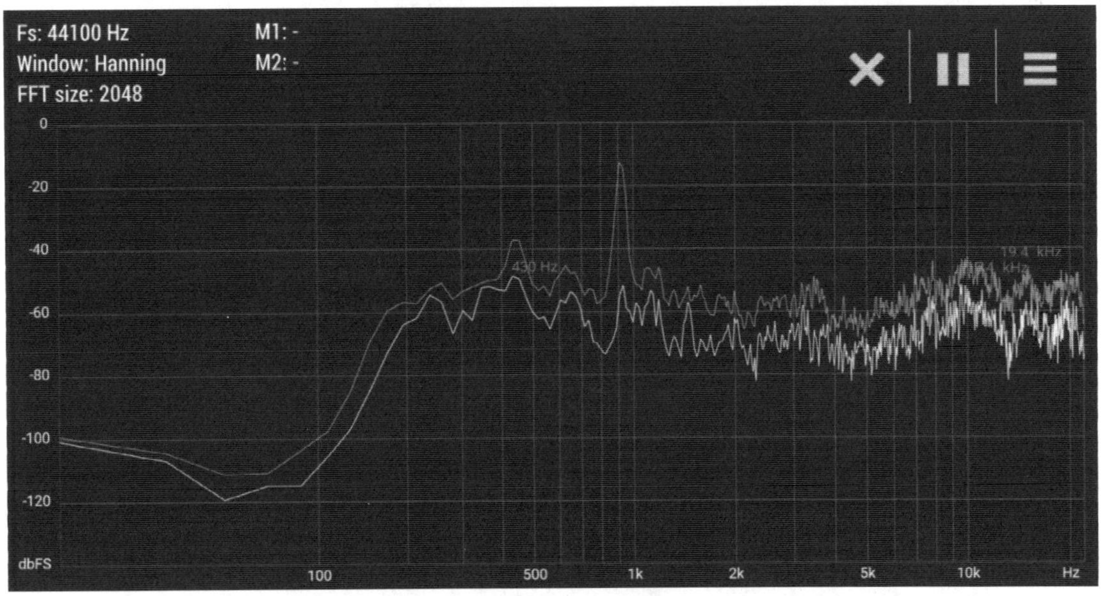

Figure 7. Noise levels measurement during the testing

After the testing, the study of paths surface was carried out by borescope and milling for opening of some path parts. Among other things, was created the fluid precipitation in the reservoir, where numerous aluminum particles up to 0,1 mm in size were found out. Signs of corrosion in the opening path of heat exchanger made from D16T [aged-hardening duralumin]are shown in figures 8 and 9, the inner surface of the same path is shown in figure 10.

IOP Conf. Series: Materials Science and Engineering **589** (2019) 012037 doi:10.1088/1757-899X/589/1/012037

Figure 8. Corrosion of the path surface

Figure 9. Open pitting corrosion on inside of heat exchangerpath

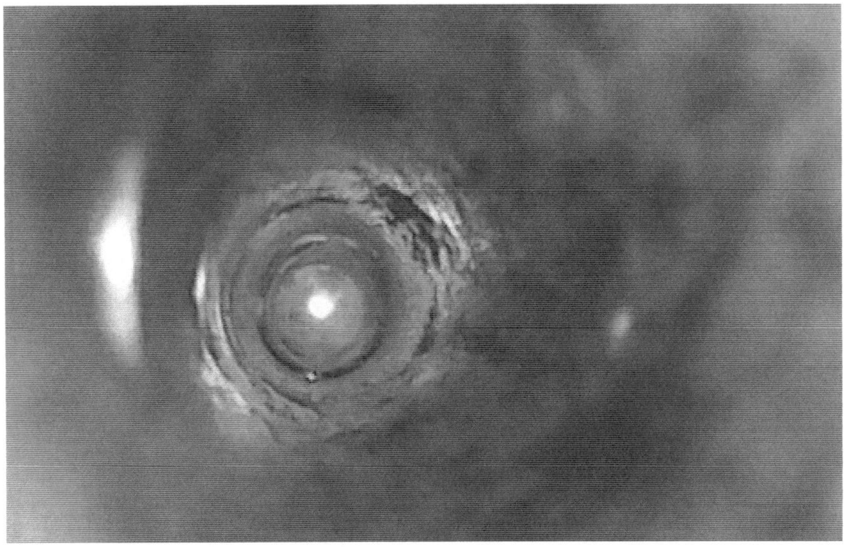

Figure 10. The path of heat exchanger made from D16T [aged-hardening duralumin], operating time is 130 hours, corrosion is visible

The effect of spray coating Sn-Bi on the corrosion developing is interesting. Zone of the path interface with and without spray coating that, perhaps, proves electrochemical type of corrosion [10,11].

Figure 11. Corrosion zones in the path with and without spray coating

The testing of heat exchanger made from AMg6 [93,5% Al, 5,5% Mg] has shown the absence of corrosion signs on the surface (figure 12):

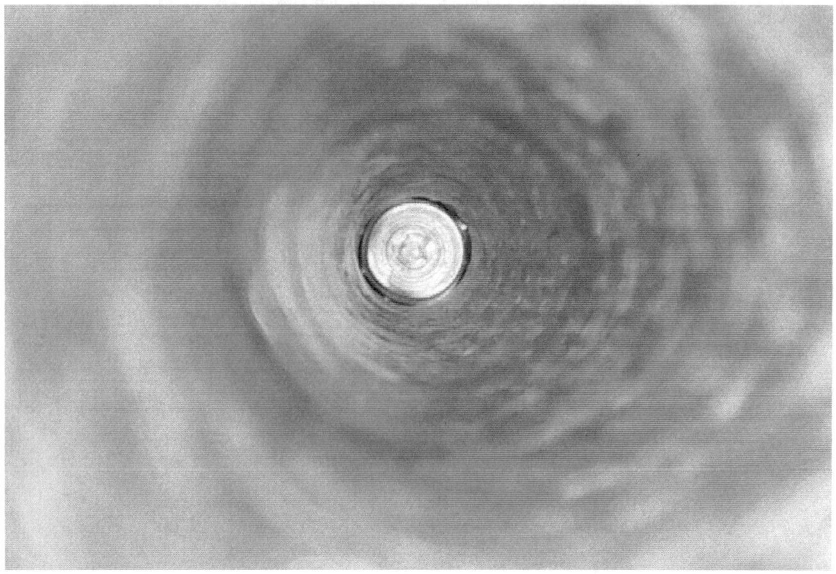

Figure 12. The path of heat exchanger made from AMg6 [93,5% Al, 5,5% Mg], operating time is 100 hours, there are no corrosion signs (there are signs of treatment)

Discussion

1. Hydrodynamic simulation and acoustic flow pattern showed that in the heat exchanger in the zone of rapid decay of pressure gas can be released. It appears to suggest that the released gas can (especially at low velocities of fluid in the paths) be accumulated in the upper parts of paths and thereby sharply accelerate the corrosion of the walls of the paths in the presence of excess oxygen;

2. The presence of copper color tone on the paths walls and a lot of corrosion rate mean the possible electrochemical corrosion in the system, especially in copper aluminum combination.

3. D16T [aged-hardening duralumin]alloy in the circumstances is extremely unstable to corrosion, in any case, without additional coating the inner surface of the paths walls;

4. D16T [aged-hardening duralumin] alloy within the time of testing, proved to be corrosion resistant;

5. The detection of a significant amount of small aluminum pieces shows the developed cavitation of the paths walls with carry-over of metal from the affected zones.

List of references

[1] Lomakin V O, Chaburko P S, Kuleshova M S 2017 Multi-criteria Optimization of the Flow of a Centrifugal Pump on Energy and Vibroacoustic Characteristics Procedia Engineering 176 pp 476-482

[2] Lomakin V O, Kuleshova M S, Kraeva E A 2015 Fluid Flow in the Throttle Channel in the Presence of Cavitation Procedia Engineering 106 pp 27-35

[3] Ionaitis, R.R., Chain, P.L. Principles for engineering active-passive hydrodynamic control-and-safety systems for nuclear reactors (2009) Atomic Energy, 106 (3), pp. 175-184.

[4] Krapivtsev, V.G. Model Studies of Coolant Flow Hydrodynamics in VVER-1000 In-Reactor Pressure Channel (2017) Atomic Energy, 122 (5), pp. 304-310.

[5] Marchevskii, I.K., Puzikova, V.V. Numerical simulation of the flow around two fixed circular airfoils positioned in tandem using the LS-STAG method (2016) Journal of Machinery Manufacture and Reliability, 45 (2), pp. 130-136.

[6] K.A. Morch. Dynamics of Cavitation bubbles and Cavitating liquids //Treatise on Materials Science and Technology, 1976. 16. P. 309-335.

[7] Kendrick H. Light. Development of cavitation erosion resistant advanced material system. Orono (Maine, USA) The University of Maine, 2005. 76 p.

[8] Patankar S. Numerical methods in heat transfer and fluid dynamics // Translated from English (1980), Energoatomizdat publishing, 1984. - 152 p.

[9] Petrov A.I., Skobelev M.M., Khanychev A.G. Research of comparative resistance to cavitation erosion of material samples and coatings of the flow section of hydraulic machines // Herald of the BMSTU. “Mechanicalengineering”. 2015. - №2

[10] Keshe G. Corrosion of metals. Physical and chemical principles and actual problems. Translated from German. / G. Keshe. Edited by IA.M. Kolotyrkina. Metallurgypublishing, 1984. – 400 p.

[11] Siniavskii V. S. Corrosion and protection of aluminum alloys. 2nd publ., translated and completed by V.S. Siniavskii, V. D. Valkov, V. D. Kalinin. Metallurgypublishing, 1986. – 368 p.

[12] N Egorkina and A Petrov 2019 *IOP Conf. Ser.: Mater. Sci. Eng.***492** 012015

[13] N Isaev 2019 *IOP Conf. Ser.: Mater. Sci. Eng.***492** 012026

[14] A Shablovskiy and E Kutovoy 2019 *IOP Conf. Ser.: Mater. Sci. Eng.***492** 012035

[15] V Cheremushkin and V Lomakin 2019 *IOP Conf. Ser.: Mater. Sci. Eng.***492** 012039

Travelling salesman problem concerning to manipulator Kawasaki FS03N trajectory formation

B Kulakov[1,2] **and D Kulakov**[1,2,3]

[1]Peoples Friendship University of Russia (RUDN University), 6 Miklukho-Maklaya St, Moscow, 117198, Russian Federation;
[2]Bauman Moscow State Technical University, 5 Second Baumanskaya Street, Moscow, 105005, Russian Federation

[3]E-mail: kulakov@bmstu.ru

Abstract: An idea of travelling salesman problem (TSP) application to an optimal manipulator path finding was implemented. A set of known poses robot should pass is given. The poses sequence may vary. A traveling time is a performance criteria to be minimized. This task was realized concerning to a special case of Kawasaki FS03N manipulator movement by dynamic programming method. A special Bellman function state notation was used that in some cases gives an ability to solve the task at industrial robot controller. The state notation rule is one-one mapping rule between TSP state and array index it is stored.

Introduction

In a number of cases a manipulator's goal is some object motion with specific poses it should be placed in a case when the poses order is not important – some workpiece several sides examining task for example. Herewith a necessary aim is to reduce some *additive* quality factor [1]. In that case the problem is the same with the travelling salesman problem (TSP) [2-4]. The aim is to find a time-optimal path-tracing [5-7]. The set of poses is given. Initial and final pose is set, intermediate poses should be passed single time only and interleaving order may differ. Sometimes classical TSP is formalized as follows: a complete weighted loop-free graph G is presented with vertex set $N = \{1, 2, ..., n\}$; any arc weight is nonnegative; a Hamilton circuit with minimal weight should be found [8].

This article contains a particular solution of such a task. A Kawasaki FS03N manipulator should place a workpiece mockup (just "workpiece" in further description) in fixed and known poses set (Fig. 1). Initial and final pose is the same and is labeled a.

The intermediate poses are grouped into an ordered set $M = \{b, c, d, e\}$ (Fig. 2). The workpiece should be examined by an instrument with fixed position in space. Thefastestpathshouldbefound. The intermediate poses sequence is not fixed (may vary). The manipulator should stay in each pose from among M for examination purpose during a fixed time.

A time of detail examination – summarized time the manipulator should stay at each pose – we assume as a constant value, neglect in discussion and exclude from the travelling time.

All Kawasaki FS03N manipulator traveling time values T_{ij} (in seconds) from pose i to pose j ($i, j = 0..4$) are presented in a Table 1 (for a definite motion speed). Nonzero i, j are the orders of the corresponding poses from the set M. $i, j = 0$ correspond to the pose a.

IOP Conf. Series: Materials Science and Engineering **589** (2019) 012038 doi:10.1088/1757-899X/589/1/012038

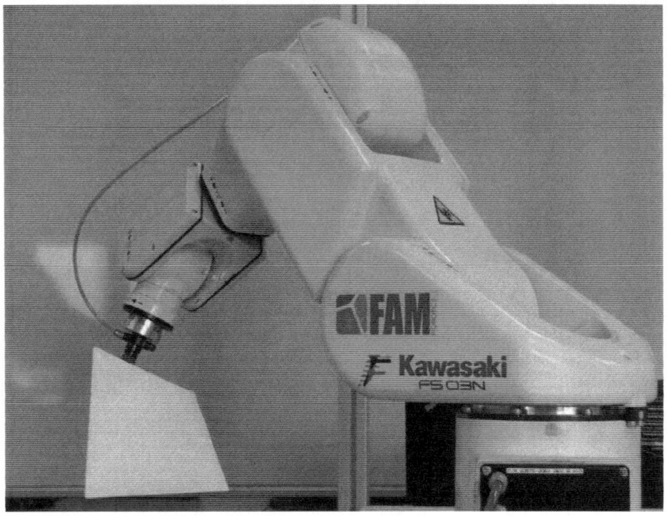

Fig. 1. Kawasaki FS03N robot with the workpiece mockup in the initial pose *a*

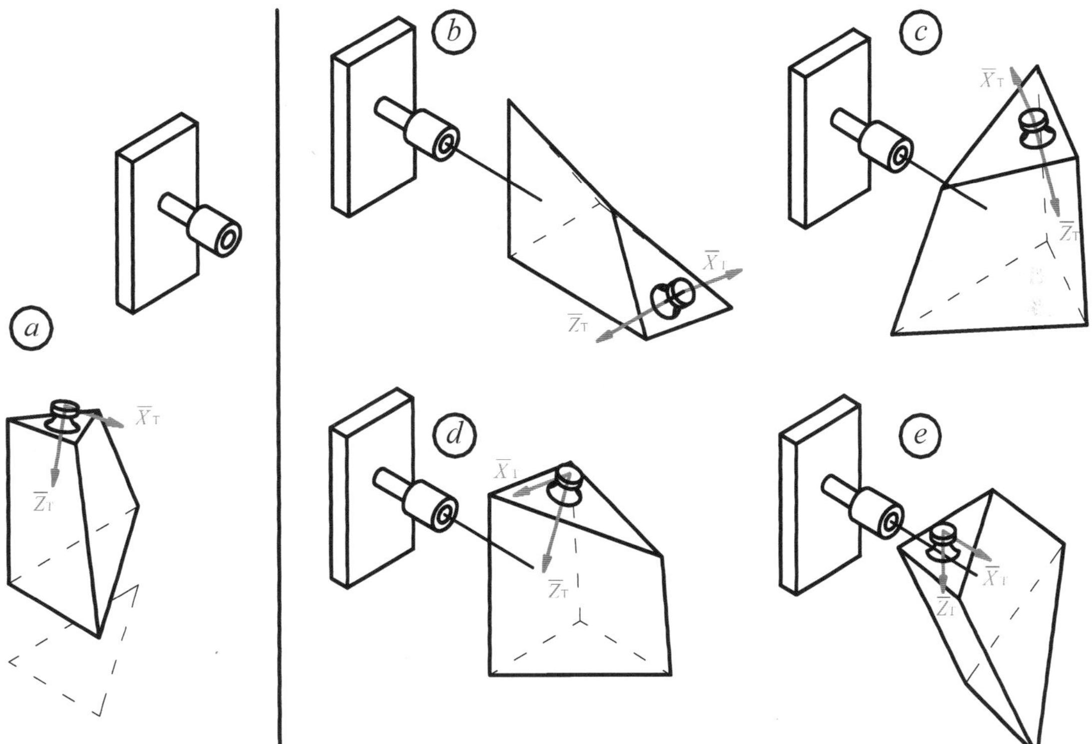

Fig. 2.A schematic view of the workpiece mockup at the set of poses M={b, c, d, e} and the instrument.

Table 1. Manipulator traveling time values T_{ij} (in seconds) from pose i to pose j

a manipulator «toward» pose, (j) / a manipulator «from» pose, (i)	a, 0	b, 1	c, 2	d, 3	e, 4
a, 0	X	5.12	3.984	2.528	2.64
b, 1	5.112	X	6.032	5.52	5.408
c, 2	3.984	6.032	X	3.488	4.4
d, 3	2.528	5.52	3.488	X	2.704
e, 4	2.64	5.408	4.4	2.704	X

All possible path variants are shown in Fig. 3.On the right of Fig. 3, opposite each terminal vertex the total traveling time is given for each case. Each vertex corresponds to some **state**: current manupulator pose and the poses it have not passed yet (a letter before the brackets and the letters in the brackets mark that correspondingly).

The task was considered and solved as the TSP with time-optimal sequence of the robot-manipulator internal poses determination. The solution we have got was realized at Kawasaki FS03N robot-manipulator. A special state notation was used to implement ease of dynamic programming realization for the assigned TSP task at the robot-manipulator.

State notation method for using dynamic programming for TSP

A calculated Bellman function value for almost any state should be stored at some ***memory address*** for further usage at dynamic programming method [9-13]. In this work the state notation is organized to compute the state memory adress and do a reverse computation – define the state by the corresponding address. We use very simple rule for that:

- the state is represented by a **binary number-word** (a "0110" as an example)
- a bit order is a manipulator pose number in the ordered set M;
- the passed poses are units, the poses are not yet passed – zeros;
- current pose is a bold unit;

For the 01**1**0 state c, d poses are passed, d pose is current, b, e poses aren't passed yet (the initial and final pose a is not used in state recording) (Fig. 3).

This state notation in Bellman function arguments gives visually quite understandable expression as follows:

$$B(0000) = \min\{T_{01} + B(\mathbf{1}000), T_{02} + B(0\mathbf{1}00), \ldots$$
$$T_{03} + B(00\mathbf{1}0), T_{04} + B(000\mathbf{1})\},$$

$$B(\mathbf{1}000) = \min\{T_{12} + B(1\mathbf{1}00), T_{13} + B(10\mathbf{1}0), T_{14} + B(100\mathbf{1})\},$$

$$B(0\mathbf{1}00) = \min\{T_{21} + B(\mathbf{1}100), T_{23} + B(01\mathbf{1}0), T_{24} + B(010\mathbf{1})\},$$

… etc.,

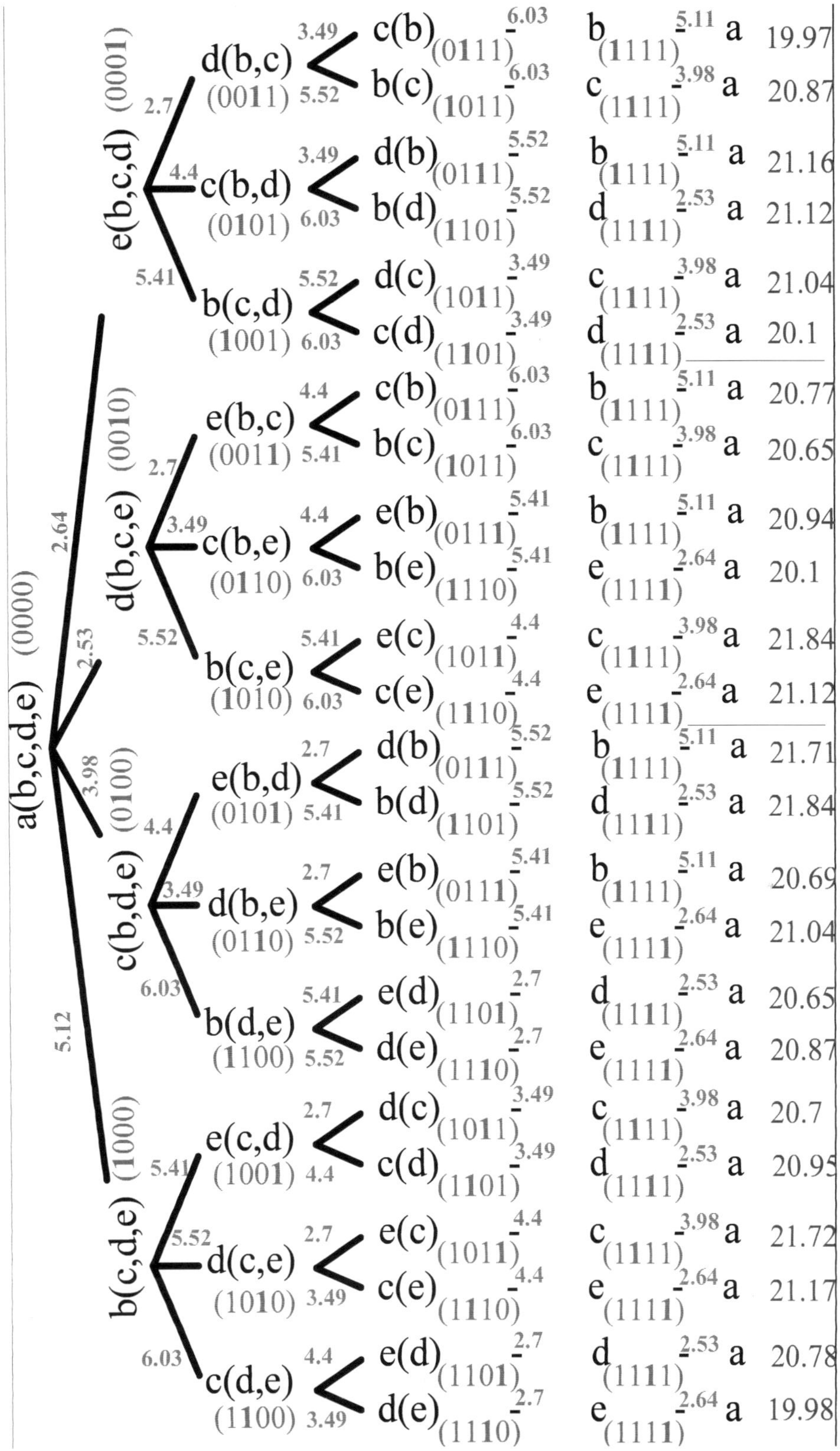

Fig. 3. Possible variants of the workpiece poses sequence during its handling time.

and also (see the Table 1):

$$B(\mathbf{11}00) = \min\{T_{23} + T_{34} + T_{40}, T_{24} + T_{43} + T_{30}\},$$

$$B(\mathbf{1}1\mathbf{0}0) = \min\{T_{13} + T_{34} + T_{40}, T_{14} + T_{43} + T_{30}\},$$

… etc.

Here T_{ij} is a manipulator traveling time from pose i to pose j according to the Table 1.

The main **features** of the ***state notation rule*** for memory adress computing and ability of programming usage are the follows:

– A state is described by the couple of numbers (p,n),, where: p is a passed pose index (decimal representation of the ***binary number-word***); n is a bit order of the current pose (unit).For example:

$B(0000) = B(0,0)$, $B(\mathbf{1}000) = B(8,1)$,

$B(0\mathbf{1}00) = B(4,2)$, $B(10\mathbf{1}0) = B(10,3)$.

etc.

– Let enter a Bc array (an array element is a traveling time from corresponding current state to final pose – i.e. a Bellman function value) and a Bs array (an array element is a next optimal pose from corresponding current state). A matrix row corresponds to passed poses, a matrix column corresponds to current pose index (a pose serial number in the set of poses M). Then let Bellman function argument (p,n) be cell adress of thearrays: p – row numer, n – column number.

– With such adressing modeapproximately half of all Bc and Bs array cells will be empty. For example, for the 10-th array row only the first element (state "**1**010") and the third element (state "10**1**0") will be filled. For reduction of unused address spaceit is rational to use a symmetry of binary notation of serial natural numbers, demonstrated in a Table 2. Asymmetrylawmathematicallyisshownbelow:

$$bin(i) = \overline{bin}(2^N - i - 1)$$

where i is a Table 2 row number in decimal representation, N is a binary digit number we use.

If only «unit cells» is filled with some information then using the described at the Table 2 symmetry we can write information to a Table 3 that don't have empty/zero cells.

The row number k of the Bc and Bs arrays will be defined as

$k = i$ at $i < 2^{N-1}$

$k = 2^N - i - 1$ at $i \geq 2^{N-1}$ (1)

IOP Conf. Series: Materials Science and Engineering **589** (2019) 012038 doi:10.1088/1757-899X/589/1/012038

Table2. Binary notation symmetry of row serial number

№ = *i*				
0	0	0	0	0
1	0	0	0	1
2	0	0	1	0
3	0	0	1	1
4	0	1	0	0
5	0	1	0	1
6	0	1	1	0
7	0	1	1	1
8	1	0	0	0
9	1	0	0	1
10	1	0	1	0
11	1	0	1	1
12	1	1	0	0
13	1	1	0	1
14	1	1	1	0
15	1	1	1	1

Table 3. Information record rule illustration for the case of data filling only at nonzero array cells in the Table 2

№ = k					
0	*15*	*1*	*1*	*1*	*1*
1	*14*	*1*	*1*	*1*	**1**
2	*13*	*1*	*1*	**1**	*1*
3	*12*	*1*	*1*	**1**	**1**
4	*11*	*1*	**1**	*1*	*1*
5	*10*	*1*	**1**	*1*	**1**
6	*9*	*1*	**1**	**1**	*1*
7	*8*	1	**1**	**1**	**1**

The *Bc* and *Bs* arrays are listed below at the Tables 4 and 5 correspondingly.

Table 4. *Bc* array – optimal traveling time from a current state to the final pose

№ *i*	*bin(i)*	№*i*	*bin(i)*	**B**	**C**	**D**	**E**
1	*0001*	**14**	1110	–	–	–	*17.34*
2	*0010*	**13**	1101	–	–	*17.57*	–
3	*0011*	**12**	**1100**	**10.64**	**8.83**	*14.63*	*15.42*
4	*0100*	**11**	1011	–	*16.67*	–	–
5	*0101*	**10**	**1010**	**13.07**	*14.08*	**10.53**	*13.34*
6	*0110*	**9**	**1001**	**12.05**	*14.08*	*13.22*	**10.18**
7	0111	**8**	**1000**	**14.86**	–	–	–

Table 5. *Bs* array – optimal further pose from a current state

№*i*	№*i*	**B**	**C**	**D**	**E**
1	**14**	–	–	–	*3*
2	**13**	–	–	*2*	–
3	**12**	**4**	**3**	*2*	*1*
4	**11**	–	*1*	–	–
5	**10**	**2**	*1*	**2**	*3*
6	**9**	**2**	*1*	*4*	**3**
7	**8**	**2**	–	–	–

The algorithm program realization

The described above memory adressing method was tested in the algorithm for optimal sequence choice of the manipulator - workpiece poses.

During dynamic programming realization at first step an algorithm consider all the states that correspondes with any two havn't passed poses and the appropriate *Bc* and *Bs* cells are filling according the rule (1). Then at the second step all the states with three not passed poses are considered, etc.

The algorithm was realized in Matlab. Some of used built-in functions are:

– dec2bin, bin2dec, str2num, num2str, strfind – for translation between binary and decimal notations;

– nchoosek – all possible combinations of not passed poses at each step.

The algorithm work result is in the Tables 4 and 5. The cells, corresponding to $i = 7,11,13,14$ of the Tables are empty, since the corresponding states don't have variants of further trajectory.

Analysis of results and discussion

For better illustration of presented work the case with limit to four intermediate manipulator - workpiece poses is explored. Maximum time saving is defined by the variantswith maximum and minimum traveling time. In this case it is 8.5%.

In general case many conditions affect on manipulator traveling time from one pose to another in the general case [14-17]. Someofthem:

– the manipulator speed parameters;

– the instrument and base coordinate systems relative position choice [18];

– required movement accuracy;
– some type of movement parameters, etc.

In spite of above listed factors we got result that could say of:

– a manipulator path choice influence on time traveling by the given example;
– principal TSP applicability to manipulator path selection at the given conditions.

The dynamic programming state memory adressing was consideredthat gives some potential benefits in industrial computer realization of the TSP.

References

[1] Domenico Spensieri, Robert Bohlin, Johan S Carlson. Coordination of robot paths for cycle time minimization. Conference: Automation Science and Engineering (CASE), 2013 IEEE International Conference on Automation Science and Engineering 6654032, 522-527 pp.

[2] Eremeev, A.V., Kovalenko, Y.V. On solving travelling salesman problem with vertex requisitions (2017) Yugoslav Journal of Operations Research, 27 (4), pp. 415-426. DOI: 10.2298/YJOR161012003E

[3] SanjoyDasgupta, Christos Papadimitriou, and UmeshVazirani Algorithms, McGraw-Hill, 2006

[4] Poznyak A.S., Najim K., Gomez-Ramirez E. Self-Learning Control of Finite Markov Chains, Marcel Dekker, Inc., New York, Basel. 2000, 298 p.

[5] Belomestnov, V.G., Pastukhovskiy, A.V., Pervushin, N.N., Serenko, V.A. Optimization of product assembly time in a robotics complex with multiple manipulators (1993) Journal of Computer and Systems Sciences International, 31 (2), pp. 163-166.

[6] Zacharia, P.T., Xidias, E.K., Aspragathos, N.A. Task scheduling and motion planning for an industrial manipulator. (2013) Robotics and Computer-Integrated Manufacturing, 29 (6), pp. 449-462. DOI: 10.1016/j.rcim.2013.05.002

[7] Sethi, S.P., Sidney, J.B., Sriskandarajah, C. Scheduling in dual gripper robotic cells for productivity gains. (2001) IEEE Transactions on Robotics and Automation, 17 (3), pp. 324-341.DOI: 10.1109/70.938389

[8] Kogan D.I. Tasks and methods of finite-dimentional optimization. Part 3. Tutorial. Dynamic programming and discrete multidimentional optimization. Publ. Nizhny Novgorod university, 2004, 157 p.

[9] Moiseev N.N. The elements of optimal systems theory, M. Publ.: Nauka, 528 p., 1975.

[10] Volkov I.K., Zagoruyko E.A. Operations research, series: Mathematics in technical university, M. Publ. house Bauman MSTU, 2004. 435 p.

[11] Bellman, Richard (1954), "The theory of dynamic programming", Bulletin of the American Mathematical Society, **60** (6): 503–516, doi:10.1090/S0002-9904-1954-09848-8.

[12] Bellman, Richard (1957), Dynamic Programming, Princeton University Press. Dover paperback edition (2003), ISBN 0-486-42809-5.

[13] R.Bellman On the theory of dynamic programming, Proc. Nat. Acad. Sci. U.S.A. vol. 38 (1952) pp. 716-719.

[14] Zenkevich, S., Maximov, A., Nazarova, A., Korshunov, A. Control of robot-based assembly cell (1993) Lecture Notes in Control and Information Sciences, 187, pp. 418-427. https://www.scopus.com/inward/record.uri?eid=2-s2.0-84946075718&partnerID=40&md5=5298f3e34b5f4f2dfe79e320cd2b551c

[15] Nosova, N.Yu., Glazunov, V.A., Misyurin, C.Yu., Filippov, D.N. Synthesis and the kinematic analysis of mechanisms of parallel structure with the outcome of progress (2015) IzvestiyaVysshikhUchebnykhZavedenii, SeriyaTeknologiyaTekstil'noiPromyshlennosti, 2015-January (2), pp. 109-113.

[16] Leskov, A., Golovin, V., Arkhipov, M., Kocherevskaya, L. Training of robot to assigned geometric and force trajectories (2016) Mechanisms and Machine Science, 39, pp. 75-84. DOI: 10.1007/978-3-319-30674-2_6

[17] Leskov, A., Illarionov, V., Zimin, A., Moroshkin, S., Kalevatykh, I. Distance robotics learning using Hybrid Simulating Testbed (2014) Proceedings of 2014 11th International Conference on Remote Engineering and Virtual Instrumentation, REV 2014, статья № 6784261, pp. 225-226. DOI: 10.1109/REV.2014.6784261.

[18] Domenico Spensieri, Johan S Carlson, Robert Bohlin, Jonas Kressin. Optimal Robot Placement for Tasks Execution. DOI: 10.1016/j.procir.2016.02.105